Peter Zörnig
Probability Theory and Statistics with Real World Applications

Also of interest

Observability and Mathematics.
Quantum Yang-Mills Theory and Modelling
Boris Khots, 2024
ISBN 978-3-11-139735-1, e-ISBN (PDF) 978-3-11-139843-3,
e-ISBN (EPUB) 978-3-11-139938-6

Business Statistics with Solutions in R
Mustapha Abiodun Akinkunmi, 2019
ISBN 978-1-5474-1746-9, e-ISBN (PDF) 978-1-5474-0135-2,
e-ISBN (EPUB) 978-1-5474-0137-6

Asymptotic Statistics.
With a View to Stochastic Processes
Reinhard Höpfner, 2014
ISBN 978-3-11-025024-4, e-ISBN (PDF) 978-3-11-025028-2,
e-ISBN (EPUB) 978-3-11-036778-2

Forum Mathematicum
Edited by: Thomas Alazard, Chantal David, Manfred Droste, Siegfried
Echterhoff, Maria Gordina, Philipp Habegger, Matthias Hieber, Clara Löh,
Karin Melnick, Anke Pohl, Maksym Radziwill, Freydoon Shahidi,
Christopher D. Sogge, Shigeharu Takayama;
Managing Editor: Jan Hendrik Bruinier
ISSN 1435-5337

Peter Zörnig

Probability Theory and Statistics with Real World Applications

Univariate and Multivariate Models Applications

2nd, Revised and Extended Edition

DE GRUYTER

Mathematics Subject Classification 2020
60-XX, 60Axx, 60E05, 62-XX, 62D05, 62Fxx,97-XX, 97K40

Author
Prof. Dr. Peter Zörnig
Institute of Exact Sciences
Statistical Department
University of Brasilia
Asa Norte
Brasil
peter@unb.br

ISBN 978-3-11-133220-8
e-ISBN (PDF) 978-3-11-133227-7
e-ISBN (EPUB) 978-3-11-133232-1

Library of Congress Control Number: 2024936933

Bibliographic information published by the Deutsche Nationalbibliothek
The Deutsche Nationalbibliothek lists this publication in the Deutsche Nationalbibliografie;
detailed bibliographic data are available on the Internet at http://dnb.dnb.de.

www.degruyter.com

Acknowledgment

I would like to thank Dr. Ranis Ibragimov and especially Ms. Melanie Götz for their support in developing the content and preparing the manuscript, respectively.

Preface to the first edition

The present book is based on my lecture notes for the course "Probability and Statistics" which I have taught a number of times at the Statistical Department of the University of Brasilia since 1998. The book is written primarily for undergraduate students from applied mathematics, statistics, computer science, engineering, and natural sciences that want to become familiar with probability theory and statistical applications. According to its title, the philosophy of the present book is to present an illustrative didactical text with profound explanations making it adequate for a self-study. The access to the discipline is facilitated as much as possible. In order to motivate the student, many examples and indications of practical applications are presented. Moreover, the book contains 326 exercises whose solutions are given at the end. The emphasis is on applications and intuitive insight. Excessive formalism is avoided without sacrificing the necessary mathematical rigor. The only prerequisite is knowledge of differentiation and integral calculus. No previous expertise of probability or statistics is presumed.

The text is subdivided into fifteen chapters that lead from an elementary discussion of probability until a somewhat advanced level. At the beginning the necessary mathematical concepts from combinatorics, special functions and bidimensional integration are reviewed. The last three chapters are devoted to the standard topics of statistical inference like sampling distributions, parameter estimation, and hypothesis tests.

I consider it appropriate to say a few words on the development of teaching mathematics during the past decades. Since my own studies from the late seventies to the early eighties, many things have changed. When I was studying mathematics myself, examples were "frowned upon" in the lecture halls, i.e., their demonstration was considered as something below the professor's level. If a student nevertheless asked for an example, it happened that the teacher wrinkled his nose and mildly smiling commented that "all and everything" is trivial . . . . These times have passed. Nowadays examples and applications are not only tolerated in mathematics, but even desirable since mathematics and its applications inspire and mutually benefit from each other.

Another profound change in education has been caused by the Internet. While it was formerly necessary to spend hours looking at libraries, today one finds information at home on the computer after a few clicks.

I would encourage the students to use the Internet in the research. However, one must be careful with regard to the reliability of the information found.

Finally, the availability of software that is capable of symbolic computation has influenced the process of learning. Beyond numerical calculation such programs also support logical reasoning since they can, for example, perform formal derivation and integration.

It is highly recommended that the students acquire software that allows symbolic calculations. For some examples and exercises in this book, it is necessary. Advanced software can also be used to evaluate complicated distributions, which formerly

https://doi.org/10.1515/9783111332277-203

could be done only with the aid of tables. But by no means one should believe that software could replace one's own reasoning since, e.g., rounding errors may distort the result completely.

One thing has not changed over the past decades: to learn mathematics it is still necessary to sit down and work seriously . . . .

Peter Zörnig, July 2016

Preface to the second edition

I was very pleased when Dr. Ibragimov from the De Gruyter publisher invited me one year ago to write a second edition of my probability book. The idea was to present a text that is, among others, useful for the practitioner who needs univariate or multivariate probabilistic models. In an attempt to meet these requirements, I have modified and expanded the first edition accordingly. Two new chapters are presented:

The relative simple classical probability distributions (like, e.g., the exponential, the normal, and the binomial distribution) are not sufficient to model all the constantly growing applications that require, e.g., asymmetrical and multimodal models. Therefore, the new Chapter 12 introduces various concepts useful to create new distributions; for example, exponentiated generalized distributions, mixture and slash distributions, among others. This is a very fertile current research area that particularly stimulates creative modeling. In my opinion, it should not be missing from any modern probability book, even if it goes beyond the level of an undergraduate course.

The second new chapter is Chapter 17, treating exploratory data analysis. After introducing the standard concepts of descriptive statistics, the shape characteristics symmetry and kurtosis are intensively studied. The reader should not be satisfied with a mere knowledge of the definitions; the examples and exercises should enable him or her to perform a proper judgment about the adequacy of the concepts. Moreover, bidimensional relations are explored in detail, introducing the sample (linear) correlation coefficient and the coefficient of determination. The nontrivial relationship between these terms is illustrated, among others, with the help of real data on the growth of the Brazilian population.

Some changes have also been made to the "old part" of the book. For example, the concepts of survival function and bivariate exponential distribution have been added.

I hope that the book will be useful for some readers. Any feedback would be appreciated . . . .

Peter Zörnig, July 2024

https://doi.org/10.1515/9783111332277-204

Contents

5.4	Continuous random variables —— **72**
5.5	Distribution and survival function —— **75**
	Exercises —— **78**
6	**Functions of random variables —— 82**
6.1	Continuous random variables —— **82**
6.2	Discrete random variables —— **88**
	Exercises —— **90**
7	**Bidimensional random variables —— 92**
7.1	Discrete random variables —— **92**
7.2	Continuous random variables —— **94**
7.3	Marginal distributions and independent variables —— **97**
7.4	Conditional distributions, distribution and survival functions —— **101**
7.5	Functions of a random variable —— **105**
	Exercises —— **108**
8	**Characteristics of random variables —— 111**
8.1	The expected value of a random variable —— **111**
8.2	Expectation of a function of a random variable —— **113**
8.3	Properties of the expected value —— **115**
8.4	Variance, moments, and shape characteristics —— **117**
8.5	Probability inequalities —— **123**
8.6	Covariance and correlation —— **126**
	Exercises —— **130**
9	**Discrete probability models —— 133**
9.1	The Poisson distribution —— **133**
9.2	The Poisson process —— **135**
9.3	The geometric distribution —— **137**
9.4	The Pascal distribution —— **139**
9.5	The hypergeometric distribution —— **141**
9.6	The multinomial distribution —— **144**
	Exercises —— **145**
10	**Continuous probability models —— 148**
10.1	The normal distribution —— **148**
10.2	The exponential distribution —— **154**
10.3	The gamma distribution —— **155**
10.4	The chi-square distribution —— **157**
10.5	The beta distribution —— **159**
10.6	The multivariate normal distribution —— **160**

1 Mathematics revision

In this chapter, we revise the necessary concepts from set theory, combinatorics, gamma function and the like.

1.1 Basic notions of sets

We start with some elementary concepts from set theory which are necessary to develop probabilistic models.

A *set* is a well-defined collection of distinct objects, usually denoted by capital letters. For example, $A = \{x \in IR | 1 \leq x \leq 4\}$ is the set of real numbers between 1 and 4 inclusive, $B = \{x^2 | x \in IN\}$ is the set of square numbers and $C = \{f : [3, 5] \rightarrow IR | f \text{ continuous}\}$ is the set of continuous real-valued functions over the interval [3, 5]. The objects that make up a set are called the *elements* of the set. We write $a \in A$ when a is an element of the set A. If a is not an element of A, we write $a \notin A$. There are two special sets of interest: the *universal set* U, denoting the set of all objects under consideration and the *empty set*, denoted by Φ.

We write $A \subset B$, when A is a subset of B, i.e., when $x \in A$ implies $x \in B$. We say that A and B are equal, if $A \subset B$ and $B \subset A$ hold. We now consider different ways of combining given sets in order to form new sets.

Given the sets A and B, the *union* of A and B is defined by $A \cup B := \{x \mid x \in A \text{ or } x \in B \text{ (or both)}\}$. The intersection of A and B is defined by $A \cap B = \{x \mid x \in A \text{ and } x \in B\}$. The sets A and B are called *disjoint*, if $A \cap B = \Phi$. Moreover, we define the (set-theoretic) *difference* of A and B by $A \backslash B = \{x \in A \mid x \notin B\}$. In particular, the *complement* of A is defined by the difference of U and A, i.e., $\bar{A} := U \setminus A$. Consider, for example, the universal set $U = \{1, \ldots, 10\}$ and the sets $A = \{1, \ldots, 6\}$ and $B = \{4,5,8,10\}$. Then $A \cup B = \{1, \ldots, 6, 8, 10\}$, $A \cap B = \{4, 5\}$, $\bar{A} = \{7, \ldots, 10\}$ and $A \backslash B = \{1,2,3,6\}$.

The above operations of sets can be geometrically illustrated by means of a *Venn diagram*, where sets are represented by regions in the plane, and the shaded portion represents the set under consideration (Fig. 1.1).

The above operations of union and intersection can be easily extended to finite numbers of sets. We define $A \cup B \cup C$ as $A \cup (B \cup C)$ or $(A \cup B) \cup C$ and $A \cap B \cap C$ as $A \cap (B \cap C)$ or $(A \cap B) \cap C$ (see the following relations (iii) and (iv)). It should be clear how these concepts can be generalized for an arbitrary number of sets. We obtain the following set identities:

(i) $A \cup B = B \cup A$
(ii) $A \cap B = B \cap A$
(iii) $A \cup (B \cup C) = (A \cup B) \cup C$
(iv) $A \cap (B \cap C) = (A \cap B) \cap C$
(v) $A \cup (B \cap C) = (A \cup B) \cap (A \cup C)$

https://doi.org/10.1515/9783111332277-001

(vi) $A \cap (B \cup C) = (A \cap B) \cup (A \cap C)$
(vii) $A \cap \Phi = \Phi$
(viii) $A \cup \Phi = A$
(ix) $\overline{(A \cup B)} = \bar{A} \cap \bar{B}$
(x) $\overline{(A \cap B)} = \bar{A} \cup \bar{B}$
(xi) $\bar{\bar{A}} = A$

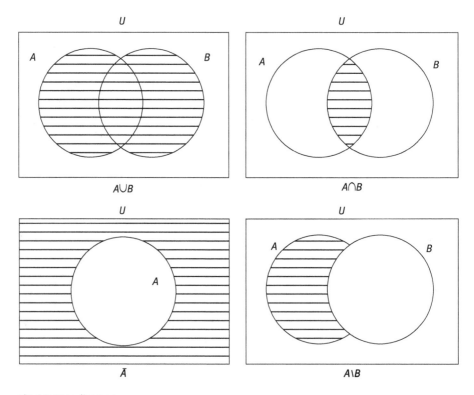

Fig. 1.1: Venn diagrams.

In particular, relations (i) and (ii) are known as the *commutative laws*; (iii) and (iv) are known as the *associative laws*; and (ix) and (x) are called *De Morgan´s laws*.

We still consider some concepts of set theory. The *Cartesian product* of the sets A and B is given by the set of pairs $A \times B = \{(a,b) \mid a \in A, b \in B\}$. If, for example, $A = \{1,2,3\}$ and $B = \{4,5\}$, then $A \times B = \{(1,4), (1,5), (2,4), (2,5), (3,4), (3,5)\}$. It is clear that in general holds $A \times B \neq B \times A$. The Cartesian product can be extended to more than two sets by $A_1 \times A_2 \times \cdots \times A_n = \{(a_1, \ldots, a_n) \mid a_i \in A_i \text{ for } i = 1, \ldots, n\}$, which is the *set of ordered n-tuples*.

Definition 1.1: For any set S, the *power set* $P(S)$ is defined as the set of all subsets of S. The set $P(S)$ contains 2^n elements if S has n elements, since each of the n elements of S can be contained in a subset or not. For example, if $S = \{1,2,4\}$, then $P(S) = \{\Phi,\{1\},\{2\}, \{4\},\{1,2\},\{1,4\},\{2,4\},\{1,2,4\}\}$.

Definition 1.2: Let A be the set. The set $\{A_1, \ldots, A_m\}$, where $A_1, \ldots, A_m$ are pairwise disjoint subsets of A, satisfying $A_1 \cup \ldots \cup A_m = A$, is called a *partition* of A (into m components).

We finally study some concepts related to the cardinality of a set. If a set A has a finite number of elements, we say that A is a *finite set*. Otherwise, A is called an *infinite set*. Recall that a *biunique function* $f\colon A \to B$ is a function such that for any $b \in B$ exists a *unique $a \in A$ with $f(a) = b$*.

Definition 1.3: An infinite set A is called *countably infinite* or *denumerable*, if there exists a biunique function $f\colon \mathbb{N} \to A$.

In simple words, A is countably infinity, if its elements may be put into a one-to-one correspondence with the natural numbers, i.e., we can identify a first element of A, namely $f(1)$, a second element of A, namely $f(2)$, etc.

Example 1.1: The set of even numbers $A = \{2, 4, 6, \ldots\}$ is countably infinite, since $f(n) = 2n$ is a biunique function from $\mathbb{N}$ into A.

Example 1.2: The Cartesian product $P = Z \times Z$, where Z is the set of integers, is countably infinite. The explicit definition of a biunique function from $\mathbb{N}$ into P is not trivial; however, one can easily construct such a function in a geometric manner (see Fig. 1.2).

The set P consists of all points with two integer components. Obviously, one obtains a biunique function from $\mathbb{N}$ into P, by defining the first element of P as $f(1) = (0,0)$, the second element as $f(2) = (1,0)$, the third element as $f(3) = (1,1)$, etc. The enumeration is continued along the "spiral" illustrated in Fig. 1.2.

Example 1.3: The set of real numbers IR and an interval $[a, b]$ where $a < b$ are examples for uncountably infinite sets.

The proof uses Cantor's diagonalization argument. The interested reader is referred to the literature cited in Wikipedia. We will make use of the following statement.

Proposition 1.1:
(i)   If B is countably infinite, then any subset $A \subset B$ is finite or countably infinite.
(ii) If A is uncountably infinite, then any set B with $A \subset B$ is noncountably infinite.

Example 1.4: The set of rational numbers $Q = \{x/y | x,y \in Z, y \neq 0\}$ is countably infinite. We first observe that this set can be also written as $Q = \{x/y | x,y \in Z, \ y > 0, \ x$

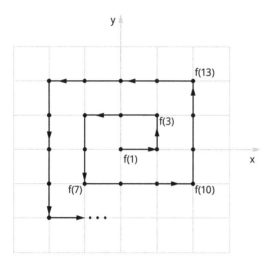

Fig. 1.2: Enumerating the elements of P.

and y are relatively prime}. (For example, $\frac{2}{-3}$ can be written as $\frac{-2}{3}$ and $\frac{15}{20}$ as $\frac{3}{4}$.) By using the second form for the definition of Q, any rational number can be uniquely written as x/y, and this rational number can be identified with the element (x, y) of the set P in Example 1.2. Since the latter is countably infinite, Proposition 1.1(a) implies that Q is also countably infinite.

Exercises

1.1.1 Illustrate the above identities (v), (vi), (ix) and (x) by means of a Venn diagram.

1.1.2 Given the universal set $U = \{1, \ldots, 10\}$ and the sets $A = \{4, 5, 6, 9\}$, $B = \{1, 2, 3, 5, 6\}$ and $C = \{6, 7, 8, 10\}$, determine
   (a) $A \cup (\bar{B} \cap \bar{C})$;
   (b) the complement of $\bar{A} \cup (B \cap C)$;
   (c) $\bar{A} \cup \bar{B} \cup \bar{C}$;
   (d) the complement of $A \cup B \cup \bar{C}$.

1.1.3 In how many ways can the sets A and B be defined such that $A \cup B = \{1, \ldots, 5\}$ and $A \cap B = \{2,3,4\}$?

1.1.4 Prove the following relations:
   (a) $(A \backslash C) \cup (B \backslash C) = (A \cup B) \backslash C$
   (b) $(A \cup B) \cap (A \cup \bar{B}) = A$

1.1.5 Let $\{A_1, \ldots, A_m\}$ be the partition of A and B a nonempty subset of A. Is $\{A_1 \cap B, A_2 \cap B, \ldots, A_m \cap B\}$ a partition of B?

1.1.6 List all partitions of the set $\{1,2,3,4\}$ into two components.

1.1.7 Given the sets $P = Z \times Z$, $A = \{x, y) \mid x^2 + y^2 \le 3\}$ and $B = \{x, y) \mid 0 \le y \le x^2, -2 \le x \le 3\}$, where Z is the set of integers, list the elements of the sets $P \cap A$ and $P \cap B$.

1.1.8 How many elements has the power set of $A \times B$, if A and B have n and m elements, respectively?

1.1.9 Let A and B denote countably infinite sets and let C and D denote uncountably infinite sets. For each of the following four statements, give a proof or a counter example:

(a) $A \cap B$ is countably infinite.
(b) $A \cup B$ is countably infinite.
(c) $C \cap D$ is uncountably infinite.
(d) $C \cup D$ is uncountably infinite.

1.1.10 Show that the Cartesian product $A_1 \times A_2 \times \cdots \times A_n$ is countably infinite, if $A_i = \mathbb{N}$ for $i = 1, \ldots, n$.

1.2 Basic concepts of combinatorics

The following notions are particularly useful to calculate the probability of events in finite sample spaces (see Section 3.1).

Given n different objects that are 1, 2, . . ., n without loss of generality, a *permutation* of n elements is obtained by writing the elements in a specific order. For example, for $n = 3$, we obtain the six permutations 123, 132, 213, 312, 231 and 321. Arranging n objects is equivalent to putting them into a box with n compartments, in a specific order (see Fig. 1.3).

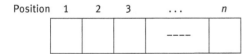

Position   1   2   3   . . .   n

Fig. 1.3: Box with n compartments.

The first compartment (position) can be filled in n ways, the second compartment in $(n - 1)$ ways, the third in $(n - 3)$ ways. etc. The number of permutations of n elements is therefore $n \cdot (n - 1) \cdot (n - 2) \cdot \cdots \cdot 1$.

Definition 1.4: For any positive integer n, we define $n! := n \cdot (n - 1) \cdot (n - 2) \cdot \cdots \cdot 1$ and call it *n-factorial*. Moreover, we define $0! := 1$.

Thus, the number of permutations of n different objects is given by $n!$.

There exist numerous approximations for the factorial in the literature. The most famous is surely Stirling's formula, given by

$$n! \approx \sqrt{2\pi n}\left(\frac{n}{e}\right)^n,$$

where $e \approx 2.71828$ denotes *Euler's number*.

We consider again n different objects, but now we choose k of them and permute them $(0 \le k \le n)$. We can imagine that we now fill a box with k compartments, using only k of the n objects. As above, the first compartment can be filled in n ways, the second in $(n - 1)$ ways and the kth in $n - (k - 1)$ ways. The number of ways to fill the box (number of *arrangements*) is therefore $n \cdot (n - 1) \ldots (n - k + 1) = n!/(n - k)!$.

We consider again n different objects. We are now interested in the number of ways we may choose k of the elements *without* regard to order.

Definition 1.5:
(i) Given the set $S = \{1, \ldots, n\}$ and a nonnegative integer $k \le n$, any subset of S with k elements is called a *k-combination* or a *k-subset* of S.

(ii) The *binomial coefficient* is defined by $\binom{n}{k} := \frac{n!}{k!\,(n-k)!}$.

Proposition 1.2: The number of k-combinations of a set with n elements is given by $\binom{n}{k}$.

Proof: We have shown that the number of arrangements with k of n elements is $n!/(n - k)!$. Any k-combination corresponds to $k!$ arrangements that are obtained by writing the k-combinations in all possible orders. Thus, the number of k-combinations is obtained, dividing $n!/(n - k)!$ by $k!$.

The statement is illustrated in Tab. 1.1 for $n = 4$ and $k = 3$. The number of arrangements there is $(4!/(4 - 3)!) = 24$ and the number of 3-combinations is $24/3! = 4$.

Tab. 1.1: Combinations and arrangements.

k-Combinations	{1,2,3}	{1,2,4}	{1,3,4}	{2,3,4}
Arrangements	1,2,3	1,2,4	1,3,4	2,3,4
	1,3,2	1,4,2	1,4,3	2,4,3
	2,1,3	2,1,4	3,1,4	3,2,4
	3,1,2	4,1,2	4,1,3	4,2,3
	2,3,1	2,4,1	3,4,1	3,4,2
	3,2,1	4,2,1	4,3,1	4,3,2

1.2.1 More about binomial coefficients

The name "binomial coefficient" is due to the fact that these numbers appear as coefficients in the expansion of the binomial expression $(a+b)^n$. For any natural number n, it holds that $(a+b)^n = (a+b) \cdot (a+b) \cdots (a+b)$ and when multiplied out, each element of the resulting sum is of the form $a^k b^{n-k}$. The number of these summands is the number of ways in which we can choose k out of n a's, disregarding order. This number is given by $\binom{n}{k}$.

We obtain the following relation, known as the *binomial theorem*:

$$(a+b)^n = \sum_{k=0}^{n} \binom{n}{k} a^k b^{n-k}. \tag{1.1}$$

For example, for $n = 3$ we obtain

$$(a+b)^3 = b^3 + 3a^2 b + 3ab^2 + b^3.$$

The numbers $\binom{n}{k}$ have many interesting properties. A variety of identities for binomial coefficients have been published. We state only three.

Proposition 1.3: For integers k, n with $0 \le k \le n$ holds:

(a) $\binom{n}{k} = \binom{n}{n-k}$,

(b) $\binom{n}{k} = \binom{n-1}{k-1} + \binom{n-1}{k}$,

(c) $2^n = \sum_{i=0}^{n} \binom{n}{i}$.

Proof: It is easy to verify the above relations algebraically:

(a) This part follows directly from the definition.
(b) The statement is equivalent to

$$\frac{n!}{k!\,(n-k)!} = \frac{(n-1)!}{(k-1)!\,(n-k)!} + \frac{(n-1)!}{k!\,(n-k-1)!}.$$

Multiplying this equation by $k!\,(n-k)!/(n-1)!$ yields the equivalent equation $n = k + n - k$ which is obviously correct.
(c) We prove this part by induction. Obviously the assertion is correct for $n = 1$. Assume that it is correct for $n = n_0 \ge 1$. Thus, part (b) implies that

$$\sum_i \binom{n_0+1}{i} = \sum_i \binom{n_0}{i-1} + \sum_i \binom{n_0}{i} = 2^n + 2^n = 2^{n+1}$$

that is, the statement holds also for $n = n_0 + 1$ (as usual, we set $\binom{n}{k} := 0$ for $k < 0$ or $k > n$).

It is very interesting and instructive to interpret the statements in Proposition 1.3 in a "combinatorial manner":

(a) In this sense $\binom{n}{k}$ and $\binom{n}{n-k}$ can be interpreted as the number of subsets of $S = \{1, \ldots, n\}$ with k and $(n-k)$ elements, respectively. Thus, $\binom{n}{k}$ counts the k-subsets of S, while $\binom{n}{n-k}$ counts their complements. These numbers must therefore be equal.

(b) Assume that an urn contains $(n-1)$ white balls and one black ball (Fig. 1.4). Now $\binom{n}{k}$ represents the number of k-combinations of balls. One can say that $\binom{n-1}{k}$ counts the k-combinations that do *not* contain the black ball, while $\binom{n-1}{k-1}$ counts the k-combinations, containing the black ball. The latter number is equal to the number of ways to choose the $(k-1)$ white balls. Hence, the statement must be true.

(c) The left side of this statement represents the *total number* of subsets of $S = \{1, \ldots, n\}$ (see Definition 1.1). Any summand $\binom{n}{i}$ of the right side counts the number of *subsets with i elements*, therefore, summing up the $\binom{n}{i}$ results in 2^n.

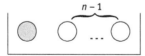

Fig. 1.4: Urn model.

We present another relation that has an interesting "combinatorial interpretation."

Proposition 1.4: $\binom{n+m}{k} = \sum_{i=0}^{n} \binom{n}{i}\binom{m}{k-i}.$

Assume that an urn contains n black balls and m white balls. The left side in Proposition 1.4 represents the number of all k-combinations of balls from this urn. Any summand $\binom{n}{i}\binom{m}{k-i}$ on the right side counts the number of k-combinations, consisting of i black balls and $(k-i)$ white balls.

If we arrange the binomial coefficients as follows, where line n contains the numbers $\binom{n}{0}, \ldots, \binom{n}{n}$, we get *Pascal's triangle*:

$$
\begin{array}{lc}
n=0 & 1 \\
n-1 & 1 \ \ 1 \\
n=2 & 1 \ \ 2 \ \ 1 \\
n=3 & 1 \ \ 3 \ \ 3 \ \ 1 \\
n=4 & 1 \ \ 4 \ \ 6 \ \ 4 \ \ 1 \\
& \vdots \quad \vdots \quad \vdots \quad \vdots
\end{array}
$$

Each number is obtained by adding two consecutive numbers of the previous line, where all lines begin and end with a 1. This regularity is a consequence of Proposition 1.3(b). Due to parts (a) and (c) of this proposition, the triangle is symmetric with respect to the vertical axis and the sum of line n is 2^n.

The binomial coefficients are usually introduced in the above combinatorial context, where n and k are nonnegative integers. However, by reducing the fraction $n!/(k!\,(n-k)!)$ in Definition 1.5(ii), we get

$$
\binom{n}{k} = \frac{n\cdot(n-1)\cdots(n-k+1)}{k!}, \tag{1.2}
$$

which is defined for any *real number* n and any natural number k. We also set $\binom{n}{0} := 1$ for every $n \in \mathrm{IR}$.

From definition (1.2) we get for example

$$
\binom{\sqrt{5}}{4} = \frac{\sqrt{5}\cdot(\sqrt{5}-1)\cdot(\sqrt{5}-2)\cdot(\sqrt{5}-3)}{4!} \approx -0.0208
$$

$$
\binom{-\pi}{3} = \frac{-\pi\cdot(-\pi-1)\cdot(-\pi-2)}{3!} \approx -11.15.
$$

It is interesting to observe that the binomial theorem remains valid for the generalization (1.2), i.e., the following holds.

Proposition 1.5: Generalized binomial theorem.

For real numbers x and $n \neq 0$, it holds:

$$(x+1)^n = \sum_{k=0}^{\infty} \binom{n}{k} x^k.$$

Note that the power series terminates with $k = n$, when n is a natural number. In the literature, the generalized binomial theorem is usually formulated in this way, but it can also be stated as in (1.1) (see Exercise 1.2.9).

By substituting specific values for n, one gets a variety of interesting expansions, e.g.:

$$n := -1 : \frac{1}{x+1} = \sum_{k=0}^{\infty} \binom{-1}{k} x^k = 1 - x + x^2 - x^3 + x^4 - \cdots,$$

$$n := -\frac{1}{2} : \frac{1}{\sqrt{1+x}} = \sum_{k=0}^{\infty} \binom{-1/2}{k} x^k = 1 - \frac{1}{2}x + \frac{3}{8}x^2 - \frac{5}{16}x^3 + \cdots.$$

Interesting relations also arise by substituting x. For $x := -1$ we obtain

$$0 = \sum_{k=0}^{\infty} \binom{n}{k} (-1)^k = \binom{n}{0} - \binom{n}{1} + \binom{n}{2} - \cdots + (-1)^n \binom{n}{n},$$

and by setting $x := -2/3$ and $n := 1/2$ we get

$$\sqrt{\frac{1}{3}} = \sum_{k=0}^{\infty} \binom{1/2}{k} \left(-\frac{2}{3}\right)^k = 1 - \frac{1}{3} - \frac{1}{18} - \frac{1}{54} - \frac{5}{648} - \cdots.$$

Finally, by using the form (1.1), we obtain, e.g.,

$$\sqrt{a+b} = \sum_{k=0}^{\infty} \binom{1/2}{k} a^k b^{1/2-k} = b^{1/2} + \frac{1}{2}ab^{-1/2} - \frac{1}{8}a^2 b^{-3/2} + \cdots.$$

The reader is suggested to verify the above relations (as well as other relations introduced later in this chapter) by means of a software that can perform symbolic computations, for example, MAPLE or MATHEMATICA.

We finally state another generalization of (1.1), which will be later used in the multinomial distribution.

Definition 1.6: For nonnegative integers $k_1, \ldots, k_r$ with $k_1 + \cdots + k_r = n$, the *multinomial coefficient* is defined by

$$\binom{n}{k_1, \ldots, k_r} := \frac{n!}{k_1! \cdot k_2! \ldots k_r!}.$$

Proposition 1.6: Multinomial theorem

For real numbers $a_1, \ldots, a_k$ and any natural number n, it holds:

$$(a_1 + a_2 + \cdots + a_k)^n = \sum_{(n_1, \ldots, n_k)} \binom{n}{n_1, \ldots, n_k} a_1^{n_1} \cdots a_k^{n_k},$$

where the summation is carried out over all nonnegative k-tuples $(n_1, .., n_k)$ with $n_1 + \cdots + n_k = n$.

If, for example, $n = 2$ and $k = 3$, we get

$$(a_1 + a_2 + a_3)^2 = \sum_{(n_1, n_2, n_3)} \binom{2}{n_1, n_2, n_3} a_1^{k_1} \cdot a_2^{k_2} \cdot a_3^{k_3}$$

$$= \binom{2}{2,0,0} a_1^2 a_2^0 a_3^0 + \binom{2}{0,2,0} a_1^0 a_2^2 a_3^0 + \binom{2}{0,0,2} a_1^0 a_2^0 a_3^2$$

$$+ \binom{2}{1,1,0} a_1^1 a_2^1 a_3^0 + \binom{2}{1,0,1} a_1^1 a_2^0 a_3^1 + \binom{2}{0,1,1} a_1^0 a_2^1 a_3^1$$

$$= a_1^2 + a_2^2 + a_3^2 + 2a_1 a_2 + 2a_1 a_3 + 2a_2 a_3.$$

1.2.2 Specific permutations and a generalization

Permutations occur in many theoretical and applied fields. Therefore, we still study the following concepts.

Definition 1.7: We say that a permutation $(p_1, \ldots, p_n)$ of the elements $1, \ldots, n$ has a *fixed point* if there is an index $i \in \{1, \ldots, n\}$ such that $p_i = i$.

A permutation without such points is called a *fixed-point-free permutation* or a *derangement*.

For example, the permutation $(p_1, \ldots, p_6) = (3,2,6,4,5,1)$ has three fixed points, since $p_i = i$ holds for $i = 2$, 4 and 5. The term "fixed point" becomes clear, if the permutation $(p_1, \ldots, p_n)$ is identified with the function $f(i) = p_i$ from the set $\{1, \ldots, n\}$ into itself. So, the above permutation $(3,2,6,4,5,1)$ is identified with the function given by $f(1) = 3$, $f(2) = 2$,

$f(3) = 6$, $f(4) = 4$, $f(5) = 5$ and $f(6) = 1$, and the numbers 2, 4 and 5, and only these remain unchanged when applying this function. The fixed points are marked by underlining in Tab. 1.2.

Tab. 1.2: Some permutations of four elements.

1, 2, 3, 4	2, 1, 3, 4
1, 2, 4, 3	2, 1, 4, 3
1, 3, 2, 4	3, 1, 2, 4
1, 4, 2, 3	4, 1, 2, 3
1, 3, 4, 2	3, 1, 4, 2
1, 4, 3, 2	4, 1, 3, 2

Proposition 1.7:

Let D_n denote the number of derangements of n objects. Then the following recurrence formula applies:

$$D_n = (n-1)(D_{n-1} + D_{n-2}) \qquad \text{for } n \geq 3.$$

Proof: In constructing a derangement of n objects, the element n can be placed on $(n-1)$ possible positions. Let x be the position of n:

$$\text{Position} \qquad 1 \ldots \underset{n}{x} \ldots n-1 \quad n.$$

Then two cases can be distinguished:

Case a): The element x is assigned to position n:

$$\text{Position:} \qquad 1 \ldots \underset{n}{x} \ldots n-1 \quad \underset{x}{n}.$$

Then exist D_{n-2} ways to continue the construction, where the element i may not be put on position i for $i \in \{1, \ldots, n-1\}\backslash\{x\}$.

Case b): The element x is *not* assigned to position n. Then exist D_{n-1} ways to continue the construction, where the element i may not be put on position i for $i \in \{1, \ldots, n-1\}\backslash\{x\}$ and x may not be put on position n. For small n the numbers D_n is presented in Tab. 1.3.

Tab. 1.3: Number of derangements.

N	D_n
2	1
3	2
4	9
5	44

Tab. 1.3 (continued)

N	D_n
6	265
7	1,854
8	14,833

Proposition 1.8:

It holds $D_n = n! \sum_{i=2}^{n} (-1)^i / i!$ for $n \geq 2$.

The statement can be proved, e.g., by using Proposition 1.7.

At the beginning of Section 1.2, we studied permutations of *distinct* elements. We now permit that the same object appears repeatedly in a sequence.

Proposition 1.9: Given n objects such that n_1 of them are of type 1, n_2 are of type 2, and n_k are of type k ($n_1 + \cdots + n_k = n$), the number of permutations of these n objects (we call them *permutations with repetitions*) is given by $\binom{n}{n_1, \ldots, n_k}$.

Proof: There are $\binom{n}{n_1}$ ways to choose the positions of the objects of type 1, then there are $\binom{n - n_1}{n_2}$ ways to choose the positions of the objects of type 2 among the $n - n_1$ unoccupied positions. The positions of the elements of type 3 can then be chosen in $\binom{n - n_1 - n_2}{n_3}$ ways. The number of positions with repetitions is therefore

$$\binom{n}{n_1} \cdot \binom{n - n_1}{n_2} \cdot \binom{n - n_1 - n_2}{n_3} \cdots \binom{n_k}{n_k} = \binom{n}{n_1, \ldots, n_k}.$$

Assume, for example, that we want to construct a word by using 2 times the letter a, 3 times the letter b, 3 times the letter c and 4 times the letter d. The number of possible words, i.e., permutations of the sequence $a\,a\,b\,b\,b\,c\,c\,c\,d\,d\,d\,d$, is therefore

$$\binom{12}{2,3,3,4} = \frac{12!}{2! \cdot 3! \cdot 3! \cdot 4!} = 277{,}200.$$

One can also derive the number of these permutations alternatively.

If the 12 elements were distinguishable, say $a_1, a_2, b_1, b_2, b_3, c_1, c_2, c_3, d_1, d_2, d_3, d_4$, there would be 12! possible permutations. If we, for example, do not distinguish between the b_i, the number of permutations reduces to $12!/3!$, since we identify the 3! permutations resulting by permuting only the b_i. Applying this reasoning to the a_i, b_i, c_i and d_i at the same time, one obtains the number $\frac{12!}{2! \cdot 3! \cdot 3! \cdot 4!}$.

Exercises

1.2.1 In how many ways 8 persons can form a waiting queue?

1.2.2 How many words of 10 letters can be formed, using each letter only once?

1.2.3 In how many ways a password can be formed, consisting of three letters, followed by five digits?

1.2.4 Calculate the factorials 10!, 15! and 20! and compare with the approximate values obtained by Stirling's formula.

1.2.5 An urn contains 8 white balls and 10 black balls.
  (a) In how many ways one can select 6 balls from the urn?
  (b) In how many ways one can select 7 balls such that 3 of them are white and four of them are black?

1.2.6 A community is composed of 200 Catholics, 300 Protestants and 400 persons with another religion. In how many ways a committee of 12 persons can be performed, consisting of 3 Catholics, 5 Protestants and 4 persons with another religion?

1.2.7 Prove Proposition 1.4 algebraically.

1.2.8 Formulate a generalization of Proposition 1.4 and interpret it in a "combinatorial manner."

1.2.9 Prove that Proposition 1.5 is equivalent to formula (1.1), where a, b and $n \neq 0$ are real numbers.

1.2.10 Show that Proposition 1.3(b) is valid for any real number $n \neq 0$ and any integer k, if the binomial coefficients are defined by (1.2).

1.2.11 Prove that the recurrence formula $D_{n+1} = (n+1) D_n + (-1)^{n+1}$ holds.

1.2.12 Show that the following two relations are valid, where the summation is defined as in Proposition 1.6:

  (a) $k^n = \displaystyle\sum_{(n_1, ..., n_k)} \binom{n}{n_1, \ldots, n_k}$.

  (b) $0 = \displaystyle\sum_{(n_1, ..., n_k)} \binom{n}{n_1, \ldots, n_k} (-1)^{n-n_k} \cdot (k-1)^{n_k}$.

1.2.13 How many words can be formed, using three times the letter a, four times the letter b, five times the letter c, and six times the letter d?

1.2.14 Determine the number of permutations of n distinct objects, having exactly k fixed points ($0 \leq k \leq n$).

1.2.15 A secretary has written letters to four different persons. The envelopes are already prepared. In how many ways the letters can be put into envelopes such that at least one letter is sent to the right address?

1.2.16 Prove the following relations for a positive integer n:

(a) $\sum_{k=0}^{n} \binom{n}{k}^2 = \binom{2n}{n}$.

(b) $\binom{-1/2}{n} = \frac{1}{(-4)^n} \binom{2n}{n}$.

1.3 Some special functions

We now consider a very intensively studied area. However, we will limit ourselves to some basic concepts. The *gamma function* is one of the most important special functions. There exist several equivalent definitions. The most important one defines this function by the *Euler integral* of the second kind

$$\Gamma(z) = \int_0^\infty e^{-t}\, t^{z-1} dt \quad \text{for Re}(z) > 0, \tag{1.3}$$

where Re(z) denotes the real part of z. By analytical continuation, the definition can be extended to the entire complex plane except for the points $z = 0, -1, -2, \ldots$. It can be easily seen that

$$\Gamma(1) = \int_0^\infty e^{-t} dt = 1. \tag{1.4}$$

Moreover, using integration by parts it can be shown that

$$\Gamma(z+1) = z\Gamma(z) \quad \text{for } z \neq 0, -1, -2, \ldots. \tag{1.5}$$

By combining (1.4) and (1.5), it follows immediately that

$$\Gamma(m+1) = m! \tag{1.6}$$

for any nonnegative integer m. The gamma function is therefore an extension of the factorial function. It is illustrated in Fig. 1.5 for real values.

A series of interesting properties can be found in the literature, e.g., *Euler's reflection formula*

$$\Gamma(1-z)\Gamma(z) = \frac{\pi}{\sin(\pi z)} \tag{1.7}$$

and the *Legendre duplication formula*

$$\Gamma(z)\Gamma\left(z + \frac{1}{2}\right) = 2^{1-2z} \sqrt{\pi}\,\Gamma(2z). \tag{1.8}$$

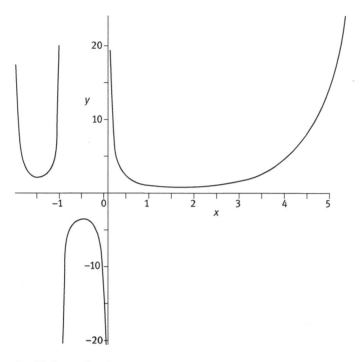

Fig. 1.5: Gamma function.

By setting $z = 1$, the latter implies in particular

$$\Gamma\left(\frac{1}{2}\right) = \sqrt{\pi}. \tag{1.9}$$

A very broad class of functions is given by the *generalized hypergeometric function*

$$_pF_q(n_1, \ldots, n_q; d_1, \ldots, d_q; z) = \sum_{k=0}^{\infty} \frac{z^k \, n_1^{(k)} \cdots \cdots n_q^{(k)}}{k! \cdot d_1^{(k)} \cdots \cdots d_q^{(k)}}, \tag{1.10}$$

where z, n_i and d_j are complex numbers and $n^{(k)}$ denotes the *Pochhammer symbol*. In the general case, the latter is defined by $n^{(k)} = \frac{\Gamma(n+k)}{\Gamma(n)}$. In particular, when n is a natural number, it holds $n^{(k)} = n \cdot (n+1) \cdots \cdots (n+k-1)$, and therefore $n^{(k)}$ is sometimes called a *rising factorial*.

An important special case is the *confluent hypergeometric function*

$$_1F_1(n; d; z) = \sum_{k=0}^{\infty} \frac{z^k \, n^{(k)}}{k! \cdot d^{(k)}} = 1 + z + \frac{z^2 n}{2! \cdot d} + \frac{z^3 n \, (n+1)}{3! \cdot d(d+1)} + \frac{z^4 n \, (n+1)(n+2)}{4! \cdot d(d+1)(d+2)} + \cdots . \tag{1.11}$$

By representing a probability density via a hypergeometric function, we access the powerful theory of these functions which has applications in diverse scientific areas.

Many elementary and special functions can be expressed as hypergeometric functions. For example, we get

$$\sin(z) = z \; _0F_1\left(-;\frac{3}{2};\frac{-z^2}{4}\right),$$

$$\text{BesselJ}(a,z) = \frac{z^a}{2^a \; \Gamma(1+a)} \; _0F_1\left(-,1+a;\frac{-z^2}{4}\right),$$

$$\text{LegendreP}(a,b,z) = \left(\frac{z+1}{z-1}\right)^{b/2} \frac{1}{\Gamma(1-b)} \; _2F_1\left(-a\,,a+1;\; 1-b;\frac{1-z}{2}\right),$$

where BesselJ denotes the Bessel function of the first kind and LegendreP the associated Legendre function of the first kind. The *incomplete gamma function* is defined by

$$\Gamma(a,z) = \Gamma(a) - \frac{z^a}{a} \; _1F_1(a;1+a;-z). \tag{1.12}$$

For Re(a) > 0, this function can also be represented by the integral

$$\Gamma(a,z) = \int_z^\infty e^{-t}t^{a-1}dt. \tag{1.13}$$

A related concept is the *exponential integral Ei(a,z)*, defined for 0 < Re(z). We limit ourselves to presenting the following definitions:

$$Ei(a,z) = z^{a-1} \; \Gamma(1-a,z), \tag{1.14}$$

$$Ei(x) = -Ei(1,-x) \text{ for } x < 0, \tag{1.15}$$

$$Ei(z) = -Ei(1,-z) + \frac{1}{2}\ln(z) + \frac{1}{2}\ln\left(\frac{1}{z}\right) - \ln(-z) \text{ for complex } z. \tag{1.16}$$

The *logarithmic derivative* of the gamma function, also known as the *Psi function*, is given by

$$\psi(z) = \frac{\partial}{\partial z}\ln\Gamma(z) = \frac{\Gamma'(z)}{\Gamma(z)}. \tag{1.17}$$

This function, illustrated in Fig. 1.6 for positive real values, has also many interesting properties, e.g.,

$$\psi(z) = -\gamma + (z-1)\sum_{k=0}^\infty \frac{1}{(z+k)\cdot(k+1)}, \tag{1.18}$$

where $\gamma \approx 0.5772$ is the *Euler constant*,

$$\psi(z+n) = \sum_{k=0}^{n-1} \frac{1}{(z+k)} + \psi(z),$$ (1.19)

where n is a natural number and

$$2\psi(2z) = \psi(z) + \psi\left(z + \frac{1}{2}\right) + 2\ln 2.$$ (1.20)

By setting $z = 1$ in (1.18) and $z = \frac{1}{2}$ in (1.20) one obtains the specific values $\psi(1) = -\gamma$ and $\psi\left(\frac{1}{2}\right) = -\gamma - 2\ln 2$.

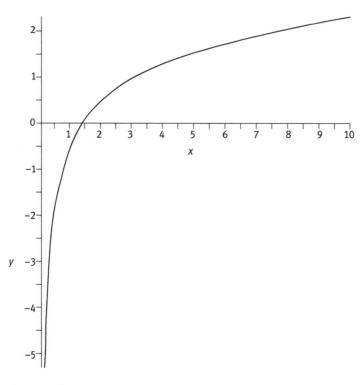

Fig. 1.6: Psi function.

If we also consider higher derivatives of $\ln \Gamma(z)$, we obtain the *polygamma function*, defined by

$$\psi(n, x) = \frac{\partial^n}{\partial x^n} \psi(x).$$ (1.21)

The *Riemann zeta function* is given by

$$\varsigma(z) = \sum_{i=1}^{\infty} \frac{1}{i^z} \quad \text{for } \text{Re}(z) > 1. \tag{1.22}$$

The *Hurwitz zeta function*, also known as generalized *Riemann zeta function*, is defined by

$$\varsigma(n, z, v) = \frac{\partial^n}{\partial z^n} \sum_{i=1}^{\infty} \frac{1}{(i+v)^z}. \tag{1.23}$$

Obviously, the ordinary Riemann zeta function is obtained for $n = v = 0$. There are several relations between the above functions, in particular

$$\psi(n-1, z) = (-1)^n (n-1)! \varsigma(0, n, z) \quad \text{for } n \geq 2. \tag{1.24}$$

Another important concept based on the gamma function is the *beta function* (see Fig. 1.7), given by

$$B(x, y) = \frac{\Gamma(x)\Gamma(y)}{\Gamma(x+y)} \quad \text{for } \text{Re}(x), \text{Re}(y) > 0, \tag{1.25}$$

which can be equivalently expressed by

$$B(x, y) = \int_0^1 t^{x-1}(1-t)^{y-1} dt. \tag{1.26}$$

Obviously it holds $B(x, y) = B(y, x)$, i.e., the graph of this bidimensional function is symmetric with respect to the plane $y = x$. A three-dimensional plot is shown in Fig. 1.7.

The *incomplete beta function* is defined by

$$I_s(x,y) = \frac{1}{B(x, y)} \int_0^s t^{x-1}(1-t)^{y-1} dt \tag{1.27}$$

for $\text{Re}(x)$, $\text{Re}(y) > 0$ and $0 \leq s \leq 1$.

The above concepts are very important for continuous probability models.

Exercises

1.3.1 Show that $\Gamma(x)$ cannot be zero.
1.3.2 Determine the maximum of $\Gamma(x)$ over the interval $]-1, 0[$ and the minimum over $]-2,-1[$.
1.3.3 Prove the relation $\Gamma\left(\frac{n}{2}\right) = \frac{2^{1-n}\sqrt{\pi}(n-1)!}{((n-1)/2)!}$ for any odd positive integer n.
1.3.4 Derive a formula for $\Gamma(-(n/2))$, where n is an odd positive integer.

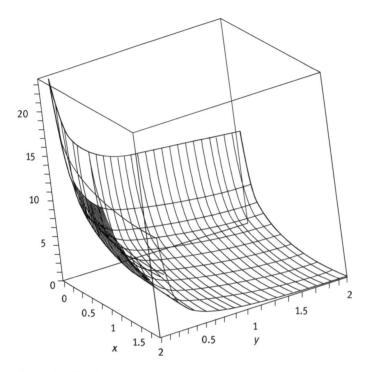

Fig. 1.7: Beta function.

1.3.5 Show that $1 + 1/2 + \cdots + 1/n = \psi(1 + n) + \gamma$ for $n \in \mathbb{N}$.

1.3.6 Show that the beta function satisfies the following relations:
    (a) $xB(x,y + 1) = yB(x + 1,y)$,
    (b) $B(x,y) = B(x + 1,y) + B(x,y + 1)$,
    (c) $(x + y) B(x,y + 1) = yB(x,y)$,
    (d) $B(x,y) B(x + y,z) = B(y,z) B(y + z,x)$.

1.3.7 Derive the following integral representations for the functions gamma and beta. Use integration by substitution:
    (a) $\int_0^1 x^a (1 - x^b)^{c-1} dx = \frac{1}{b} B\left(\frac{a+1}{b}, c\right)$ for Re$(a + 1)$, Re(b), Re$(c) > 0$,
    (b) $\int_0^\infty \frac{x^a}{(1+x)^b} dx = B(a + 1, b - a - 1)$ for Re$(a + 1)$, Re$(b{-}a{-}1) > 0$,
    (c) $\int_0^\infty x^{a-1} e^{-bx^c} dx = \frac{\Gamma(a/c)}{c\, b^{a/c}}$ for Re(a), Re(b), Re$(c) > 0$.

1.3.8 Express $B(x,y)$ by means of a binomial coefficient.

1.4 Integration of bidimensional functions

A double integral is of the form

$$\iint_R f(x,y)\,dy\,dx. \tag{1.28}$$

With regard to applications we will assume that $f(x,y)$ is a continuous function and that the region $R \subset IR^2$ of integration can be expressed as

$$R = \{(x,y)\mid a \le x \le b,\ g_1(x) \le y \le g_2(x)\} \tag{1.29}$$

(see Fig. 1.8) or as

$$R = \{(x,y)\mid c \le y \le d,\ h_1(y) \le x \le h_2(y)\} \tag{1.30}$$

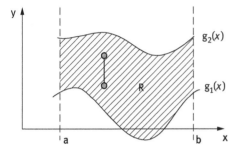

Fig. 1.8: x-Representable region.

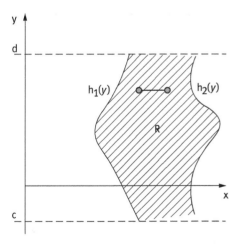

Fig. 1.9: y-Representable region.

(see Fig. 1.9). This means that the boundary of that region can be expressed by means of functions g_1, g_2 of x or functions h_1, h_2 of y. We will say that R is x-representable when it has the form (1.29) and y-representable when it has the form (1.30). Figure 1.8 illustrates that R is x-representable, if for any two points of R with equal x-components the connecting line segment is entirely contained in R. An analog relation holds when R is y-representable (see Fig. 1.9). From this follows that a convex set R is both x-representable and y-representable.

If f is a positive function, then the double integral (1.28) can be interpreted as the volume of the three-dimensional region M between the surface of $f(x,y)$ and the xy-plane (see Fig. 1.10).

If R is x-representable, the double integral (1.28) can be calculated by means of two one-dimensional integrations:

$$\iint\limits_{R} f(x,\ y)\ dy\ dx = \int\limits_{b}^{b} \int\limits_{g_1(x)}^{g_2(x)} f(x,\ y)\ dy\ dx. \tag{1.31}$$

The inner integral of the last expression is a function of x. We will denote it as $s(x) = \int_{g_1(x)}^{g_2(x)} f(x,y)dy$. One can interpret $s(x_0)$ as the area of the intersection of M and the plane $E(x_0) = \{(x,y,z)|x = x_0\}$. Therefore, the volume of M is obtained as $\int_a^b s(x)dx$.

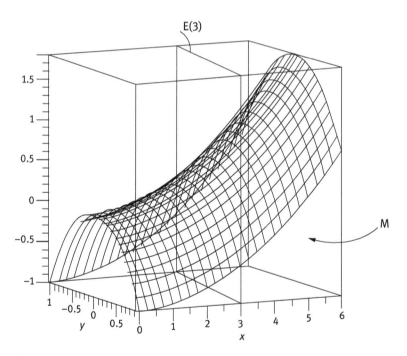

Fig. 1.10: Geometrical interpretation of the double integral for a specific function.

Example 1.5: For the double integral $\iint_R (5x^2 - 2y^3)\,dy\,dx$ with $R = \{(x,y)\mid -1 \le x \le 2,\ 0 \le y \le x^2\}$ we obtain

$$\iint_R (5x^2 - 2y^3)\,dy\,dx = \int_{-1}^{2}\int_{0}^{x^2} (5x^2 - 2y^3)\,dy\,dx = \int_{-1}^{2}\left(5x^4 - \frac{x^8}{2}\right)dx = 9/2.$$

For any y-representable integration domain we obtain the analogous formula

$$\iint_R f(x,\ y)\,dy\,dx = \int_{c}^{d}\int_{h_1(y)}^{h_2(y)} f(x,\ y)\,dx\,dy. \tag{1.32}$$

As already mentioned, a convex domain R is both x-representable and y-representable. In this case, both formulas (1.28) and (1.32) can be applied, yielding of course the same result.

Example 1.6: We calculate the double integral $V = \iint_R (x^2 + 3xy)\,dy\,dx$ with $R = \{(x,y)\mid -1 \le x \le 1,\ x^2 \le y \le 1\}$; see Fig. 1.11(a). The convex domain R may also be presented as $R = \{(x,y)\mid 0 \le y \le 1,\ -\sqrt{y} \le x \le \sqrt{y}\}$; see Fig. 1.11(b)).

From (1.31) we obtain

$$V = \int_{-1}^{1}\int_{x^2}^{1} (x^2 + 3xy)\,dy\,dx = \int_{-1}^{1}\left(-\frac{3}{2}x^5 - x^4 + x^2 + \frac{3}{2}x\right)dx = \frac{4}{15},$$

and using (1.32) yields

$$V = \int_{0}^{1}\int_{-\sqrt{y}}^{\sqrt{y}} (x^2 + 3xy)\,dx\,dy = \int_{0}^{1}\frac{2}{3}y^{3/2}\,dy = \frac{4}{15}.$$

One can easily check that the function $f(x,y)$ in Example 1.6 is not positive, in particular $f(-1,1) = -2$. But the formulas (1.31) and (1.32) also apply in this case. However, in the geometrical interpretation of the double integral, one must bear in mind that the area below the xy-plane has a negative sign. Whenever both formulas can be used, the reader should check in advance, which of them can be calculated more easily.

A very useful statement for the calculation of double integrals is the following substitution rule.

Theorem 1.1: Assume that the variables x, y of f can be expressed by new variables u and v such that any pair (x,y) corresponds biuniquely to a pair (u,v), where

$$x = x(u, v), \quad y = y(u, v) \quad \text{and} \quad u = u(x,y), \quad v = v(x,y) \tag{1.33}$$

are continuously differentiable functions. Then it holds

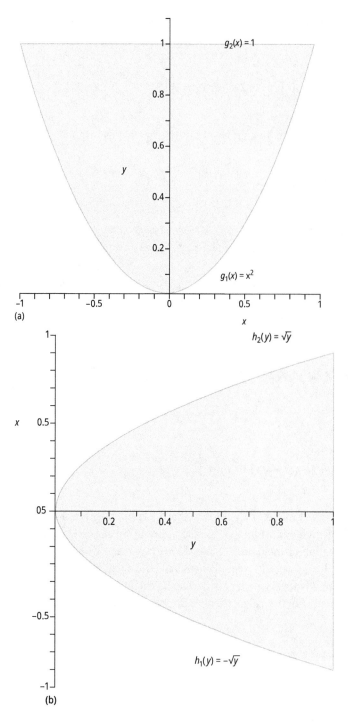

Fig. 1.11: Alternative representations of the integration domain.

$$\iint_R f(x, y)\, dy\, dx = \iint_G f(x(u, v), y(u, v))\, |J|\, dv\, du, \qquad (1.34)$$

where G is the integration domain in terms of u and v and J is the Jacobian of the transformation (1.30), i.e., the determinant

$$J = \begin{vmatrix} \dfrac{\partial x}{\partial u} & \dfrac{\partial x}{\partial v} \\[2ex] \dfrac{\partial y}{\partial u} & \dfrac{\partial y}{\partial v} \end{vmatrix}.$$

We point out that $|J|$ in (1.34) denotes the absolute value of the determinant J.

Example 1.7: Consider the double integral $\iint_R (x + 2xy)\, dy\, dx$ where $R = \{(x,y)\mid 0 \le x \le 1, 0 \le y \le 1-x\}$, see Fig. 1.12. From (1.31) we get

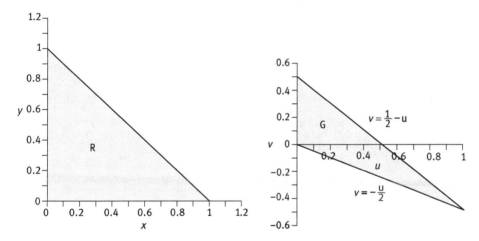

Fig. 1.12: Transformation of the integration domain.

$$\iint_R (x + 2xy)\, dy\, dx = \int_0^1 \int_0^{1-x} (x + 2xy)\, dy\, dx = \int_0^1 (x^3 - 3x^2 + 2x)\, dx = \frac{1}{4}.$$

Consider the linear transformation of variables

$$\begin{pmatrix} x \\ y \end{pmatrix} = \begin{pmatrix} 1 & 2 \\ 1 & 0 \end{pmatrix} \begin{pmatrix} u \\ v \end{pmatrix}$$

with inverse function

$$\begin{pmatrix} u \\ v \end{pmatrix} = \begin{pmatrix} 0 & 1 \\ 1/2 & -1/2 \end{pmatrix} \begin{pmatrix} x \\ y \end{pmatrix},$$

mapping the region R into G (Fig. 1.12). In particular, the vertices $\begin{pmatrix} 0 \\ 0 \end{pmatrix}$, $\begin{pmatrix} 1 \\ 0 \end{pmatrix}$ and $\begin{pmatrix} 0 \\ 1 \end{pmatrix}$ of R are mapped to the vertices $\begin{pmatrix} 0 \\ 0 \end{pmatrix}$, $\begin{pmatrix} 0 \\ 1/2 \end{pmatrix}$ and $\begin{pmatrix} 1 \\ -1/2 \end{pmatrix}$ of G. The Jacobian is $J = \begin{vmatrix} 1 & 2 \\ 1 & 0 \end{vmatrix} = -2$. The integral over the variables u and v is now

$$\iint_G f(x(u,\,v),y(u,\,v))\,|J|\,dv\,du = \int_0^1 \int_{-u/2}^{1/2-u} (u + 2v + 2(u+2v)u)| - 2|\,dv\,du$$

$$= 2 \int_0^1 \int_{-u/2}^{1/2-u} (u + 2v + 2u^2 + 4uv)\,dv\,du = 2 \int_0^1 \left(\frac{1}{4} - \frac{3}{4}u^2 + \frac{1}{2}u^3 \right) du = \frac{1}{4},$$

which is identical to the previous calculation.

Of special importance is the substitution of Cartesian coordinates by polar coordinates:

Example 1.8: If r and Φ denote radius and angle, relation (1.33) receives the form $x = r \cos\Phi$, $y = r \sin\Phi$ and $r = \sqrt{x^2 + y^2}$, $\Phi = \arctan(y/x)$ and the Jacobian is

$$J = \begin{vmatrix} \frac{\partial x}{\partial r} & \frac{\partial x}{\partial \Phi} \\ \frac{\partial y}{\partial r} & \frac{\partial y}{\partial \Phi} \end{vmatrix} = \begin{vmatrix} \cos\Phi & -r\sin\Phi \\ \sin\Phi & r\cos\Phi \end{vmatrix} = r,$$

yielding the following important special case of (1.34):

$$\iint_R f(x,y)\,dy\,dx = \iint_G f(r\cos\Phi, r\sin\Phi)\,r\,d\Phi\,dr. \tag{1.35}$$

We still consider the following numerical example in which R is the unit circle, i.e., $R = \{(x,y)|\ 0 \le x^2 + y^2 \le 1\}$. Then for $f(x, y) = x(x + y)$ relatively complex computations for the left side of (1.35) yield

$$V = \int_{-1}^1 \int_{-\sqrt{1-x^2}}^{\sqrt{1-x^2}} (x^2 + xy)\,dy\,dx = \int_{-1}^1 2x^2\sqrt{1-x^2}\,dx = \frac{\pi}{4}.$$

The right side of (1.35) is

$$V = \iint_G r^3 \cos \Phi (\cos \Phi + \sin \Phi) d\Phi dr,$$

where $G = \{(r, \Phi)| \ 0 \le r \le 1, 0 \le \Phi \le 2\pi\}$ is a simple rectangular region, implying

$$V = \int_0^1 r^3 \int_0^{2\pi} (\cos^2\Phi + \cos\Phi\sin\Phi) d\Phi dr = \int_0^1 r^3 \pi \, dr = \frac{\pi}{4}.$$

These calculations are considerably easier.

Exercises

1.4.1 (a) Calculate the double integral $\iint_R x \ln(xy) dy \, dx$ over the square $R = \{(x, y)|0 \le x, \ y \le 1\}$.

(b) Determine the inner integrals arising when (1.31) and (1.32) are applied.

1.4.2 Calculate $\iint_R x^2 y \, e^{xy} dy dx$

(make use of a suitable software, if possible)

(a) for $R = \{(x,y) \ | \ 0 \le x \le 2, -x \le y \le x\}$.

(b) for $R = \{(x,y) \ | \ 0 \le x \le \pi, 0 \le y \le \sin(x)\}$.

(c) Illustrate the integration domains geometrically.

1.4.3 (a) Calculate $\iint_R 3ye^x dy \, dx$ by means of (1.31), where R is the square with vertices

$$\begin{pmatrix} 0 \\ 0 \end{pmatrix}, \begin{pmatrix} 1 \\ 1 \end{pmatrix}, \begin{pmatrix} 2 \\ 0 \end{pmatrix}, \begin{pmatrix} 1 \\ -1 \end{pmatrix}.$$

(b) Consider the quadratic transformation $x = u^2 + v$, $y = -u^2 + v$. Formulate the transformed integral in (1.31) and calculate it. What are the region G and the Jacobian J?

1.4.4 Calculate the integral $\iint_R (x^2 + 5xy^2) dy \, dx$ over the ellipse $R = \{(x, y)| \ x^2 + 2y^2 \le 2\}$.

(a) Calculate it directly, using one of the formulas (1.31) or (1.32).

(b) Introduce polar coordinates and use Theorem 1.1. Determine the region G and illustrate it geometrically.

1.4.5 Calculate the integral

$$\int_8^{10} \int_2^5 y^{3/2} \ln(x^2) dy \, dx.$$

1.4.6 Calculate

$$\int\limits_{a}^{b}\int\limits_{c}^{d} y^3 x^2 dy \; dx.$$

1.4.7 Calculate

$$\int\limits_{a}^{b}\int\limits_{c}^{d} \sqrt{y}\ln(x)dy \; dx \quad \text{for } a, b, c, d > 0.$$

1.4.8 Derive the formula for the volume of the ellipsoid. Use the equation

$$\frac{x^2}{a^2} + \frac{y^2}{b^2} + \frac{z^2}{c^2} + \leq 1.$$

Express z as a function of x and y and integrate this bidimensional function.

1.4.9 Calculate

$$\iint\limits_{E} \left(x^2 + y^2\right) dy \; dx,$$

where E is the ellipse given by

$$\frac{x^2}{a^2} + \frac{y^2}{b^2} \leq 1.$$

Determine the integral
(a) directly using Cartesian coordinates, or
(b) using polar coordinates.

2 Introduction to probability

2.1 Mathematical models

At first we illustrate the type of phenomenon which will be the subject of all further investigations of this book and lay the foundations for developing appropriate models. Basically, mathematical models can be subdivided into *deterministic* and *nondeterministic* models. The former is characterized by the fact that the conditions under which an experiment is performed determine the outcome. For example, if we insert a power source into a simple electrical circuit, the flow of current will be given by Ohm's law $I = E/R$. Keeping the voltage E and the resistance R unchanged, the experiment has the same outcome for every repetition of the experiment (or at least possible deviations are negligible). For many real-life situations, deterministic models are sufficient; for example, the gravitational law describes precisely the movement of a falling object in a vacuum, and Kepler's laws give us the motion of planets. However, there are many other situations that require a nondeterministic model, also known as *probabilistic, stochastic* or *random model*. Suppose that a piece of a radioactive material is emitting α-particles. The exact number of particles emitted in a certain interval of time cannot be predicted, even if all characteristics of the source material like shape, mass and chemical composition were known. Thus the conditions under which the experiment is carried out do not determine the result. Other phenomena that require nondeterministic models are meteorological observations. For example, the exact rainfall in a given area during a certain period cannot be accurately predicted, even if comprehensive meteorological data like barometric pressure and wind velocity in several locations are known. From now on, we restrict ourselves to the study of nondeterministic phenomena.

2.2 Further examples of random experiments

The following terms are of particular importance.

Definition 2.1: An experiment is called a *random experiment*, if the result is unknown but the whole of possible outcomes is known.

For a given random experiment E, the *sample space*, denoted as S, is the set of possible outcomes of E. Any subset of S is called an *event*. An event with a single element is called an *elementary event*.

Of course, a nondeterministic experiment in the broadest sense may also yield unexpected or unknown outcomes, but this situation will be excluded from consideration, since no probabilistic modeling is possible under these circumstances. In Tab. 2.1, the above concepts are illustrated by means of diverse examples.

https://doi.org/10.1515/9783111332277-002

Tab. 2.1: Examples of random experiments.

No.	Experiment	Observation/result	Sample space	Example of event	
				Formal	**In words**
E_1	A die is tossed	Number on the top	{1,2,3,4,5,6}	{1,3,5}	Number is odd
E_2	A coin is tossed three times	Sequence of heads and tails obtained	{HHH,HHT,HTH, THH,HTT,THT,TTH, TTT}	{HHT,HTH,THH}	Exactly two heads are observed
E_3	A coin is tossed three times	Number of heads obtained	{0,1,2,3}	{0,1}	Less than two heads are observed
E_4	Given an urn with identical balls, numbered by 1, .., n, select k balls randomly at once	Set of balls drawn from the urn	Set of k-subsets of {1, .., n}	set of k-subsets of {2, .., n}	The selection of balls does not contain the ball with number 1
E_5	Items are produced until the 10th perfect item is obtained	Total number of items produced	{10,11,12, .. .}	{10, .. ., 15}	At most 15 items are produced
E_6	A light bulb is inserted into a socket	Time elapsed until the bulb burns out (in hours)	[0,200] or [0, ∞[	[120,180]	Lifetime is between 120 and 180 h
E_7	A long-distance bus crosses Europe	Coordinates (x,y) at a certain time	IR^2	[1,000,2,000] × [5,000,7,000]	The bus is located within a particular rectangle
E_8	A thermograph records the temperature continuously over a time period [0,t]	Change of temperature during the period [0,t] (in degree Celsius)	{f: [0,t] → IR, f continuous}	{f: [0,t] → IR, f continuous, f(x) ≤ 20 for $0 \leq x \leq t$}	The temperature was 20 °C at most

The experiment E_1 is one of the simplest examples. The outcomes are obviously 1, . . ., 6. Thus, the sample space is {1, . . ., 6}. Any subset of S represents an event for this random experiment. An event can be described formally in set-theoretic terms but also in words (which is important for the communication with users of probability theory from various applied areas). There are often several formulations possible, for example, the event E_3 in Tab. 2.1 can be described by "less than two heads" or by "at most one head."

A comparison between E_2 and E_3 shows that for a complete description of a random experiment, it is necessary to define the observation clearly. In spite of the simplicity of Definition 2.1, some caution is necessary for a correct modeling of the sample space. Note that, e.g., in case of E_2, the outcomes are sequences of the symbols H and T and *not* simply H and T. The observation of E_5 is the total number of items produced, which is of course at least 10. In practice, there are often several possibilities to define the sample space. For example, in the case of E_6, the set of possible outcomes can be modeled as [0, ∞[or as [0,200] if it is (almost) impossible that the light bulb works for more than 200 h. There are infinite results related with conducting the experiments E_7 and E_8. The respective sample spaces have dimensions 2 and ∞.

It can be easily checked that the sample spaces are finite for experiments E_1, . . ., E_4. The random experiment E_5 has a countably infinite sample space and E_6, . . ., E_8 have uncountably infinite sample spaces. According to these characteristics, we will design different types of probabilistic models in the further course of the text. In principle, all the random experiments considered so far can be repeated. Considering a few realizations of an experiment, the individual outcomes may seem to appear in a haphazard manner. However, as an experiment is repeated a *large* number of times, a definite regularity appears. This regularity allows us to construct a precise mathematical model to analyze the experiment.

2.3 Assigning probabilities to events

The modern probability theory has its roots in attempts to analyze games of chance in the sixteenth and seventeenth centuries. The history is closely associated with names like Fermat, Pascal and Huygens. Exact probability theory makes use of a very complex formalism; however, we will restrict ourselves to the essential concepts that are sufficient to describe most situations occurring in real life. An intuitive approach to associate probabilities to events is to make use of "symmetry" in a broad sense. It is reasonable to assume that head and tail appear "equally likely" when a fair coin is tossed. So it makes sense to assign the same probability ½ to both of the possible outcomes. With a similar reasoning one can assign the probability 1/6 to any of the results 1,. . ., 6, when a fair die is tossed (see Fig. 2.1(a)). If one selects randomly one of n identical balls of an urn, it is useful to assign the probability $1/n$ to each possible outcome (Fig. 2.1(b)). Similarly, when a pointer is set in motion around the center of a smooth circular surface, one can assume that it stops moving with equal probability in any of the eight sectors (Fig. 2.1(c)). Of course, the same reasoning applies for any other number of sectors of the same size.

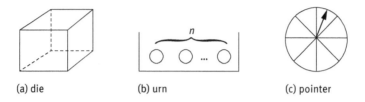

(a) die   (b) urn   (c) pointer

Fig. 2.1: The symmetry approach.

The symmetry approach is important for specific situations and we will apply it soon (see Section 3.1); however, it is obvious that the approach works only under ideal conditions, rarely encountered in practice. If the die is not a perfect cube, if the balls in the urns differ slightly in weight or size or if the circular surface is not uniformly smooth, the method does not apply. Moreover, in defining, e.g., probabilities in various other areas like meteorology, finances or economics, the considered approach is in general completely inappropriate.

Another idea of assigning probabilities – the so-called *frequentist approach* – is based on the following definition.

Definition 2.2: Given the random experiment E and an event A, associated with E, if E is realized n times, then the *relative frequency* of A is defined as $f_A = n_A/n$, where n_A is the number of occurrences of A.

Example 2.1: Suppose that in constructing a certain product we distinguish between the results "perfect (p)," "with minor defects (d)" and "with major defects (D)." Consider the hypothetical relative frequencies in Tab. 2.2, associated with different production quantities.

Tab. 2.2: Relative frequencies.

Number of items produces, n	Perfect items	Items with minor defects	Items with major defects	Relative frequency of major defects
100	65	20	15	0.15
1,000	680	209	111	0.111
10,000	6,952	2,153	895	0.0895
100,000	69,708	19,802	10,490	0.1049

Studying the last column of Tab. 2.2, we can observe that the relative frequency of items with major defects tends to "stabilize" near some definite value in turn of 0.1. We emphasize that the data are merely hypothetical. But they illustrate a phenomenon which is in fact observed in practice: As the number of repetitions of an experiment increases, the relative frequency of some event "converges" in some intuitive sense to a certain value. This characteristic is known as *statistical regularity*.

We now study some further properties of the relative frequency.

Definition 2.3: Two events A and B are called *mutually exclusive*, if $A \cap B = \Phi$, i.e., A and B cannot occur together.

For instance, the events "the product is perfect" and "the product has a defect" in Example 2.1 are mutually exclusive.

One can easily verify that relative frequencies (Definition 2.2) have the following properties:

– $0 \le f_A \le 1$ for any event A.
– $f_A = 1$ if and only if A occurs in all realizations of the random experiment.
– $f_A = 0$ if and only if A never occurs in the realizations of the random experiment.
– If A and B are mutually exclusive events, then $f_{A \cup B} = f_A + f_B$.

The statistical regularity suggests to define the probability of an event as the relative frequency f_A, obtained for a "large" number of repetitions of the experiment. However, there are two objections to this idea. At first, it is not clear how many realizations of the random experiment are necessary to be sure that the relative frequency is a good approximation for the probability. Secondly, the relative frequency may depend on chance or on the experimenter. What we need is a number without having to resort to experimentation.

2.4 Basic notions of probability

The fundamental idea is to derive a measure for probability, "by abstracting" the concept of relative frequency in a certain sense, e.g., we define a probability measure as a quantity that has essentially the properties of the relative frequency.

Definition 2.4: Let E be a random experiment with sample space S. To each event A, we associate a real number $P(A)$, called the *probability* of A, satisfying the following properties:

(1) $0 \le P(A) \le 1$.
(2) $P(S) = 1$.
(3) If A and B are mutually exclusive events, then $P(A \cup B) = P(A) + P(B)$.
(4) If $A_1, A_2, \ldots$ are pairwise mutually exclusive events, then $P\left(\bigcup_{i=1}^{\infty} A_i\right) = \sum_{i=1}^{\infty} P(A_i)$.

Note that condition (4) does not directly follow from (3) but is included among the axioms of an "idealized" sample space. Though Definition 2.4 does not yet enable us to compute probabilities (which will be done in the next chapter), the axioms have useful consequences, formulated in the following as theorems.

Theorem 2.1: If Φ is the empty set, then $P(\Phi) = 0$.

Proof: For any event A, property (2) implies
$P(A) = P(A \cup \Phi) = P(A) + P(\Phi)$, thus $P(\Phi) = 0$.

The converse of this statement is not true. We will see later that $P(A) = 0$ does not imply that A is empty.

Theorem 2.2: If $\bar{A}$ is the complement of A, then $P(A) = 1 - P(\bar{A})$.

Proof: The sample space S can be written as $S = A \cup \bar{A}$, Thus, the properties (2) and (3) imply $1 = P(S) = P(A \cup \bar{A}) = P(A) + P(\bar{A})$ from which the statement follows immediately.

This statement is very intuitive. If a taxi driver has an accident during the day with probability 0.01, he has no accident with probability 0.99. This result is very useful, since in many problems it is easier to compute the probability of the complement than that of the given event.

Theorem 2.3: If A and B are events with $A \subset B$, then $P(A) \leq P(B)$.

Proof: The event B can be written as $B = A \cup (B \backslash A)$, which is a union of mutually exclusive events (see Fig. 2.2). Thus, property (3) implies $P(B) = P(A \cup (B \backslash A)) = P(A) + P(B \backslash A)$. Since $P(B \backslash A)$ is nonnegative by property (1), it follows the theorem.

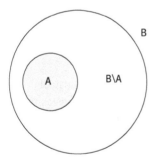

Fig. 2.2: Partition of event B.

Theorem 2.4: If A and B are *any* two events, then

$$P(A \cup B) = P(A) + P(B) - P(A \cap B)$$

Proof: We decompose both $A \cup B$ and B into mutually exclusive events, yielding (see Fig. 2.3)

$$A \cup B = A \cup (B \backslash A),$$
$$B = (A \cap B) \cup (B \backslash A).$$

Thus,

$$P(A \cup B) = P(A) + P(B \backslash A),$$
$$P(B) = P(A \cap B) + P(B \backslash A).$$

Subtracting the second equation from the first yields

$$P(A \cup B) - P(B) = P(A) - P(A \cap B),$$

implying the theorem.

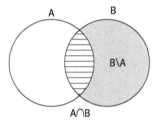

Fig. 2.3: Partition of $A \cup B$.

Theorem 2.4 is an obvious generalization of the above property (3). We now generalize the statement for an arbitrary number of events. For illustrative purposes, we first
restrict ourselves to three events.

Theorem 2.5: If A, B and C are any three events, then

$$P(A \cup B \cup C) = P(A) + P(B) + P(C) - P(A \cap B) - P(A \cap C) - P(B \cap C) + P(A \cap B \cap C).$$

Proof: To make the previous theorem applicable, we write $A \cup B \cup C$ as a union of *two*
events, yielding

$$P(A \cup B \cup C) = P((A \cup B) \cup C) = P(A \cup B) + P(C) - P((A \cup B) \cap C)$$
$$= P(A) + P(B) - P(A \cap B) + P(C) - P((A \cup B) \cap C). \tag{2.1}$$

Applying the law (vi) of Section 1.1 to the last term, we get

$$P((A \cup B) \cap C) = P((A \cap C) \cup (B \cap C)) = P(A \cap C) + P(B \cap C) - P(A \cap B \cap C). \tag{2.2}$$

Substituting the last term of (2.1) for the right side of (2.2) completes the proof.

Example 2.2: Assume that a person is selected randomly from a population. We consider the events

A = {the person is a foreigner},
B = {the person is poor},
C = {the person has a certain disease}.

Given the probabilities $P(A) = 0.3$, $P(B) = 0.4$, $P(C) = 0.35$, $P(A \cap B) = 0.11$, $P(A \cap C) = 0.15$, $P(B \cap C) = 0.1$ and $P(A \cap B \cap C) = 0.05$, from Theorem 2.5, we obtain $P(A \cup B \cup C) = 0.3 + 0.4 + 0.35 - 0.11 - 0.15 - 0.1 + 0.05 = 0.74$. For the specific case with three events, it might be more illustrative to calculate the probabilities of the $2^3 = 8$ regions $A \cap B \cap C$, $A \cap B \cap \bar{C}$, $A \cap \bar{B} \cap C$, ..., $\bar{A} \cap \bar{B} \cap \bar{C}$, in a Venn diagram (see Fig. 2.4) and add the "inner probabilities" to determine $P(A \cup B \cup C)$. For example, the region (event) given by $A \cap B \cap \bar{C} = (A \cap B) \setminus (A \cap B \cap C)$ has the probability $P(A \cap B) - P(A \cap B \cap C) = 0.11 - 0.05 = 0.06$. We obtain $P(A \cup B \cup C) = 0.09 + 0.06 + 0.05 + 0.1 + 0.24 + 0.05 + 0.15 = 0.74$.

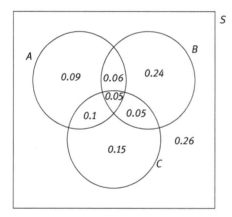

Fig. 2.4: Venn diagram.

By using mathematical induction, one can prove the following general result, known as *inclusion-exclusion principle*.

Theorem 2.6: For any events $A_1, A_2, \ldots, A_k$ it holds

$$P(A_1 \cup A_2 \cup \ldots \cup A_k) = \sum_{i=1}^{k} P(A_i) - \sum_{1 \le i < j \le k} P(A_i \cap A_j)$$
$$+ \sum_{1 \le i < j < l \le k} P(A_i \cap A_j \cap A_l) + \ldots + (-1)^{k-1} P(A_1 \cap A_2 \cap \ldots \cap A_k).$$

For the sake of completeness, we mention that in addition to the above-studied approaches based on symmetry or frequencies, there exists another important method to assign probabilities to events. In the so-called *Bayesian approach*, the probability is closely related to the information we have about an event. It represents the "state of knowledge" which is updated when new relevant data are obtained. The concept of probability is also extended to cover degrees of certainty about statements.

Exercises

2.1 Describe the following events by means of words, where the event A_i is related to the experiment E_i in Tab. 2.1: $A_1 = \{4,5,6\}$, $A_2 = \{HHH,HHT,HTH, THH\}$, $A_3 = \{1,3\}$, $A_5 = \{20,21,22, \ldots \}$, $A_6 = [10,100]$, $A_7 = [0,1000] \times [0,2000]$, $A_8 = \{f\colon [0,t] \to IR \mid f$ continuous, $f(x) \le f(y)$ for $0 \le x \le y \le t\}$.

2.2 The weights of a certain group of persons have been determined (in kg). Consider the sample space $[0, \infty[$ and the events $A = [60,100]$, $B = [80,120]$, $C = [110,150]$. Describe verbally the events $A \cap B, B \cap C, A \cap B \cap C, A \cup C, A \cup B \cup C, C \backslash B$.

2.3 An urn contains 6 red, 10 blue and 8 green balls. The balls are drawn at random, one after the other and without replacement, until the last green ball is encountered. The observation is the number of balls, chosen from the urn. What is the sample space?

2.4 Assume that a random permutation with the elements $1, \ldots, 4$ is determined. Consider the events $A = \{1$ is in the first position$\}$ and $B = \{2$ is in the second position$\}$. List all the elements of the sample space. List the elements of the events $A \cap B$ and $B \backslash A$.

2.5 Let A, B and C be three events, associated with a random experiment. Express the following verbal statements in set notations, using Venn diagrams:
(i)   At least one of the events occurs.
(ii) Exactly two of the events occur.
(iii) At least one of the events A and B occurs, while C does not occur.
(iv) At least one of the events A, B, C does not occur.

2.6 Describe the sample space in set notation and determine the number of outcomes.
(i)   A die is tossed two times.
      Two out of five fishes in an aquarium are randomly selected, one after the other
(ii) without replacement and
(iii) with replacement.

2.7 Consider the experiment in part (i) of the previous exercise. List the elements of the following events:
(i)   The sum of the results is 10.
(ii) The sum of the results is at most 5.
(iii) The sum of the results is 13.

2.8 A family is randomly chosen from a certain district, and the number of male and female children is observed. Model the sample space. Express the following event formally: A family has at most three male children and less than two female children.

2.9 The (simplified) patient record of a physician contains data of four characteristics: A, age; W, weight; B, blood pressure; and G, glucose values. Model the sample space for the random selection of a patient, where the mentioned characteristics are observed. Assume that a person is classified as a high-risk patient, if he is at least 70 years old, weighting at least 100 kg, and at least one of the conditions $B \geq 170$ and $G \geq 160$ is satisfied. Describe formally the event that a randomly chosen patient from the record is a high-risk patient.

2.10 Any of two neighboring branches of a bank has six cash machines. At a specific time, the respective numbers of machines in operation are observed.
(i) Determine the sample space.
(ii) List the elements of the following events:
(a) In the first branch are less machines in use than in the second.
(b) In each branch at most two machines are in use.
(c) In each branch at least one machine is in use.
(iii) Describe the complement of the event (c) verbally.

2.11 A social worker reports that the probability to encounter a foreigner in a certain community is 80%, the probability to encounter a person with a certain disease is 70% and a foreigner with that disease is encountered with probability 40%. Is there a reason to doubt the statement?

2.12 Let A, B and C denote events such that $P(A) = P(B) = P(C) = 1/5$, $P(A \cap B) = P(B \cap C) = 0$, $P(A \cap C) = 1/10$.
Calculate the probability that none of the events A, B, C occurs.

2.13 Let A and B denote events such that $P(A) = x$, $P(B) = y$ and $P(A \cap B) = z$. Express the following probabilities in terms of x, y and z:
(a) $P(\bar{A} \cup \bar{B})$ (b) $P(A \cap \vec{B})$.
(Use Venn diagrams.)

2.14 Is there a geometric approach to assign probabilities to the outcomes, when a "die" is not a cube but an irregular convex hexahedron?

2.15 Prove that $P(A_1 \cup A_2 \cup \cdots \cup A_n) \leq P(A_1) + \cdots + P(A_n)$ holds for any events $A_1, A_2, \ldots, A_n$. The relation is known as *Boole's inequality*.

2.16 Show the relations $P(\bar{A} \cap B) = P(B) - P(A \cap B)$ and $P(\bar{A} \cap \bar{B}) = 1 - P(A) - P(B) + P(\cap B)$

2.17 Let A, B and C be three events such that $P(A) = 0.5$, $P(A \cap B) = 0.25$, $P(A \cap C) = 0.2$ and $P(A \cap B \cap C) = 0.15$. What is the probability of the event "A occurs and B and C do not occur"?

2.18 Write the formula in Theorem 2.6 for the cases $k = 4$ and $k = 5$.

3 Finite sample spaces

In this chapter, we study exclusively experiments, for which the sample space has a finite number of elements (see the experiments $E_1, \ldots, E_4$ in Tab. 2.1). In this case, the sample space can be written as $S = \{a_1, \ldots, a_k\}$. We associate probabilities to events in a very intuitive manner. To each elementary event $\{a_i\}$, we assign a probability p_i such that

$$
\begin{array}{ll}
\text{(a)} & p_i \geq 0 \text{ for } i = 1, \ldots, k, \\
\text{(b)} & p_1 + \cdots + p_k = 1.
\end{array}
\tag{3.1}
$$

For the general event $A = \{a_{j_1}, a_{j_2}, \ldots, a_{j_r}\}$, where $j_1, \ldots, j_r$ represent any r indices with $1 \leq j_1 \leq \cdots \leq j_r \leq k$ we define the probability as $P(A) = p_{j_1} + \cdots + p_{j_r}$. It can be easily checked that this definition satisfies the axioms (1)–(4) in Definition 2.4.

Example 3.1: A hypothetical city in South America is composed of 25% Argentines, 15% of Brazilians, 30% of Columbians, 15% of Spanish, 8% of French and 7% of Portuguese. We now select a person at random from the city. The sample space can be modeled as $S = \{a, b, c, s, f, p\}$, where the elements correspond to the six possible nationalities. It is reasonable to define the probabilities of the outcomes (elementary events) by the corresponding proportions in the city, i.e., $p_a = 0.25$, $p_b = 0.15$, $p_c = 0.3$, $p_s = 0.15$, $p_f = 0.08$ and $p_p = 0.07$. These probabilities are motivated by the urn model in Fig. 2.1(b). As we assumed that all the balls of the urn are selected with equal probabilities, we may assume that all persons of the city are selected with equal probabilities.

We consider the events $A = \{$the person is South American$\}$ and $B = \{$the person is native Portuguese speaker$\}$, which are formally expressed as $A = \{a,b,c\}$ and $B = \{b,p\}$. The probabilities of these events are therefore $P(A) = 0.25 + 0.15 + 0.3 = 0.7$ and $P(B) = 0.15 + 0.07 = 0.22$.

3.1 Equally likely outcomes

We now study the important specific case of the above model in which all outcomes occur with the same probability. Then it follows immediately from eq. (3.1) that $p_i = 1/k$ for all $i = 1, \ldots, k$. In consequence, any event $A = \{a_{j_1}, a_{j_2}, \ldots, a_{j_r}\}$ with r elements has the probability $P(A) = p_{j_1} + \cdots + p_{j_r} = r/k$. Denoting by $|A|$ the number of elements of an event or set A, we obtain the simple and intuitive formula

$$
P(A) = \frac{|A|}{|S|}.
\tag{3.2}
$$

https://doi.org/10.1515/9783111332277-003

In many situations, the model with equally likely outcomes is applicable. We present several examples in the following.

Example 3.2:
(a) A coin is tossed three times. What is the probability of observing exactly two heads? The sample space and the event are given by $S = \{$HHH,HHT,HTH,THH, HTT,THT,TTH,TTT$\}$ and $A = \{$HHT,HTH,THH $\}$. Hence, $P(A) = \frac{|A|}{|S|} = \frac{3}{8}$.
(b) An urn contains 30 white balls and 70 black balls. If two balls are drawn randomly at once, what is the probability that both are white?

The sample space S consists of all 2-subsets of the 100 balls, thus $|S| = \binom{100}{2}$,

and the event A consists of all the 2-subsets of the 30 white balls, thus $|A| = \binom{30}{2}$

(see Proposition 1.2). The probability of the considered event is therefore
$$P(A) = \binom{30}{2} / \binom{100}{2} = \frac{29}{330} \approx 0.0879.$$

Example 3.3: Given an urn, containing 8 white balls numbered by 1, . . ., 8 and 41 black balls numbered by 9, . . ., 49. We select 6 balls randomly at once. Determine the probability of the following events:
(a) The balls with numbers 3, 7, 12, 15, 29 and 49 are selected.
(b) The selection is composed of two white balls and four black balls.

Solutions:
(a) The event A consists of a unique outcome, hence $|A| = 1$. The sample space S consists of the 6-subsets of $\{1, . . ., 49\}$; thus, $|S| = \binom{49}{6} = 13{,}983{,}816$. Therefore, the event has probability $P(A) = 1/13{,}983{,}816$ and every other outcome has the same probability.
(b) The sample space S consists of the 6-subsets of $\{1, . . ., 49\}$, composed by 2 white balls and 4 black balls. There are $\binom{8}{2}$ ways to select the white balls and $\binom{41}{4}$ ways to select the black balls. Hence, the event has $|A| = \binom{8}{2}\binom{41}{4}$ elements all of which are obtained by combining any 2-set of the 8 white balls with any 4-set of the 41 black balls. The sample space is the same as in part (a), yielding $P(A) = \dfrac{\binom{8}{2} \cdot \binom{41}{4}}{\binom{49}{6}} = \dfrac{28 \cdot 101{,}270}{13{,}983{,}816} \approx 0.2028.$

We will later see that we have calculated probabilities of the hypergeometric distribution (see Section 9.5). The following example is of the same form as the previous one; however it is presented to demonstrate that the probabilistic applications of this section are by no means restricted to games of change and the like.

Example 3.4: To introduce first ideas related to the testing of hypotheses, we consider the following hypothetical situation:

The department of a university is composed of 8 women and 10 men. The university administration informs that four persons who will receive a special Christmas bonus have been randomly selected from the 18 department members. It turns out that only men have been selected and the women's representative complains, arguing that women are discriminated because this result could not have occurred in a true random selection. Is the complaint justified?

In order to decide this, we calculate the probability of the event A = {the selection contains only men} in case of a random selection.

Similar to the previous example, we obtain the probability

$$P(A) = \frac{\binom{8}{0} \cdot \binom{10}{4}}{\binom{18}{4}} = \frac{1 \cdot 210}{3{,}060} \approx 0.0686.$$

At a significance level of $\alpha = 5\%$, the result is not "improbable," since $P(A) > \alpha$. This means that the hypothesis

H_0: "the department members have been randomly selected" cannot be rejected.

Assume now that all selected persons are women. Would there be a reason to complain? Consider the event B = {the selection contains only women}. In perfect analogy to the previous calculation, we obtain

$$P(B) = \frac{\binom{8}{4} \cdot \binom{10}{0}}{\binom{18}{4}} = \frac{70 \cdot 1}{3{,}060} \approx 0.0228 < 0.05$$

in the case of a random selection. Now in fact there is a reason to reject the hypothesis H_0 that the persons have been randomly selected.

Example 3.5: An urn contains five red balls, four blue balls and six green balls (Fig. 3.1). We select five balls randomly at once. Determine the probabilities of the following events:

(a) A = {the selection contains two red, two blue and one green ball}.
(b) B = {all selected balls have the same color}.
(c) C = {the selection contains at least one ball of each color}.

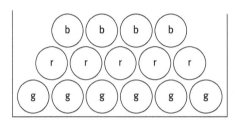

Fig. 3.1: Urn with balls of different colors.

Solutions:

(a) Similar to the previous examples, the sample space S consists of all 5-subsets of
the set of 15 balls in the urn, thus $|S| = \binom{15}{5}$. A selection related to the event A
consists of two red balls, two blue balls and one green ball. The number of ele-
ments in A is therefore $|A| = \binom{5}{2} \cdot \binom{4}{2} \cdot \binom{6}{1}$, since there are $\binom{5}{2}$ ways to se-
lect two red balls, $\binom{4}{2}$ ways to select two blue balls and $\binom{6}{1}$ ways to select one
red ball (compare with Example 3.3(b)). Hence $P(A) = \dfrac{\binom{5}{2} \cdot \binom{4}{2} \cdot \binom{6}{1}}{\binom{15}{5}} =$

$\dfrac{10 \cdot 6 \cdot 6}{3{,}003} \approx 0.1199.$

(b) The required probability of the respective event is $P(B) = P(\text{all balls are red}) +$
$P(\text{all balls are blue}) + P(\text{all balls are green}) = \binom{5}{5} / \binom{15}{5} + 0 + \binom{6}{5} / \binom{15}{5} = \dfrac{1+6}{3{,}003}$
$\approx 0.0023.$

(c) Consider the events
R = {the selection does *not* contain a red ball},
B = {the selection does *not* contain a blue ball},
G = {the selection does *not* contain a green ball}.
The event in question is the complement of the event $R \cup B \cup G$. Using the inclusion-exclusion principle, we obtain

$$P(R \cup B \cup G) = P(R) + P(B) + P(G) - P(R \cap B) - R(R \cap G) - P(B \cap G) + P(R \cap B \cap G)$$

$$= \frac{\binom{15}{5} + \binom{11}{5} + \binom{9}{5} - \binom{6}{5} - 0 - \binom{5}{5} + 0}{\binom{15}{5}}$$

$$= \frac{252 + 462 + 126 - 6 - 1}{3{,}003} = \frac{833}{3{,}003}.$$

Hence, the required probability is $P(C) = 1 - \dfrac{833}{3{,}003} \approx 0.7226$.

Example 3.6: Consider the experiment of the previous example. Let N_r, N_b, N_g denote the numbers of red, blue and green balls in the selection, respectively. What is the probability that $N_r \leq N_b \leq N_g$?

Tab. 3.1: Probabilities of outcomes.

N_r	N_b	N_g	Probability
0	0	5	$\binom{5}{0} \cdot \binom{4}{0} \cdot \binom{6}{5} / \binom{15}{5}$
0	1	4	$\binom{5}{0} \cdot \binom{4}{1} \cdot \binom{6}{4} / \binom{15}{5}$
0	2	3	$\binom{5}{0} \cdot \binom{4}{2} \cdot \binom{6}{3} / \binom{15}{5}$
1	1	3	$\binom{5}{1} \cdot \binom{4}{1} \cdot \binom{6}{3} / \binom{15}{5}$
1	2	2	$\binom{5}{1} \cdot \binom{4}{2} \cdot \binom{6}{2} / \binom{15}{5}$

Table 3.1 gives the outcomes related to the event $N_r \leq N_b \leq N_g$ and the corresponding probabilities.

The required probability is therefore the sum of the values in the right column of Tab. 3.1, i.e.,

$$\frac{6 + 4 \cdot 15 + 6 \cdot 20 + 5 \cdot 4 \cdot 20 + 5 \cdot 6 \cdot 15}{\binom{15}{5}} = \frac{1{,}036}{3{,}003} \approx 0.3450.$$

Example 3.7: An urn contains 12 red, 13 blue and 15 green balls, and 12 balls are selected randomly at once. What is the probability that exactly two colors are represented in the selection?

Solution: The probability p_{rb} that the colors red and blue are represented is

$$p_{rb} = P(\text{no ball is green}) - P(\text{all balls are red}) - P(\text{all balls are blue}).$$

By using the notations in Example 3.5(c), we can write this as

$$p_{rb} = P(G) - P(G \cap B) - P(G \cap R) = \frac{\binom{25}{12}}{\binom{40}{12}} - \frac{\binom{12}{12}}{\binom{40}{12}} - \frac{\binom{13}{12}}{\binom{40}{12}}$$

$$= \frac{\binom{25}{12} - \binom{12}{12} - \binom{13}{12}}{\binom{40}{12}} \approx 0.0009.$$

In the same way, we determine the probability that the colors red and green or blue and green are presented, respectively, as

$$p_{rg} = \frac{\binom{27}{12} - \binom{12}{12} - \binom{15}{12}}{\binom{40}{12}} \approx 0.0031 \text{ and } p_{bg} = \frac{\binom{28}{12} - \binom{13}{12} - \binom{15}{12}}{\binom{40}{12}} \approx 0.0054.$$

The probability in question is $p_{rb} + p_{rg} + p_{bg} \approx 0.0095$.

3.2 Variants of a random experiment

We have seen above that many random experiments with a finite number of equally likely outcomes can be formally interpreted as a selection of balls drawn from an urn.

We may assume that the balls are numbered from 1 to N and as a second characteristic we may assign a color to each ball. Now the selection can be obtained in different manners. If there are selected n balls from the urn ($n < N$), there exist three ways to select them:

(i) If the n balls are selected at once, then the outcomes are the *n-subsets* of the set $\{1, \ldots, N\}$.

(ii) If the n balls are selected successively (one after the other) *without replacement*, then the outcomes are the *arrangements* with n elements (see Section 1.2).

(iii) If the n balls are selected successively *with replacement*, then the outcomes are the *sequences* of length n with elements from the set $\{1, \ldots, N\}$.

It may be intuitively clear that the corresponding outcomes of the procedures (i) and (ii) occur with the same probability, but the probabilities of events are calculated in different manners. Therefore, it seems appropriate to study the relationship between the three methods in detail.

Given the urn in Fig. 3.2, containing r black balls and $N-r$ white balls ($r < N$). We select n balls at random in different manners (see above). We study the events $E_k =$ {the selection contains exactly k black balls} for each selection procedure.

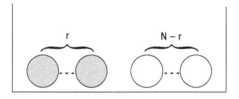

Fig. 3.2: Urn with balls of two colors.

We can now state that the selection procedures (i) and (ii) are equivalent in the sense that the events E_k have the same probability for both procedures (and any k). However, procedure (iii) is not equivalent to the others (see Fig. 3.3).

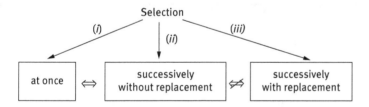

Fig. 3.3: Selection procedures.

We illustrate the relations by means of a numerical example.

Example 3.8: Assume that the urn contains four black and six white balls, and we select five balls (i.e., $N = 10$, $r = 4$ and $n = 5$). We are interested in the probability that $k = 2$ white balls are selected.

(i)   The balls are selected at once and in analogy to the previous calculations the event has probability $\dfrac{\dbinom{4}{2} \cdot \dbinom{6}{3}}{\dbinom{10}{5}} = \dfrac{6 \cdot 20}{252} = \dfrac{10}{21} \approx 0.4762.$

(ii) Now the 5 balls are selected one after the other *without* replacement. The event $E_2 = \{$the selection contains exactly 2 black balls$\}$ can occur in different ways, corresponding to the lines in Tab. 3.2.

Tab. 3.2: Probabilities of outcomes.

Sequence	Probability
bbwww	$\dfrac{4}{10} \cdot \dfrac{3}{9} \cdot \dfrac{6}{8} \cdot \dfrac{5}{7} \cdot \dfrac{4}{6} = \dfrac{1}{21}$
bwbww	$\dfrac{4}{10} \cdot \dfrac{6}{9} \cdot \dfrac{3}{8} \cdot \dfrac{5}{7} \cdot \dfrac{4}{6} = \dfrac{1}{21}$
bwwbw	$\dfrac{1}{21}$
$\vdots$	$\vdots$
wwwbb	$\dfrac{1}{21}$

The first column lists the sequences of *individual* selections of balls. For example, the first line corresponds to the case in which the black balls were chosen in the first and second selection of a ball, while the white balls have been encountered in the selections 3 to 5. The corresponding probability is obtained as follows: The first ball is drawn with probability 4/10, since 4 of the 10 balls in the urn are black. After this, the urn contains 3 black and 6 white balls. The second selected ball is therefore black with probability 3/9. Next, the urn contains 2 black and 6 white balls. The probability to select now a white ball is therefore 6/8. With this argumentation, one can verify that any sequence of balls in Tab. 3.2 has the *same* probability $\frac{1}{21}$.

The required probability is now the sum of all the probabilities in the second column. But all these probabilities are equal and the number of lines is $\binom{5}{2}$, corresponding to the number of ways to select the two positions of the black balls among five possible positions. The required probability is therefore $\binom{5}{2} \cdot \dfrac{1}{21} = \dfrac{10}{21}$, which is identical to the result in case (i).

(iii) The five balls are selected one after the other *with* replacement. The event $E_2 = \{$the selection contains exactly 2 black balls$\}$ can occur in the same way as in the previous case (Tab. 3.2). But now all the probabilities are equal to $\frac{4}{10} \cdot \frac{4}{10} \cdot \frac{6}{10} \cdot \frac{6}{10} \cdot \frac{6}{10} = 0.03456$, since in any individual selection of a ball, the urn contains 4 black and 6 white balls. The required probability is therefore $0.03456 \cdot 10 = 0.3456$ which differs from the previously calculated probabilities.

Exercises

3.1 An urn contains five white and four black balls. If there are chosen randomly four balls at once, what is the probability that the selection contains at least three white balls?

3.2 Solve Example 3.5(c) with the method of Example 3.6

3.3 An urn contains seven red balls, six blue and five green balls. We select eight balls randomly at once. Determine the probability that the selection contains at least two balls of each color:

   (i) Use the approach of Example 3.6.

   (ii) Use the approach of Example 3.5(c).

       Which of these methods is more elegant?

3.4 A lot consists of 200 defective and 700 perfect articles. If 300 articles are chosen randomly at once, what is the probability that

   (i) the selection contains exactly 140 perfect articles, and

   (ii) at least 70 of the chosen articles are defective. (Make use of an adequate computer program.)

3.5 Express the number of derangements (see Proposition 1.8) in terms of the incomplete gamma function.

3.6

   (a) What is the probability that a random permutation of n elements has exactly k fixed points?

       Calculate the numerical values for $n = 8$ and all $k = 0, \ldots, 8$.

   (b) Calculate the probability that a random permutation of 100 elements has more than 3 fixed points.

3.7 Calculate the limits of the probabilities in part (a) of the previous exercise for $n \to \infty$.

3.8 Perform the calculations of Example 3.8 for the case $N = 12$, $r = 5$, $n = 8$ and $k = 3$.

3.9 Generalize the relations in Section 3.2 for the case in which the urn contains balls with three different colors.

3.10 The persons in a room consist of 10 men over 40 years, 15 men under 40, 18 women over 40 and 17 women under 40. One person is chosen at random. We consider the events $J = \{$the person is under 40$\}$ and $M = \{$the person is a man$\}$. Calculate the probabilities

   (i) $P(J \cup \bar{M})$,

   (ii) $P(\bar{J} \cup M)$.

3.11 A shipment of 1,400 air conditioners contains 300 defective and 1,100 perfect items. Two hundred air conditioners are chosen at random without replacement and classified. What is the probability

   (i) that exactly 38 defective devices are found,

   (ii) that at least 50 defective devices are found?

3.12 Suppose that from 100 objects are chosen 40 at random, *with* replacement. What is the probability that no object is chosen more than once?

3.13 How many 5-letter code words may be formed, using the letters a, b, c, d, e, f and g

   (i)   if any letter may be used any number of times and

   (ii) if no letter may be repeated?

3.14 An urn contains balls marked 1, 2, . . ., n. Two balls are chosen at random, one after the other. What is the probability that the numbers are consecutive integers, if

   (i)   the balls are chosen with replacement and

   (ii) the balls are chosen without replacement?

3.15 How many subsets can be formed from a set of 30 elements, containing at least 6 elements?

3.16 (i)   An urn contains n balls, where n_i are of type i for $i = 1, . . ., r$ ($n_1 + \cdots + n_r = n$). If k balls are selected randomly at once, what is the probability that the selection contains at least m balls of each type ($m \leq n_i$ for $i = 1, . . ., r$)?

   (ii) Use an adequate software to calculate the above probability for $r = 4$, $(n_1, . . ., n_4) = (8, 10, 12, 15)$, $k = 40$ and $m = 6$.

4 Conditional probability and independence

4.1 Conditional probability

In the present section, we study the question in how far the occurrence of an event influences the probability of another event. In Definition 2.3 we already introduced the concept of mutually exclusive events A and B. In this case the occurrence of one of the events implies that the other cannot occur. Another extreme situation happens if $A \subset B$, then the realization of A implies that B must occur. In general, the occurrence of an event A may influence more or less the probability that an event B occurs. For example, in road traffic the event of rain affects the probability that a driver has an accident. In the following, we study the conditional probability of B, given that A has occurred. We denote this probability by $P(B|A)$ and define it formally below. We first motivate the concept by an example with a finite sample space. Consider the urn in Fig. 4.1, containing 12 balls numbered by 1, . . ., 12. The balls with numbers 1, . . ., 5 are black and the others are white.

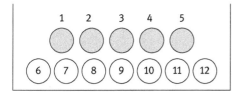

Fig. 4.1: Urn model.

We choose one ball from the urn at random and consider the events $A = \{$the ball is white$\}$ and $B = \{$the ball has an odd number$\}$. Assume that a ball has been chosen and we discover that it is white. What is the conditional probability $P(B|A)$ that the number is odd? Since any of the seven white balls is chosen with the same probability and three of these balls have an odd number we obtain intuitively

$$P(B|A) = \frac{|A \cap B|}{|A|} = \frac{3}{7}. \tag{4.1}$$

With the knowledge that the event A has occurred, we can interpret A as the reduced sample space and the observed event B can be interpreted as $A \cap B$. Formula (4.1) corresponds therefore to the reasoning in Section 3.1 (see eq. (3.2)). Now it is a simple observation that probability (4.1) can be written as

$$P(B|A) = \frac{|A \cap B|/|S|}{|A|/|S|} = \frac{P(A \cap B)}{P(A)}, \tag{4.2}$$

where S is the sample space. We have now motivated the following general concept.

https://doi.org/10.1515/9783111332277-004

Definition 4.1: If $P(A) > 0$, the conditional probability of B, given that A has occurred, is

$$P(B|A) = \frac{P(A \cap B)}{P(A)}. \tag{4.3}$$

Applying (4.3) we obtain also

$$P(B|\bar{A}) = \frac{P(\bar{A} \cap B)}{P(\bar{A})} = \frac{3}{5}$$

and the (unconditional) probability of B can be interpreted as $P(B) = P(B|S)$. One can show that for a given event A, the conditional probability $P(B|A)$ satisfies the axioms of Definition 2.4. We will illustrate some interesting applications.

Example 4.1: The composition of the employees of a fictive company is given by the following contingency table:

	Men	Women	Total
Smokers	20	10	30
Nonsmokers	40	30	70
Total	60	40	100

We select a person at random from the 100 employees and consider the events $M = \{$the person is a man$\}$ and $S = \{$the person is smoker$\}$. For example, the conditional probability that the selected person smokes, given that it is a man, is

$$P(S|M) = \frac{P(S \cap M)}{P(M)} = \frac{0.2}{0.6} = 1/3.$$

It should be easy for the reader to interpret and calculate $P(\bar{S}|M)$, $P(S|\bar{M})$ and $P(\bar{S}|\bar{M})$.

In many situations, the conditional probability $P(B|A)$ can be more easily calculated as $P(A \cap B)$, i.e., instead of (4.3) one applies the equivalent relation

$$P(A \cap B) = P(B|A)P(A), \tag{4.4}$$

which is sometimes called the *multiplication theorem of probability*. This relation can be easily extended to a case with more than two events, for example

$$P(A \cap B \cap C) = P(A|B \cap C)P(B \cap C)$$
$$= P(A|B \cap C)P(B|C)P(C). \tag{4.5}$$

Example 4.2: Given the urn in Fig. 4.1.
(a) If we chose two balls successively at random, without replacement. What is the probability that both balls are black?

Consider the events A = {the first chosen ball is black} and B = {the second ball is black}. From (4.4) we obtain the desired probability as

$$P(A \cap B) = P(B|A)P(A) = \frac{4}{11} \cdot \frac{5}{12} = \frac{5}{33}.$$

(b) If we chose 8 balls successively in the above manner. What is the probability that the 8th ball is the 6th white ball chosen from the urn?

Let A be the event that there are exactly five white balls among the first seven balls and B the event that the eighth ball is black. Then the desired probability is given as

$$P(A \cap B) = P(B|A)P(A) = \frac{5}{7} \cdot \frac{\binom{7}{5} \cdot \binom{5}{2}}{\binom{12}{7}} = \frac{5}{7} \cdot \frac{21 \cdot 10}{792} \approx 0.1894.$$

Note that we have made use of the fact that the two variants (i) and (ii) of urn selections in Fig. 3.3 are equivalent. To formulate further applications of the multiplication theorem, we need the following concept.

Definition 4.2: Given a sample space S, we say that the events $B_1, \ldots, B_m$, represent a *partition of S*, if
(i) $B_i \cap B_j$ for all $i \neq j$,
(ii) $B_1 \cup \ldots \cup B_m = S$ and
(iii) $P(B_i) > 0$ for all i.

The definition is identical with Definition 1.2 except for the additional condition (iii). The latter guarantees that exactly one of the events B_i occurs, when the random experiment is performed.

For example, in the case of tossing a die, B_1 = {1}, B_2 = {3,5} and B_3 = {2,4,6} represent a partition of the sample space. Exactly one of these events occurs in each tossing of the die.

We now present the *total probability theorem*.

Theorem 4.1: Given a partition $B_1, B_2, \ldots, B_m$ of the sample space S and an event A, (Fig. 4.2) then the probability of A can be expressed as

$$P(A) = \sum_{i=1}^{m} P(A|B_i)P(B_i).$$

Proof: The set A can be written as $A = A \cap S = A \cap (B_1 \cup \cdots \cup B_m) = (A \cap B_1) \cup \cdots (A \cap B_m)$, where the sets $A \cap B_i$ are mutually exclusive. Hence, from the multiplication theorem, it follows that

$$P(A) = \sum_{i=1}^{m} P(A \cap B_i) = \sum_{i=1}^{m} P(A|B_i)P(B_i).$$

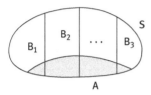

A

Fig. 4.2: Partition of a sample space.

Example 4.3: Assume that a specific article is produced by the factories F_1, F_2 and F_3. The production quantities are 10,000, 8,000 and 7,000 items and the probabilities of producing a defective item are 0.02, 0.05 and 0.03, respectively. The total quantity of articles produced by the three factories is stored in a warehouse and an article is chosen at random. What is the probability that the selected article is defective?

The sample space S consists of all the 25,000 articles in the warehouse. Let A be the event that the selected article is defective. The partition of the sample space is naturally given by the sets B_i of articles produced by the factory F_i ($i = 1, 2, 3$). From Theorem 4.1 we obtain the desired probability as

$$P(A) = P(A|B_1)P(B_1) + P(A|B_2)P(B_1) + P(A|B_3)P(B_3)$$

$$= 0.02 \cdot \frac{10}{25} + 0.05 \cdot \frac{8}{25} + 0.03 \cdot \frac{7}{25} = 0.0324.$$

The calculation of the total probability can be well illustrated by a tree diagram in Fig. 4.3, which shows the different ways the event A may occur. The terms $P(A|B_i)P(B_i)$ in the above formula correspond to the different paths from the starting point 0 to the event A.

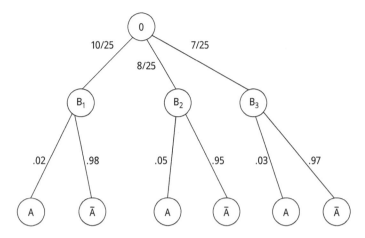

Fig. 4.3: Tree diagram.

4.2 Bayes' theorem

As an immediate consequence of Theorem 4.1, we obtain the following interesting result, known as Bayes' theorem.

Theorem 4.2: Given a partition $B_1, B_2, \ldots, B_m$ of the sample space S and an event A, then the conditional probability of B_i, given that A has occurred, can be expressed as

$$P(B_i|A) = \frac{P(A|B_i)P(B_i)}{\sum_{j=1}^{m} P(A|B_j)P(B_i)} \qquad \text{for } i = 1, \ldots, m.$$

Proof: From the definition of the conditional probability, it follows that

$$P(B_i|A) = \frac{P(A \cap B_i)}{P(A)} = \frac{P(A|B_i)\, P(B_i)}{P(A)}$$

and substituting the denominator according to Theorem 4.1 yields the statement.

In continuation of Example 4.3, we could ask the following. If the selected article is defective, what is the conditional probability that it has been produced by factory 1? From Theorem 4.2 we get

$$P(B_1|A) = \frac{P(A|B_1)\, P(B_1)}{P(A)} = \frac{0.02 \cdot \frac{10}{25}}{0.0324} \approx 0.2469.$$

Bayes' theorem has been controversially discussed since its publication. It is mathematically correct; however, its application is limited by the fact that the probabilities $P(B_i)$ are often not known. Anyway the statement is very interesting since it enables to determine with which probability the event B_i has caused the event A. Therefore, it is also known as the "formula for the probability of causes."

We illustrate another potential application.

Example 4.4: The patients of the infirmary of a cardiology clinic have been operated by three physicians P_1, P_2 and P_3. Assume that these physicians have operated 50%, 30% and 20% of all the patients of the infirmary and that they commit a malpractice with probability 0.04, 0.05 and 0.02, respectively. If a patient of the infirmary is chosen at random and if he/she is victim of a malpractice, what is the probability that the physician P_3 has caused the problem?

Consider the events A = {the patient is victim of a malpractice}and B_i = {the patient has been operated by P_i}, i = 1,2,3. The total probability is given by $P(A) = 0.5 \cdot 0.04 + 0.3 \cdot 0.05 + 0.2 \cdot 0.02 = 0.039$. Hence, from Theorem 4.2 we get the desired probability as

$$P(B_3|A) = \frac{P(A|B_3)\,P(B_3)}{P(A)} = \frac{0.02 \cdot 0.2}{0.039} \approx 0.1026.$$

Motivated by the COVID-19 pandemic, we finally present a fictive multistage problem.

Example 4.5: Suppose that a recently developed vaccine against an infective disease is given to patients in two doses. Assume that 90% of the population takes the first dose, and 80% of these people take also the second dose. Moreover, 20% of the people that missed the first dose take the second one. The probability of an outbreak of the disease depends on the vaccination status. A fully vaccinated person has 1% chance of becoming infected. If only the first or only the second dose was taken, the probability of becoming infected is 5% or 10%, respectively. An unvaccinated person eventually becomes infected with a 20% chance. The situation is illustrated in Fig. 4.4, where the following events are defined:

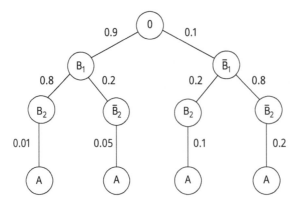

Fig. 4.4: Multistage tree diagram.

A = {a person is infected with the disease},
B_1 = {the person has taken the first dose of the vaccine},
B_2 = {the person has taken the second dose of the vaccine}.

We calculate the probabilities along the four paths from 0 to A. For the path $0{\rightarrow}B_1{\rightarrow}B_2{\rightarrow}A$, we obtain the probability

$p_1 = P(B_1 \cap B_2 \cap A) = P(B_1)\,P(B_2|B_1)\,P(A|B_1 \cap B_2) = 0.9\cdot0.8\cdot0.01 = 0.0072.$ Analogously
we get
$p_2 = P(B_1)\,P(\bar{B}_2|B_1)P(A|\bar{B}_2\cap B_1) = 0.9\cdot0.2\cdot0.05 = 0.009$ for the path $0 \rightarrow B_1 \rightarrow \bar{B}_2 \rightarrow A$
$p_3 = P(\bar{B}_1)\,P(B_2|\bar{B}_1)P(A|\bar{B}_1 \cap B_2) = 0.1\cdot0.2\cdot0.1 = 0.002$ for the path $0 \rightarrow \bar{B}_1 \rightarrow B_2 \rightarrow A$ and
$p_4 = P(\bar{B}_1)\,P(\bar{B}_2|\bar{B}_1)P(A|\bar{B}_1\cap\bar{B}_2) = 0.1\cdot0.8\cdot0.2 = 0.016$ for the path $0 \rightarrow \bar{B}_1 \rightarrow \bar{B}2 \rightarrow A.$

Thus, the total probability of the event A is $P(A) = p_1 + \cdots + p_4 = 0.0342$, resulting in the conditional probabilities

$$P(\text{no dose taken}|A) = \frac{p_4}{P(A)} = \frac{0.016}{0.0342} = 0.4678,$$

$$P(\text{only second dose taken}|A) = \frac{p_3}{P(A)} = \frac{0.002}{0.0342} = 0.0585,$$

$$P(\text{only first dose taken}|A) = \frac{p_2}{P(A)} = \frac{0.009}{0.0342} = 0.2632,$$

$$P(\text{fully vaccinated}|A) = \frac{p_1}{P(A)} = \frac{0.0072}{0.0342} = 0.2105.$$

In summary, one can say that an infection can be attributed to incomplete vaccination with a probability of 79%. It would be very interesting to perform such a study based on real data.

4.3 Independent events

In the intuitive sense, two events A and B are independent, if the occurrence of one of them in no way influences the occurrence or nonoccurrence of the other. In terms of conditional probability, it makes sense to define the independence by $P(B|A) = P(B)$ and $P(A|B) = P(A)$. But if $P(A) > 0$, the first condition is equivalent to

$$\frac{P(A \cap B)}{P(A)} = P(B)$$

or to

$$P(A \cap B) = P(A)P(B). \qquad (4.6)$$

In the same way, one can show that $P(A|B) = P(A)$ is equivalent to (4.6), if $P(B) > 0$. Clearly, (4.6) also holds if any of the events A, B has probability 0. Therefore, we obtain the following convenient definition.

Definition 4.3: The events A and B are *independent*, if and only if (4.6) holds.

To illustrate the concept, consider an urn containing four black and six white balls. We select two balls successively and consider the events $A = \{$the first ball is black$\}$ and $B = \{$the second ball is black$\}$. If the selection is *with* replacement, A and B are obviously independent. Formally it holds $P(A) = P(B) = 0.4$ and $P(A \cap B) = P(A) P(B) = 0.4^2 = 0.16$. But for a selection *without* replacement we obtain $P(B|A) = 3/9 \neq P(B|\bar{A}) = 4/9$; hence, A and B cannot be independent.

Example 4.6: Consider the company in Example 4.1. When an employee is selected randomly, then the events $M = \{$the person is a man$\}$ and $S = \{$the person is a smoker$\}$ are not independent, since $P(S|M) = 1/3 \neq P(S|\bar{M}) = 1/4$. This indicates that men and women have different proportions of smokers. The reader should verify that for the following modified contingency table, the considered events would be independent:

	Men	Women	Total
Smokers	18	12	30
Nonsmokers	42	28	70
Total	60	40	100

We now state some characteristics of independence that are intuitively clear.

Theorem 4.3: If A and B are independent events, then
(i) A and $\bar{B}$ are independent,
(ii) $\bar{A}$ and B are independent and
(iii) $\bar{A}$ and $\bar{B}$ are independent.

Proof of part (i):
Since $A = (A \cap B) \cup (A \cap \bar{B})$, where $(A \cap B)$ and $(A \cap \bar{B})$ are mutually exclusive, we get

$$P(A) = P(A \cap B) + P(A \cap \bar{B}) = P(A)P(B) + P(A \cap \bar{B}),$$

implying

$$P(A \cap \bar{B}) = P(A) - P(A)P(B) = P(A)(1 - P(B)) = P(A)P(\bar{B}).$$

Hence, $\bar{A}$ and $\bar{B}$ are independent.

The proof of part (ii) is analogous and part (iii) is left to the reader.

We will now define independence for more than two events. For reasons of illustration, we start with the case of three events.

Definition 4.4: The three events A, B and C are called (mutually) *independent*, if and only if the following conditions hold:

$$P(A \cap B) = P(A)P(B),$$

$$P(A \cap C) = P(A)P(C),$$

$$P(B \cap C) = P(B)P(C),$$

$$P(A \cap B \cap C) = P(A)P(B)P(C).$$

For example, when three dice are thrown, then the events {the first die shows number x}, {the second die shows number y} and {the third die shows number z} are independent for any $x, y, z \in \{1, \ldots, 6\}$.

Example 4.7: Select a number at random from the set $\{1, \ldots, 8\}$. Then the events $A = \{1,2,4,5\}$, $B = \{2,3,5,6\}$ and $C = \{4,5,6,7\}$ are independent. It holds $P(A) = P(B) = P(C) = 1/2$, $P(A \cap B) = P(A \cap C) = P(B \cap C) = 1/4$ and $P(A \cap B \cap C) = 1/8$. Hence, the conditions in Definition 4.4 are satisfied.

Example 4.8: Assume that two coins are thrown. The sample space is illustrated in Fig. 4.5. Consider the events $A = \{$the first coin shows a head$\}$, $B = \{$the second coin shows a tail$\}$ and $C = \{$both coins have equal outcomes$\}$. Similar to the previous example, it holds $P(A) = P(B) = P(C) = 1/2$, $P(A \cap B) = P(A \cap C) = P(B \cap C) = 1/4$; thus, the first three conditions in Definition 4.4 are satisfied, but $P(A \cap B \cap C) = 0 \neq P(A)P(B)P(C) = 1/8$. Thus, the three events are *dependent*, though any two of them are independent.

In the case of an arbitrary number of events we obtain the following.

Definition 4.5: The events $A_1, A_2, \ldots, A_n$ are called (mutually) *independent*, if and only if,

$$P\left(A_{i_1} \cap \ldots \cap A_{i_k}\right) = P\left(A_{i_1}\right) \cdot \ldots \cdot P\left(A_{i_k}\right)$$

holds for any $k = 2, \ldots, n$ and indices $1 \leq i_1 < \cdots, i_k \leq n$.

In generalization of Theorem 4.3, the independence of the events $A_1, A_2, \ldots, A_n$ is preserved, when any of these events are substituted by their complements. In most applications, we need not check all the conditions of Definition 4.4 since we usually assume independence based on our knowledge of the experiment.

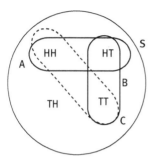

Fig. 4.5: Venn diagram.

Example 4.9: If a die is thrown 10 times, what is the probability that
(i) the result is always 4 and
(ii) the die shows the number 6 at least once.

Solution:
(i) Consider the independent events A_i = {the die shows the number 4 in the ith throw}, $i = 1, \ldots, 10$. The desired probability is therefore

$$P(A_1) \cdots P(A_{10}) = \left(\frac{1}{6}\right)^{10} \approx 1.6 \cdot 10^{-8}.$$

(ii) We now consider the events B_i = {the die does *not* show the number 6 in the ith throw}, $i = 1, \ldots, 10$. Thus, the probability that the die *never* shows the number 6 in the 10 throws is $P(B_1) \cdot \cdots \cdot P(B_{10}) = \left(\frac{5}{6}\right)^{10}$. Hence, the desired probability is $1 - \left(\frac{5}{6}\right)^{10} \approx 0.8385$.

Example 4.10: Assume that the three nurses A, B and C of the intensive care appear independently for duty with probabilities p_a, p_b, and p_c.
(i) What is the probability p_i that exactly i nurses appear for duty ($i = 0,1,2,3$)?
(ii) Calculate the probabilities p_i for $p_a = 0.94$, $p_a = 0.95$ and $p_c = 0.97$.

Solution:
(i) Now we use explicitly Theorem 4.3 in generalized form and obtain

$$p_0 = P(A \text{ does not appear}) \cdot P(B \text{ does not appear}) \cdot P(C \text{ does not appear})$$
$$= (1 - p_a)(1 - p_b)(1 - p_c),$$
$$p_0 = P(\text{only } A \text{ appears}) + P(\text{only } B \text{ appears}) + P(\text{only } C \text{ appears})$$
$$= p_a(1 - p_b)(1 - p_c) + (1 - p_a)p_b(1 - p_c) + (1 - p_a)(1 - p_b)p_c$$

Similarly we get

$$p_2 = p_a p_b(1-p_c) + p_a(1-p_b)p_c + (1-p_a)p_b p_c,$$
$$p_3 = p_a p_b p_c.$$

(ii) We obtain the probabilities

$$p_0 = 0.06 \cdot 0.05 \cdot 0.03 = 0.00009,$$

$$p_1 = 0.94 \cdot 0.05 \cdot 0.03 + 0.06 \cdot 0.95 \cdot 0.03 + 0.06 \cdot 0.05 \cdot 0.97 = 0.00603,$$

$$p_2 = 0.94 \cdot 0.95 \cdot 0.03 + 0.94 \cdot 0.05 \cdot 0.97 + 0.06 \cdot 0.95 \cdot 0.97 = 0.12767,$$

$$p_3 = 0.94 \cdot 0.95 \cdot 0.97 = 0.86621,$$

summing up to 1.

A systematic study of the probabilities p_i for an arbitrary number of nurses leads to the so-called Poisson's binomial distribution.

Further applications of the independence concept arise in electronic circuits with relays. These components were already used in the early days of computer history, but still today relays play an important role in logic circuits.

Example 4.11: The probability of the closing of each relay of the circuit in Fig. 4.6 is given by p (we say that a relay is closed, when current can flow). If all relays function independently, what is the probability that the lamp lights?

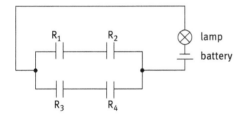

Fig. 4.6: Electronic circuit.

Solution:
Consider the events $A_i = \{$relay i is closed$\}$, $i = 1, \ldots, 4$ and $E = \{$the lamp lights$\}$. The relation between these events is given by $E = (A_1 \cap A_2) \cup (A_3 \cap A_4)$, because the electric circuit can be closed via two paths: via relays 1 and 2, and via relays 3 and 4. (Note that $A_1 \cap A_2$ and $A_3 \cap A_4$ are *not* mutually exclusive.) Thus,

$$P(E) = P(A_1 \cap A_2) + P(A_3 \cap A_4) - P(A_1 \cap A_2 \cap A_3 \cap A_4) = p^2 + p^2 - p^4.$$

Example 4.12: Consider now the circuit in Fig. 4.7.Assume again that the probability that each relay being closed is p and that all relays function independently. With the same notation, as in the previous example,

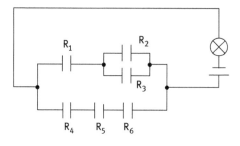

Fig. 4.7: Modified circuit.

we get $E = (A_1 \cap A_2) \cup (A_1 \cap A_3) \cup (A_4 \cap A_5 \cap A_6)$. Thus, using the inclusion-exclusion principle we obtain

$$P(E) = P(A_1 \cap A_2) + P(A_1 \cap A_3) + P(A_4 \cap A_5 \cap A_6) - P(A_1 \cap A_2 \cap A_3)$$
$$- P(A_1 \cap A_2 \cap A_4 \cap A_5 \cap A_6) - P(A_1 \cap A_3 \cap A_4 \cap A_5 \cap A_6) + P(A_1 \cap \ldots \cap A_6)$$
$$= p^2 + p^2 + p^3 - p^3 - p^5 - p^5 + p^6 = p^6 - 2p^5 + 2p^2.$$

If the circuit is obtained as a combination of series and parallel connections, the identification of circuit-closing paths is not difficult in principle. Otherwise, the problem is more complex.

Example 4.13: We still consider the circuit in Fig. 4.8.

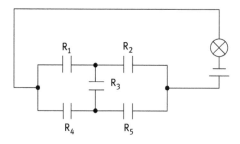

Fig. 4.8: Circuit with five relays.

Now one obtains four circuit-closing paths: via relays 1,2, via 1,3,5, via 4,3,2 and via 4, 5. The relation between the events A_i and E is now given by $E = (A_1 \cap A_2) \cup (A_1 \cap A_3 \cap A_5) \cup (A_4 \cap A_3 \cap A_2) \cup (A_4 \cap A_5)$. Similar to the previous examples we can show that $P(E) = 2p^5 - 5p^4 + 2p^3 + 2p^2$. (The reader is advised to verify this.) However, we now present an alternative solution method (see Tab. 4.1).

The symbol × in the first five columns indicates that the respective relay is closed. The last column contains the probability of the given state of the system. The desired probability $P(E)$ is obtained by adding the probabilities of the states with the lamp on. In this way one obtains again

$$P(E) = p^5 + 5p^4(1-p) + 8p^3(1-p)^2 + 2p^2(1-p)^3 = 2p^5 - 5p^4 + 2p^3 + 2p^2.$$

Tab. 4.1: States of the circuit.

Relays					Lamp on?	Probability
1	2	3	4	5		
×	×	×	×	×	Yes	p^5
×	×	×	×		Yes	$p^4(1-p)$
×	×	×		×	Yes	$p^4(1-p)$
×	×		×	×	Yes	$p^4(1-p)$
×		×	×	×	Yes	$p^4(1-p)$
	×	×	×	×	Yes	$p^4(1-p)$
×	×	×			Yes	$p^3(1-p)^2$
×	×		×		Yes	$p^3(1-p)^2$
×	×			×	Yes	$p^3(1-p)^2$
×		×	×		No	
×		×		×	Yes	$p^3(1-p)^2$
×			×	×	Yes	$p^3(1-p)^2$
	×	×	×		Yes	$p^3(1-p)^2$
	×	×		×	No	
	×		×	×	Yes	$p^3(1-p)^2$
		×	×	×	Yes	$p^3(1-p)^2$
×	×				Yes	$p^2(1-p)^3$
		⋮	⋮		No	
			×	×	Yes	$p^2(1-p)^3$

The foregoing examples provide excellent training opportunities for the probabilistic logical reasoning. Moreover, as the following example shows, some basic ideas may be also transferred to areas that have apparently nothing to do with electronic circuits.

Example 4.14: The four physicians $P_1, \ldots, P_4$ of the emergency department of a hospital are very qualified and have skills in several medical areas, as given in Tab. 4.2. Each physician appears for service with probability p and service failures occur independently. The department functions properly, if for any medical discipline in Tab. 4.2 at least one physician trained in this area has appeared for duty. What is the probability that this happens?

Tab. 4.2: Skills of physicians.

	Orthopedics	General medicine	Surgery	Ophthalmology	Urology	Pediatrics
P_1	×	×		×		
P_2		×	×	×	×	×
P_3	×	×			×	×
P_4		×	×			×

Solution: The basic idea to solve the problem is to express the relation between the events $A_i = \{$physician i is present at service$\}$, $i = 1, \ldots, 4$ and $E = \{$the department functions properly$\}$.

It is useful to introduce the following definition which can be easily adapted to another context:

Definition 4.6: A subset $S \subset \{P_1, \ldots, P_4\}$ of physicians is called a *minimum working set*, if for any medical discipline in Tab. 4.2 at least one physician of S is trained in this discipline and if no proper subset of S has this characteristic.

One can easily verify that the minimum working sets for Example 4.13 are $\{P_1, P_2\}, \{P_2, P_3\}$ and $\{P_1, P_3, P_4\}$. Now the emergency department functions properly, if and only if the set of physicians present at work contains at least one of the minimum working sets. Hence, $E = (A_1 \cap A_2) \cup (A_2 \cap A_3) \cup (A_1 \cap A_3 \cap A_4)$. Similar to the previous examples, we obtain $P(E) = p^2 + p^2 + p^3 - p^3 - p^4 - p^4 + p^4 = 2p^2 - p^4$.

It is interesting to observe that Definition 4.6 generalizes the concept of circuit-closing paths in electronic circuits. Such a path is a minimum set of closed relays that turns the lamp on, i.e., a minimum working set in the (generalized) sense of Definition 4.6.

Exercises

4.1 A box contains 10 good and 5 bad tubes. There are two tubes drawn at once. One of them is tested and found to be good. What is the probability if the other one is also good?

4.2 Urn 1 contains 5 white and 10 black balls. Urn 2 contains 6 white and 7 black balls. A ball is chosen at random from urn 1 and put into urn 2. After this, a ball is chosen at random from urn 2. What is the probability that this ball is black?

4.3 If A and B are independent events such that $P(A)$ and $P(A \cup B)$ are known. Represent $P(B)$ in dependence of the known probabilities.

4.4 From a set of 30 items (18 perfect and 12 defective), 3 items are chosen at random, one after the other and without replacement. What is the probability that
   (i) all selected items are defective and
   (ii) at most one is perfect?

4.5 A bag contains three coins, one of which is coined with two tails, while the other two are normal and not biased. A coin is chosen at random from the bag and tossed three times. If a tail turns up each time, what is the probability that this is the coin with two tails?

4.6 Let A and B be two events with $P(A) = 0.5$ and $P(A \cup B) = 0.8$. Let $P(B) = p$.
   (i) For what choice of p are A and B mutually exclusive?
   (ii) For what choice of p are A and B independent?

4.7 Each of two persons tosses three fair coins. What is the probability that they obtain the same number of tails?

4.8 Two dice are rolled. Given that the outcomes are different, what is the probability that one of them is 3?

4.9 The entries of a second-order determinant assume the values zero or 1 with equal probability 1/2. What is the probability that the value of the determinant is negative?

4.10 Prove that $P(A|B) > P(A)$ implies $P(B|A) > P(B)$.

4.11 A box contains 10 red and 15 green balls. The balls are chosen at random from the box, one after the other, without replacement. What is the probability that
   (i) the first three balls are green,
   (ii) exactly two of the first four balls are red and
   (iii) the third ball is green, given that the first two were red.

4.12 Define partitions for the sample spaces of the random experiments in Tab. 2.1.

4.13 If 4 balls are successively chosen from the urn in Fig. 4.1, what is the probability that the fourth ball is the third black ball chosen from the urn? Calculate the desired probability in two ways:
   (i) as in Example 4.2(b) and
   (ii) using a tree diagram.

4.14 A school class consists of 25 boys and 20 girls. Three pupils are chosen successively at random, without replacement. What is the probability that the second pupil is a girl, while the others are boys? Make use of the multiplication theorem.

4.15 Let X, Y and Z denote the outcomes of three dice. Calculate the conditional probabilities:
   (i) $P(X + Y = 10 | Y > X)$,
   (ii) $P(X + Y + Z > 12 | X, Y > 3)$,
   (iii) $P(Z > X | X < Y)$.

4.16 Consider Example 4.4.

  (i) Interpret and calculate the probabilities $P(B_1|A)$, $P(B_2|A)$ and $P(B_i|\bar{A})$ for $i = 1,2,3$. Make use of tree diagrams.

  (ii) Assume that the percentages of operated patients for the physicians P_3 and P_2 are unknown, i.e., P_3 and P_2 realized $a\%$ and $(50 - a)\%$ of the patients, respectively ($0 \leq a \leq 50$). Express $P(A)$ and $P(B_3|A)$ as functions of a.

4.17 An urn contains five red balls, six blue balls and eight green balls. Two balls are chosen successively at random, without replacement. What is the probability that the first ball was red, given that the second is blue?

4.18 In a fictive highly developed country live immigrants from three different nations: 60% of them come from nation A, 30% from B and 10% from nation C. The immigrants of nations A, B and C have no valid identification with probabilities 0.2, 0.3 and 0.15, respectively.

  (i) What is the probability that an immigrant chosen at random has no valid identification?

  (ii) If a randomly chosen immigrant has no valid identification, what is the probability that he/she is from nation A?

4.19 Assume that an architect commits a mistake in the planning with probability 0.05. In this case, a construction collapses with probability 0.2. In case of a perfect planning the construction breaks down with probability 0.001. If a construction collapses, what is the probability that the architect has caused the damage?

4.20 One of three identical urns contains three coins of 10 cents, the second contains two coins of 10 cents and one of 50 cents, and the third contains three coins of 50 cents. One urn is selected at random and then a coin is selected at random from it. If the coin is of 50 cents, what is the probability that the remaining two coins are of 10 cents?

4.21 A box contains 10 red and 15 green balls. Three balls are selected one after the other without replacement.

  (i) What is the probability that the two first chosen balls were green, given that the third is red?

  (ii) What is the probability that the third ball is red, given that at least one of the first two balls were red.

4.22 Consider the following contingency table that resulted from a consumer research of 200 customers of a supermarket:

	Drink alcohol rarely or never	Drink alcohol frequently	Total
Age < 40 years	$f_{1,1} = 80$	$f_{1,2} = 40$	120
Age ≥ 40 years	$f_{2,1} = 60$	$f_{2,2} = 20$	80
Total	140	60	200

A customer is randomly chosen from this sample. Consider the events $J = \{$the costumer is less than 40 years old$\}$ and $A = \{$the costumer drinks alcohol frequently$\}$. Are these events independent? If not, for what observed frequencies $f_{i,j}$ would they be independent (for unchanged totals)?

4.23 Determine four events A, B, C and D over a finite sample space with equally likely outcomes such that any three of these events are independent, and A, B, C and D are *not* independent.

4.24 In any show a trapeze artist suffers a fatal accident with probability $p = 0.001$.
   (i) What is the probability to survive 500 shows?
   (ii) For which value of p he survives 500 shows with probability 0.99?

4.25 Assume that the five nurses A, B, C, D and E of an intensive care appear independently for duty with probabilities $a = 0.98$, $b = 0.95$, $c = 0.9$, $d = 0.85$ and $e = 0.9$. The intensive care functions properly if at least three nurses are present. What is the probability of this event?

4.26 Consider the circuit in Example 4.12. Assume that the relays 1, . . ., 6 close with probabilities $p_1 = 0.5$, $p_2 = 0.7$, $p_3 = 0.3$, $p_4 = 0.4$, $p_5 = 0.8$ and $p_6 = 0.9$, respectively.
   (i) What is the probability that the lamp is on?
   (ii) If the lamp is on, what is the conditional probability that relay 2 is closed?

4.27 Given the following electronic circuit, express E in terms of A_i (see Examples 4.11–4.13).

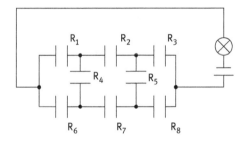

4.28 Solve Example 4.14 alternatively by the method suggested in Tab. 4.1.

4.29 The organizing committee of an international political conference employs five multilingual secretaries $S_1, . . ., S_5$. The seven languages German (G), French (F), English (E), Portuguese (P), Spanish (S), Russian (R) and Chinese (C) are spoken at the conference. The secretaries speak the languages indicated in the following table:

	G	F	E	P	S	R	C
S_1	×			×		×	
S_2		×	×				×
S_3	×			×	×		
S_4		×		×	×		
S_5	×		×		×	×	

Each secretary appears for service with probability p and they are present independently of each other. What is the probability that any of the seven languages is spoken by at least one of the secretaries present at the congress?

Solve the problem in two ways:

(i) Consider a table similar to Tab. 4.1.

(ii) Determine the minimum working sets as in Examples 4.11–4.13 and use the inclusion-exclusion principle. Which solution method is more elegant?

4.30 Realize that the calculation of the conditional probabilities in Example 4.5 corresponds to the application of Bayes' theorem. What is the partition of the sample space?

4.31 Given an urn with seven red, five blue and three green balls, three balls are randomly chosen one after the other without replacement. Consider the events A = {the last selected ball is red}, B = {the second selected ball is blue}, C = {the first two selected balls have the same color}. Make use of a tree diagram and determine the probabilities:

(i) $P(A)$, (ii) $P(B|A)$ and (iii) $P(C|A)$.

5 One-dimensional random variables

5.1 The concept of a random variable

As the examples in the previous chapters show, the outcomes of a random experiment need not be numbers. For example, in classifying a manufactured item we might simply use the categories "perfect" and "defective", and when tossing a coin a number of times, the outcomes may be the sequences of heads and tails (see Tab. 2.1). But since we are interested in measuring any characteristic of an experiment, we must associate a number with each outcome. For instance, we can assign the values 1 and 0 to a perfect and a defective manufactured item, and to a sequence of heads and tails we can assign the number of heads observed. We obtain the following concept which is fundamental to all fields of probability theory and statistics.

Definition 5.1: Consider a random experiment E with sample space S. A *random variable* (r.v.) is a function $X: S \to IR$ such that the probability $P(X \le c)$ is defined for any real number c (Fig. 5.1). The image of the function X, i.e., $R_X := \{X(s) | s \in S\}$, is called the *range space* of X.

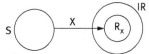

Fig. 5.1: Random variable.

We have already used this simple concept in an intuitive sense (see, e.g., the numbers N_r, N_b and N_g in Example 3.6). The terminology is somewhat unfortunate, since in fact a random variable is a function and not a variable. But the naming "variable" is universally accepted throughout the literature because of historical reasons and therefore we will not alter it. The range space can be interpreted as another sample space, when $X(s)$ is considered as the result of the experiment. The requirement that $P(X \le c)$ must be defined for any c is a "technical" condition whose meaning will become clear later. There exist innumerous examples for random variables. For instance, age, weight, height, income, number of children and number of cars are possible random variables associated with a randomly chosen person. The numbers of balls of a given color are random variables associated with a random selection of balls from an urn, or the sum of the outcomes is a random variable associated with the experiment of tossing two dice.

In the pertinent literature, usually capital letters such as X, Y and Z are used to denote random variables. However, when speaking of the *value* of these variables, in general, lowercase letters such as x, y and z are used.

https://doi.org/10.1515/9783111332277-005

5.2 Discrete random variables

Definition 5.2: Given a random variable X, if the range space R_X is finite or countably infinite, X is called a *discrete random variable*.

The range space of such a variable can be written as $R_X = \{x_1, x_2, \ldots, x_n, \ldots\}$, i.e., in the finite case, the list of values terminates and in the countably infinite case it continues indefinitely.

Similar to the beginning of Chapter 3, we associate a probability $p(x_i)$ with each element x_i of R_X, such that

$$p(x_i) \geq 0 \text{ for all } i \text{ and } p(x_1) + p(x_2) + \cdots = 1. \tag{5.1}$$

(The specific probabilities are given by an adequate discrete model.) The probability of the event $B = \{x_{j_1}, x_{j_2}, \ldots\}$ related to the sample space R_X is defined as

$$P(B) = p\left(x_{j_1}\right) + p\left(x_{j_2}\right) + \cdots. \tag{5.2}$$

The function p is called the *probability function* of the random variable X and the collection of pairs $(x_i, p(x_i))$ is called the *probability distribution* of X.

Example 5.1: If the random experiment consists of tossing two dice, one can model the sample space as $S = \{(i, j) \mid 1 \leq i, j \leq 6\}$. The values of the random variable $X = \{$sum of the outcomes$\}$ are indicated in Tab. 5.1(a). The probability distribution of X is given in part (b). It follows for example that the value of X is 4, if and only if one of the outcomes (1, 3), (2, 2), (3, 1) occurred (see the bold diagonal in Tab. 5.1(a)). Therefore, the elementary event $B_1 = \{4\}$ of R_X has the probability 3/36. From (5.2) it follows, for instance, that the event $B_2 = \{10, 11, 12\}$ of R_X has the probability $p(10) + p(11) + p(12) = 3/36 + 2/36 + 1/36 = 6/36$. We express these facts by the convenient notations $P(X = 4) = 3/36$ and $P(X \geq 10) = 6/36$.

Tab. 5.1: Tossing two dies.
(a) Sample space S and values of X

Second die	1	2	3	4	5	6
First die						
1	2	3	4	5	6	7
2	3	4	5	6	7	8
3	4	5	6	7	8	9
4	5	6	7	8	9	10
5	6	7	8	9	10	11
6	7	8	9	10	11	12

(b) Probability distribution of X

x_i	$p(x_i)$
2	1/36
3	2/36
4	3/36
5	4/36
6	5/36
7	6/36
8	5/36
9	4/36
10	3/36
11	2/36
12	1/36

In order to assign probabilities to the values of random variables, the following fact is also useful: to each event B related to the range space R_X of a random variable X corresponds to the event $A = X^{-1}(B) = \{s \in S \,|\, X(s) \in B\}$ related to S. The events A and B occur simultaneously and are therefore called *equivalent* events. Clearly, equivalent events have the same probability; thus, $P(A) = P(B)$. For instance, the above-considered event $B_2 = \{10, 11, 12\}$ of R_X is equivalent to $A_2 = \{(4,6), (5,5), (6,4), (5,6), (6,5), (6,6)\}$ of S. Thus $P(B_2) = P(A_2) = 6/36$.

5.3 The binomial distribution and extensions

This is surely one of the most important discrete probability models and the reader who studied carefully Section 3.2 will consider its derivation as a simple task. Nevertheless, we start with a typical example for the binomial distribution.

Example 5.2: Suppose that an item is perfect with probability $p = 0.8$ and defective with probability $1 - p = 0.2$. What is the probability that there are exactly two defective pieces among five manufactured items, produced in sequence?

Assume that the probability p is the same for each item throughout the duration of the study and that the results of the individual productions are *independent* events.

We consider the random variable $X = \{$number of defective items$\}$. The required probability is $P(X = 2)$. The event "$X = 2$" can occur in 10 different ways, corresponding to the sequences in Tab. 5.2.

One can easily see that all the probabilities in the right column are equal, and the number of sequences is 10, since there are $\binom{5}{2} = 10$ ways to choose the positions of the 2 defective items among 5 possible positions. Since the sequences represent mutu-

Tab. 5.2: Outcomes and probabilities.

Sequence	Probability
ddppp	$0.2 \cdot 0.2 \cdot 0.8 \cdot 0.8 \cdot 0.8$
dpdpp	$0.2 \cdot 0.8 \cdot 0.2 \cdot 0.8 \cdot 0.8$
$\vdots$	$\vdots$
pppdd	$0.8 \cdot 0.8 \cdot 0.8 \cdot 0.2 \cdot 0.2$

ally exclusive outcomes, the desired probability is now obtained by adding the equal probabilities of all the sequences, i.e., $P(X = 2) = \binom{5}{2} 0.8^3 \cdot 0.2^2 = 0.2048$.

Definition 5.3: Given an (individual) random experiment E and an event A associated with E. Suppose that $P(A) = p$ and hence $P(\bar{A}) = 1 - p$. Consider n *independent* realizations of E and assume that the probability p remains *unaltered*. The sample space of the overall experiment consists of the 2^n sequences of length n with elements A and $\bar{A}$. Define the random variable X as the number of times that A occurred in the n realizations of E. We call X the *binomial random variable* with parameters n and p. Equivalently, we say that X has a *binomial distribution*. The range space is obviously $\{0, 1, \ldots, n\}$. The individual realizations of E are called *Bernoulli's trials*.

Theorem 5.1: For the binomial variable X defined above, it holds

$$P(X = k) = \binom{n}{k} p^k (1-p)^{n-k} \qquad \text{for } k = 0, \ 1, \ \ldots, \ n. \tag{5.3}$$

Proof: As illustrated in Example 5.2, there are $\binom{n}{k}$ sequences corresponding to the event "$X = k$," namely, the permutations of the sequence $A, \ldots, A, \bar{A}, \ldots, \bar{A}$ (where A and $\bar{A}$ occur k and $n - k$ times, respectively). Due to the independence of the individual experiments, any of these sequences appears with probability $p^k(1 - p)^{n-k}$. Thus it follows the assertion.

We can easily verify that (5.3) defines a legitimate probability distributions, i.e., that (5.1) holds. Obviously, each summand in (5.3) is nonnegative and from the binomial theorem (1.1), it follows by setting $a = p$ and $b = 1 - p$:

$$\sum_{k=0}^{n} \binom{n}{k} p^k (1-p)^{n-k} = (p + 1 - p)^n = 1^n = 1.$$

The binomial distribution can be often used when independent repetitions of an experiment with dichotomous outcomes are performed.

Example 5.3: A certain type of lamp works more than 600 h with probability $p = 0.1$. If we test 30 lamps, what is the probability that exactly k of them work more than 600 h, $k = 0, 1, \ldots, 30$?

From Theorem 5.1, we obtain the required probability as $p(k) = P(X = k) = \binom{30}{k} 0.1^k \cdot 0.9^{30-k}$.

The distribution is plotted in Fig. 5.2.

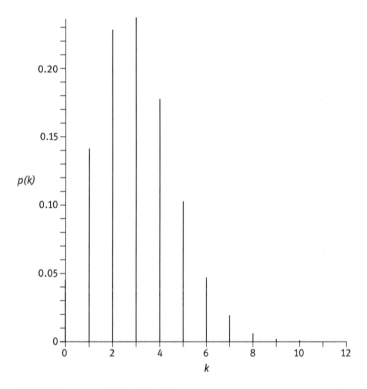

Fig. 5.2: Discrete probability distribution.

The probabilities increase monotonically until a maximum value is reached and then they decrease monotonically. In the given example, the first five numerical values are $p(0) \approx 0.042$, $p(1) \approx 0.141$, $p(2) \approx 0.228$, $p(3) \approx 0.236$, $p(4) \approx 0.177$. In the general case, there is an index $k_0 \in \{0, 1, \ldots, n\}$ such that the binomial probabilities $p(k)$ satisfy $p(0) < p(1) < \cdots < p(k_0) \geq p(k_0 + 1) > \cdots > p(n)$ (see Exercise 5.11).

The binomial distribution can be directly applied only when the probability of the event A is the same for all repetitions of E (see Definition 5.3). If we allow different probabilities, we obtain the Poisson's binomial distribution already mentioned in Chapter 4 (see Example 4.9). However, if the probability of event A assumes only *a few different values*, an approach of the following type is practicable.

Example 5.4: Suppose that 25 articles of a certain type are produced by the two machines A and B. The items are classified as perfect or defective. An article produced by A and B is defective with probability $p = 0.1$ and $q = 0.05$, respectively. The machines A and B produce 10 and 15 articles, respectively. What is the probability that exactly k among the 25 produced articles are defective ($k = 0, 1, \ldots, 25$)?

Let X and Y denote the numbers of defective articles produced by A and B. Thus, the total number of defective articles produced is $Z = X + Y$, which is a sum of two binomial random variables. Evidently, we obtain

$$P(Z = k) = P(X + Y = k) = \sum_{i=0}^{k} P(X = i)P(Y = k - i)$$

$$= \sum_{i=0}^{\min\{k,10\}} \binom{10}{i} 0.1^i \cdot 0.9^{10-I} \binom{15}{k-i} 0.05^{k-i} \cdot 0.95^{15-k+i}.$$

$$(5.4)$$

This is *not* a binomial distribution, but the pattern of the graphical representation of these probabilities is similar to that of Fig. 5.2. In principle, this approach applies also for more than two machines, but the calculations become very expensive when the number of machines increases.

5.4 Continuous random variables

In the preceding section we have assigned a probability $p(x_i)$ to each value x_i of a random variable X, satisfying conditions (5.1) . However, if a random variable is not discrete (see Definition 5.2), the sample space is uncountably infinite and we cannot speak of the ith value x_i of X, hence $p(x_i)$ is meaningless.

How can we define a concept for continuous random variables that replaces appropriately the previous concept of the probability function? This concept must enable us to assign probabilities to outcomes and events of X, where events are usually of the form $\{a \le X \le b\}$. We proceed formally as follows.

Definition 5.4: X is called a *continuous random variable*, if there is a function f, called the *probability density function* (pdf), satisfying the conditions
(i) $f(x) \ge 0$ for all x,
(ii) $\int_{-\infty}^{\infty} f(x)dx = 1$,
    and the probability of an event $\{a \le X \le b\}$ is given by
(iii) $P(a \le X \le b) = \int_{a}^{b} f(x)dx$ for any a, b with $-\infty \le a < b \le \infty$.

We are essentially saying that X is a continuous random variable if it may assume all values in some interval $[a, b]$. The intuitively reasonable assumption that a pdf exists allows us to dispense with a complicated formalism (although we do not achieve the greatest possible generality). Figure 5.3 shows an example of a pdf. The probability $P(a \leq X \leq b)$ represents the area under the graph between a and b. The conditions (i) and (ii) correspond to conditions (5.1) and guarantee that the probability of any event is a value between 0 and 1.

As a consequence of Definition 5.4(iii), the probability of any single outcome x_0 of a continuous random variable is zero, since $P(x_0) = P(x_0 \leq X \leq x_0) = \int_{x_0}^{x_0} f(x)dx = 0$. It might be difficult to accept this intuitively; however, the following idea should help to become familiar with this fact. Let us consider the probability $p_\varepsilon = P(75 - \varepsilon \leq X \leq 75 + \varepsilon)$ that the weight (in kg) of a randomly selected person is in the interval $[75 - \varepsilon, 75 + \varepsilon]$. Independent of a specific probabilistic model, the probability decreases if ε decreases, and it is consistent with our intuition that p_ε tends to zero as ε tends to zero. Hence, the weight equals 75 kg with probability zero. Moreover, probability zero is not equivalent to impossibility, since always any specific value must occur as an outcome of the random experiment. As a further consequence of $P(x_0) = 0$, we obtain that the probabilities $P(c \leq X \leq d)$, $P(c \leq X < d)$, $P(c < X \leq d)$ and $P(c < X < d)$ are all equal in the continuous case.

The development of an adequate probabilistic model is now reduced to the construction of a pdf, i.e., a function satisfying conditions (i) and (ii) in Definition 5.4. We consider some examples of pdfs.

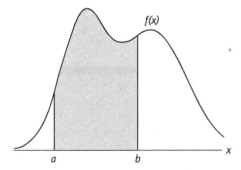

Fig. 5.3: Continuous distribution.

Example 5.5: The variable X has the pdf (see Fig. 5.4):

$$f(x) = \begin{cases} \dfrac{4}{65}x^3 & \text{for } 2 \leq x \leq 3, \\ 0 & \text{elsewhere.} \end{cases}$$

(i) Show that f is a pdf.
(ii) Determine the probability $P(1.5 \leq X \leq 2.5)$.

Solution:

(i) Obviously $f(x) \geq 0$ for all x. Moreover,

$$\int_{-\infty}^{\infty} f(x)dx = \int_{2}^{3} \frac{4}{65} x^3 dx = \left[\frac{x^4}{65}\right]_{2}^{3} = 1.$$

(ii) $P(1.5 \leq X \leq 2.5) = \int_{2}^{2.5} \frac{4}{65} x^3 dx = \left[\frac{x^4}{65}\right]_{2}^{2.5} = \frac{2.5^4 - 2^4}{65} \approx 0.355.$

If the pdf of a random variable is given, we can identify the sample space as the region of the real axis where the pdf has positive values. In particular, the sample space of Example 5.5 is the interval [2, 3].

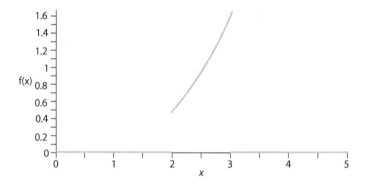

Fig. 5.4: Bounded range space.

If a function f is positive over an interval, one can easily obtain a pdf, multiplying f by an adequate constant C.

Example 5.6: Let X be the lifetime of a certain electronic component (in hours). Suppose that the pdf is given by

$$f(x) = \begin{cases} \frac{C}{x^2} & \text{for } 1{,}000 \leq x \leq 2{,}000, \\ 0 & \text{elsewhere.} \end{cases}$$

The pdf implies that we are assigning probability zero to the events $\{X < 1{,}000\}$ and $\{X > 2{,}000\}$.

Solution: The value of C follows from condition (ii) in Definition 5.4 as

$$\int_{1{,}000}^{2{,}000} \frac{C}{x^2} dx = 1 \Rightarrow \frac{1}{C} = \int_{1{,}000}^{2{,}000} \frac{1}{x^2} dx = \left[\frac{-1}{x}\right]_{1{,}000}^{2{,}000} = \frac{1}{2{,}000} \Rightarrow C = 2{,}000.$$

The constant C is called a *normalizing constant*. A further example for a simple type of pdf is given as follows.

Definition 5.5: Assume that X is a continuous random variable assuming values in the interval $[a, b]$, where a and b are finite $(a < b)$. If the pdf is given by

$$f(x) = \begin{cases} \frac{1}{b-a} & \text{for } a \le x \le b, \\ 0 & \text{elsewhere.} \end{cases}$$

We say that X is *uniformly distributed* over the interval $[a, b]$ (see Fig. 5.5).
In particular, for values c and d with $a \le c < d \le b$, we obtain

$$P(c \le X \le d) = \int_{c}^{d} \frac{1}{b-a} dx = \frac{d-c}{b-a}.$$

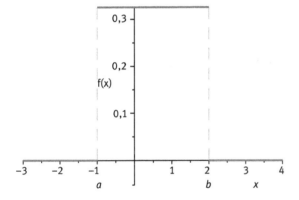

Fig. 5.5: Continuous uniform distribution for $a = -1$, $b = 2$.

5.5 Distribution and survival function

We now consider another concept to describe a distribution which will be particularly useful in the study of functions of random variables (see Chapter 6).

Definition 5.6: Let X be a discrete or continuous random variable. The *(cumulative) distribution function* of X (abbreviated as cdf) is defined by $F(x) = P(X \le x)$.

From Definition 5.1, it follows that $F(x)$ is defined on the entire real line. We obtain immediately

$$F(x) = \int_{-\infty}^{x} f(s) \, ds \tag{5.5}$$

for continuous and

$$F(x) = \sum_{j} p(x_j) \tag{5.6}$$

for discrete variables, where the sum is taken over all indices j satisfying $x_j \le x$. We study an example for both cases.

Example 5.7: To determine $F(x)$ for the density of Example 5.5, we have to distinguish the three cases $x \le 2$, $2 < x < 3$ and $x \ge 3$, yielding (see Fig. 5.6)

$$F(x) = \begin{cases} 0 & \text{for } x \le 2, \\ \int_{2}^{x} \frac{4}{65} s^3 ds = \frac{x^4 - 16}{65} & \text{for } 2 < x < 3, \\ 1 & \text{for } x \ge 3. \end{cases}$$

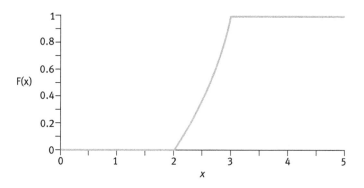

Fig. 5.6: Distribution function.

Example 5.8: Let X be the result of tossing a die, then

$$F(x) = \begin{cases} 0 & \text{for } x < 1 \\ 1/6 & \text{for } 1 \le x < 2 \\ 2/6 & \text{for } 2 \le x < 3 \\ \vdots & \vdots \\ 5/6 & \text{for } 5 \le x < 6 \\ 1 & \text{for } x \ge 6 \end{cases}$$

(see Fig. 5.7). It is very important to indicate the inclusion or exclusion of the end-points in describing the intervals. For example, for any x from the right half-open interval $[2, 3[= \{x \mid 2 \le x < 3\}$, it holds that $F(x) = P(X \le x) = F(2) = 2/6$, since for all these values x there exist two outcomes, namely 1 and 2, which are smaller or equal than x.

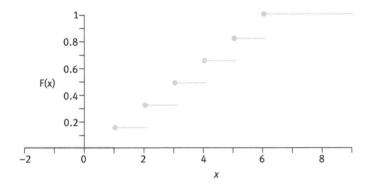

Fig. 5.7: Step function.

The graphs for the cdfs in Figs. 5.6 and 5.7 are quite typical. If X is a continuous random variable, F is continuous for all x. If X is discrete with a finite number of values, the graph is a step function. The points of discontinuity are the possible values $x_1, \ldots, x_n$ of X. At the points x_i, the function is continuous from the right and the graph shows a "jump" of magnitude $p(x_j) = P(X = x_j)$.

In general, a cdf has the properties summarized in the following statement.

Theorem 5.2: A function $F(x)$ is a distribution function if and only if it has the following properties:

(i) F is nondecreasing, i.e., if $x_1 \le x_2$, then $F(x_1) \le F(x_2)$.

(ii) $\lim\limits_{x \to -\infty} F(x) = 0$ and $\lim\limits_{x \to \infty} F(x) = 1$
(sometimes simply written as $F(-\infty) = 0$ and $F(\infty) = 1$).

(iii) In all points, F is continuous from the right, that is, $\lim\limits_{x \to x_0 +} F(x) = F(x_0)$ for all x_0. From the above definitions and considerations, it follows immediately that every cdf has the properties of the theorem. Vice versa any function with these properties is a cdf. The proof of the latter requires concepts of measure theory and will not be given here.

From a cdf, we can also recover the corresponding pdf or probability function as follows.

Theorem 5.3:
(i) If F is the cdf of a continuous random variable, then

$$f(x) = \frac{d}{dx} F(x)$$

for all x at which F is differentiable.
(ii) If the discrete variable X has values $x_1, x_2, \ldots$, then

$$p(x_1) = F(x_1) \text{ and } p(x_j) = F(x_j) - F(x_{j-1}) \text{ for } j = 2, 3, \ldots.$$

Example 5.9: The function $F(x) = 1 - (1/x)$ for $x \geq 1$, $F(x) = 0$ for $x < 1$ satisfies the conditions of Theorem 5.2 and is therefore a cdf. The function is differentiable for $x \neq 1$, hence differentiating results in the density $f(x) = 1/x^2$ for $x \geq 1$ and $f(x) = 0$ for $x < 1$.

Example 5.10: Consider the step function $F(x) = 0$ for $x < -1$, $F(x) = 1/4$ for $-1 \leq x < 1$, $F(x) = 1/2$ for $1 \leq x < 4$ and $F(x) = 1$ for $x \geq 4$ which is continuous from the right. Then F is the cdf of the discrete variable with values $x_1 = -1$, $x_2 = 1$ and $x_3 = 4$ (discontinuities of F). From Theorem 5.3(ii), we obtain the probabilities $p(x_1) = F(x_1) = 1/4$, $p(x_2) = F(x_2) - F(x_1) = 1/2 - 1/4 = 1/4$ and $p(x_3) = F(x_3) - F(x_2) = 1 - 1/2 = 1/2$.

Definition 5.7: Let X be a discrete or continuous random variable. The *survival function* or complementary distribution function is defined as $S(x) = P(X > x) = 1 - F(x)$.

When T represents, e.g., the lifetime of a person or device, $S(t) = P(T > t)$ is the probability that the person or device survives the time t.

Exercises

5.1 In tossing a coin, a head comes up with probability 0.6 and a tail with probability 0.4. This coin is tossed three times and X denotes the number of heads that appear. Determine the probability distribution and the cdf of X. Illustrate both graphically.

5.2 From a lot containing 30 items, 7 of which are defective, 5 are chosen at random. Let X denote the number of defectives found. Determine the probability distribution of X if the items are chosen
  (i) without replacement and
  (ii) with replacement.

5.3 A discrete variable assumes positive integers such that $P(X = k) = 1/2^k$ for $k = 1$, 2, ... Calculate the probabilities
  (i) $P(X$ is odd),
  (ii) $P(X \leq 8)$,
  (iii) $P(X$ is divisible by 5).

5.4 Let X denote the number of girls among 10 children born in a hospital. Determine the probability function of X if both sexes are equally likely.

5.5 Consider the probability function

$$f(x) = \begin{cases} 1/10 & \text{for } x = -2, \\ 2/5 & \text{for } x = -1, \\ 1/10 & \text{for } x = 0, \\ 2/5 & \text{for } x = 1, \\ 0 & \text{elsewhere.} \end{cases}$$

Calculate
  (i) $P(|X| \leq 1)$,
  (ii) $P(X > -1)$,
  (iii) $P(-3 < X < 1)$.

5.6 Tom has invited 10 of his friends to his birthday party and all have confirmed that they will come. But from experience, Tom knows that his friends are not very reliable. Assuming that they appear independently of each other and that each friend appears with probability 0.8, what is the probability that at least eight friends will come to the birthday party?

5.7 What is the probability that at least k times a six appears when a fair die is tossed n times?

5.8 Consider a one-dimensional random walk on the number line. A marker is initially placed at the point zero. In each step, it is moved one unit to the right with probability p and to the left with probability $1 - p$. Assume that n steps have been made.
  (i) Determine the probability that the marker is located at position k.
  (ii) Assume that $n = 100$ and $p = 1/2$. What is the probability that the marker is at least 30 units away from the origin?

5.9 One of the participants of a game must correctly answer at least four out of five questions. To each question are given four alternative answers, where exactly one is correct. If the participant cannot answer any question and chooses the alternatives at random, what is the possibility that he wins the game?

5.10 Formulate a recurrence formula for the binomial probabilities $p_n(k) = \binom{n}{k} p^k$ $(1-p)^{n-k}$.

5.11: (i) Show that the probabilities $p_n(k)$ defined in the previous exercise satisfy the relation $p_n(0) < p_n(1) < \cdots < p_n(k_0) \geq p_n(k_0 + 1) > \cdots > p_n(n)$ for an index $k_0 \in \{0, \ldots, n\}$.
(ii) Verify that two cases are possible:
(a) $p_n(k_0) > p_n(k_0 + 1)$, i.e., $p_n(k)$ is maximal for $k = k_0$,
(b) $p_n(k_0) = p_n(k_0 + 1)$, i.e., $p_n(k)$ assumes its maximum value for the *two values* $k = k_0$ and $k = k_0 + 1$.

5.12 Illustrate the probabilities $P(Z = k)$ in (5.4) graphically.

5.13 Generalize formula (5.4) for the case with three machines.

5.14 Assume that in Definition 5.3 the probabilities of the event A may be different for all realizations of E. Generalize formula (5.3) for this case.

5.15 (i) Determine the cdf for the density

$$f(x) = \begin{cases} (1-x)/4 & \text{for } 0 \leq x < 1, \\ (x-1)/2 & \text{for } 1 \leq x < 2, \\ 1/2 & \text{for } 2 \leq x < 3, \\ 7/2 - x & \text{for } 3 \leq x < 3.5, \\ 0 & \text{elsewhere.} \end{cases}$$

(ii) Calculate the probabilities $P(1.5 \leq X \leq 2.5)$ and $P(2 \leq X \leq 3.5)$.

5.16 Given the density

$$f(x) = \begin{cases} C\sqrt{x} & \text{for } 0 \leq x < 1, \\ C\left(1-(x-1)^2\right) & \text{for } 1 \leq x < 2, \\ 0 & \text{elsewhere.} \end{cases}$$

Determine (i) the value of C and (ii) the cdf.

5.17 Given the density $f(x) = 3x^2$ for $0 \leq x \leq 1$. Calculate $P(X \geq 3/4 \mid X \geq 1/2)$.

5.18 Let $f_1, \ldots, f_n$ be densities over an interval $[a,b]$ and $\lambda_1, \ldots, \lambda_n$ be nonnegative numbers such that $\lambda_1 + \cdots + \lambda_n = 1$. Show that $\lambda_1 f_1(x) + \cdots + \lambda_n f_n(x)$ is a pdf. (This pdf is called a *convex combination* of f_i or a *mixture distribution*.)

5.19 Suppose that the lifetime (in hours) of a certain electronic component is a continuous variable with pdf $f(x) = 150/x^2$ for $x > 150$ and 0 for $x \leq 150$.
(i) What is the probability that a component will last less than 250 h if it is known that it is still functioning after 200 h of service?
(ii) If three such components are in use, what is the probability that exactly one must be replaced after 200 h of service?

5.20 Construct a cdf based on the arctangent function (see Theorem 5.2) and determine the corresponding pdf.

5.21 Determine the cdf for the probability function in Exercise 5.5.

5.22 (i) Determine the cdf and the survival function for the density $f(x) = 3e^{-3x}$ for $x \geq 0$ and 0 elsewhere.

(ii) Calculate $P(2 \leq X \leq 5)$, $P(X = 1)$ and $P(-3 \leq X \leq 10)$.

5.23 Determine the probability function corresponding to the cdf

$$F(x) = \begin{cases} 0 & \text{for } x < 2 \\ 1/5 & \text{for } 2 \leq x < 5, \\ 3/5 & \text{for } 5 \leq x < 10, \\ 1 & \text{for } x \geq 10 \end{cases}$$

5.24 Determine the pdf corresponding to the cdf

$$F(x) = \begin{cases} 0 & \text{for } x < -1 \\ (1+x^3)/4 & \text{for } -1 \leq x < 1, \\ x/2 & \text{for } 1 \leq x < 2, \\ 1 & \text{for } x \geq 2. \end{cases}$$

5.25 Given the pdf $f(x) = 3/x^2$ for $2 \leq x \leq b$, determine the corresponding cdf and the survival function.

5.26 (i) Show that for given parameters n and p, the cdf $F(k) = \sum_{j=0}^{k} \binom{n}{j} p^j (1-p)^{n-j}$ of the binomial distribution is identical to the function $G(k) = (n-k)\binom{n}{k}$ $\int_0^{1-p} y^{n-k-1}(1-y)^k dy$ for $k = 0, 1, \ldots, n$.

(ii) Express $F(k)$ in terms of the incomplete beta function.

6 Functions of random variables

There are many practical situations imaginable in which the distribution of a random variable X may be known and one would like to determine the distribution of another random variable Y, which can be expressed as a function of X. For example, if a physicist has determined the distribution of the velocity of a particle, he/she may want to determine the distribution of the kinetic energy (see Example 6.1). Or if an economist has found the distribution of sales for a given product, he/she might want to derive the distribution of the profit. In the following, we will express the distribution of the function $Y = \psi(X)$ in terms of the distribution of X.

6.1 Continuous random variables

In this section, we assume that X and $Y = \psi(X)$ are continuous random variables and that ψ is a continuous function. By F and G, we denote the cdfs of X and Y, and f and g represent the pdfs of X and Y. We first consider the case in which ψ is monotone increasing or decreasing, respectively (see Fig. 6.1).

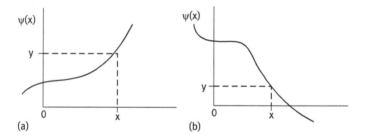

Fig. 6.1: Monotone functions of a r.v.: (a) increasing and (b) decreasing.

If ψ is monotone increasing, the events $\{X \leq x\}$ and $\{Y \leq y\}$ are equivalent (see Section 5.2), where x and y are values of X and Y, related by $y = \psi(x)$ or $x = \psi^{-1}(y)$, respectively. It follows immediately that $G(y) = P(Y \leq y) = P(X \leq x) = F(x)$. Hence, we can write

$$G(\psi(x)) = F(x) \tag{6.1}$$

or

$$G(y) = F\big(\psi^{-1}(y)\big). \tag{6.2}$$

Differentiating these equations by means of the chain rule, we get the following relations between the densities of X and Y:

https://doi.org/10.1515/9783111332277-006

$$f(x) = g(\psi(x))\psi'(x) \tag{6.3}$$

$$g(y) = f(\psi^{-1}(y))(\psi^{-1})'(y). \tag{6.4}$$

Example 6.1: Let X denote the velocity of a particle. From classical mechanics, it is known that the kinetic energy is given by the well-known relation $Y = \psi(X) = (m/2)X^2$, where X assumes nonnegative values and m is the mass of the particle. Now the event $\{X \le x\}$ related to the velocity is equivalent to the event $\{Y \le y\}$ related to the energy, where $y = \psi(x) = (m/2)x^2$. For example, the events $\{X \le 3\}$ and $\{Y \le (m/2)3^2\}$ are equivalent, and from relation (6.2) we obtain the cdf of the kinetic energy as $G(y) = F(\psi^{-1}(y)) = F(\sqrt{2y/m})$. Differentiating yields the pdf of the energy: $g(y) = f(\sqrt{2y/m})(1/\sqrt{2ym})$, see (6.4).

When the function ψ is monotone decreasing (see Fig. 6.1(b)), then the events $\{X \le x\}$ and $\{Y \ge y\}$ are equivalent. We get $P(Y \ge y) = P(X \le x)$, hence

$$1 - G(y) = F(x).$$

Expressing x in terms of y, we obtain

$$G(y) = 1 - F(\psi^{-1}(y)). \tag{6.5}$$

Differentiating yields

$$g(y) = -f(\psi^{-1}(y))(\psi^{-1})'(y) = f(\psi^{-1}(y))[-(\psi^{-1})'(y)]. \tag{6.6}$$

The main results are given by (6.4) and (6.6). The two cases can be unified as follows.

Theorem 6.1: Let X and Y be continuous random variables such that $Y = \psi(X)$, where ψ is differentiable and monotone increasing or monotone decreasing for all x. Then the pdf of Y can be written as

$$g(y) = f(\psi^{-1}(y)) \, |(\psi^{-1})'(y)|.$$

Proof: We observe that if ψ is monotone increasing, then the inverse function ψ^{-1} is also monotone increasing, thus $(\psi^{-1})'(y) \ge 0$ for all y. Therefore the statement follows from (6.4). Similarly, if ψ is monotone decreasing, then ψ^{-1} is monotone decreasing, thus $(\psi^{-1})'(y) \le 0$ for all y. The term in brackets of the relation (6.6) can therefore be written as $|(\psi^{-1})'(y)|$.

Example 6.2: Assume that the r.v. X has the pdf $f(x) = \frac{2}{3}x$ for $1 \le x \le 2$ and 0 elsewhere. Determine the pdf of the variable $Y = \psi(X) = -3X + 8$.

Solution: From $y = -3x + 8$, it follows that $x = \frac{8-y}{3} = \psi^{-1}(y)$ and $(\psi^{-1})'(y) = -\frac{1}{3}$. From Theorem 6.1, we obtain

$$g(y) = f\left(\frac{8-y}{3}\right)\left|-\frac{1}{3}\right| \qquad \text{for } 1 \le \frac{8-y}{3} \le 2,$$

$$= \frac{2}{3}\frac{8-y}{3}\frac{1}{3} \qquad \text{for } 3 \le 8-y \le 6,$$

$$= \frac{2}{27}(8-y) \qquad \text{for } 2 \le y \le 5.$$

One can easily verify that $g(y)$ is a pdf. In particular, the last example illustrates the general relation

$$R_Y = \psi(R_X) \tag{6.7}$$

between the range spaces of X and Y. In the example, we have $R_X = [1,2]$ and $\psi(R_X) = \psi([1,2]) = [2,5] = R_Y$.

Example 6.3: Assume that X is exponentially distributed, i.e., $f(x) = ae^{-ax}x$ for $x \ge 0$ and 0 for $x < 0$ ($a > 0$). Determine the pdf of the variable $Y = \psi(X) = 3/(X+1)^2$.

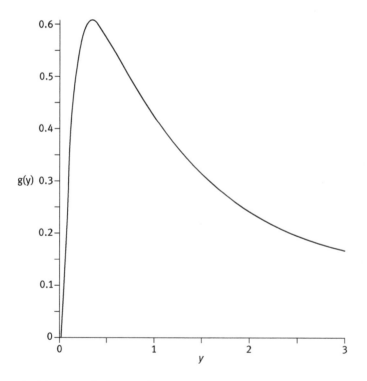

Fig. 6.2: Density of a function of a r.v.

Solution: We get $\psi^{-1}(y) = \sqrt{\frac{3}{y}} - 1 = \sqrt{3}\, y^{-1/2} - 1$ and $(\psi^{-1})'(y) = -\frac{\sqrt{3}}{2} y^{-3/2}$, hence

$$g(y) = f\left(\sqrt{\frac{3}{y}} - 1\right) \frac{\sqrt{3}}{2} y^{-3/2}$$

$$= ae^{-a(\sqrt{3/y}-1)} \frac{\sqrt{3}}{2} y^{-3/2} \qquad \text{for } \sqrt{\frac{3}{y}} - 1 \geq 0$$

$$= a\frac{\sqrt{3}}{2} e^{-a(\sqrt{3/y}-1)} y^{-3/2} \qquad \text{for } 0 < y \leq 3.$$

The density is illustrated in Fig. 6.2 for $a = 1$.

We now illustrate two cases in which the function ψ is not monotone. In the above situations, the determination of the density of Y was based on the idea to iden- tify an event related to X which is equivalent to the event $\{Y \leq y\}$ or to its complement. We will see that this procedure may be complicated, when the graph of ψ performs many alternating up and down movements. In such a case, the equivalent event may be a complicated disconnected set.

Example 6.4: Assume that a point P moves at random on the horizontal axes such that its position X is a continuous r.v. Let $Y = \psi(X) = \sqrt{1 + X^2}$ denote the distance be- tween P and the point $Q = (0, -1)$, see Fig. 6.3(a).

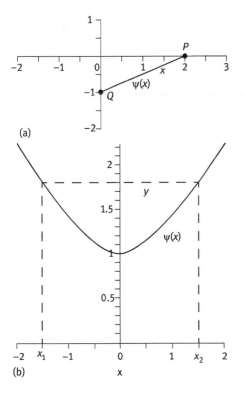

(a)

(b)

Fig. 6.3: Random movement.

Figure 6.3(b) illustrates that for $y > 1$ the event $\{Y \le y\}$ is equivalent to $\{x_1 \le X \le x_2\}$, where $x_1 = -\sqrt{y^2 - 1}$ and $x_2 = \sqrt{y^2 - 1}$ denote the two elements of the inverse image $\psi^{-1}(y)$ of y. Hence, the relation between the cdfs of X and Y is given by

$$G(y) = P(Y \le y) = P(x_1 \le X \le x_2) = F(x_2) - F(x_1) = F(\sqrt{y^2 - 1}) - F(-\sqrt{y^2 - 1}). \qquad (6.8)$$

Differentiating with respect to y gives the density of Y as

$$g(y) = f\left(\sqrt{y^2 - 1}\right) \frac{y}{\sqrt{y^2 - 1}} - f\left(-\sqrt{y^2 - 1}\right) \frac{-y}{\sqrt{y^2 - 1}}$$

$$= \frac{y}{\sqrt{y^2 - 1}} \left(f\left(\sqrt{y^2 - 1}\right) + f\left(-\sqrt{y^2 - 1}\right) \right) \text{ for } y \ge 1. \qquad (6.9)$$

If X has, for example, the "linear density" $f(x) = \dfrac{1}{4} + \dfrac{x}{8}$ for $-2 \le x \le 2$ and $f(x) = 0$ elsewhere, then

$$g(y) = \frac{y}{\sqrt{y^2 - 1}} \left(\frac{1}{4} + \frac{\sqrt{y^2 - 1}}{8} + \frac{1}{4} - \frac{\sqrt{y^2 - 1}}{8} \right) = \frac{y}{2\sqrt{y^2 - 1}} \text{ for } 1 \le y \le \sqrt{5}.$$

Note that the range space of the distance Y is $[1, \sqrt{5}]$, since X assumes values between -2 and 2.

Example 6.5: Assume that the continuous variable X has the density $f(x) = x^2/18$ for $-3 \le x \le 3$ and $f(x) = 0$ elsewhere. What is the density of the variable $Y = \psi(X) = (X^2 - 4)^2$?

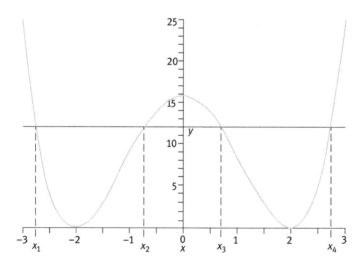

Fig. 6.4: Oscillating function ψ.

For $0 \le y \le 16$, the inverse image $\psi^{-1}(y)$ consists of the four real numbers $x_1 = -\sqrt{4 + \sqrt{y}}$, $x_2 = -\sqrt{4 - \sqrt{y}}$, $x_3 = \sqrt{4 - \sqrt{y}}$ and $x_4 = \sqrt{4 + \sqrt{y}}$. For $16 < y \le 25$, only x_1 and x_4 are real. Similar to the previous example, Fig. 6.4 shows that the event $\{Y \le y\}$ is equivalent to $\{x_1 \le X \le x_2\}$ or $\{x_3 \le X \le x_4\}$ for $0 \le y \le 16$ and to $\{x_1 \le X \le x_4\}$ for $16 < y \le 25$. Hence, the cdf of Y is given by

$$G(y) = \begin{cases} F\left(-\sqrt{4 - \sqrt{y}}\right) - F\left(-\sqrt{4 + \sqrt{y}}\right) + F\left(\sqrt{4 + \sqrt{y}}\right) - F\left(\sqrt{4 - \sqrt{y}}\right) & \text{for } 0 \le y \le 16, \\ F\left(\sqrt{4 + \sqrt{y}}\right) - F\left(-\sqrt{4 + \sqrt{y}}\right) & \text{for } 16 < y \le 25. \end{cases}$$

Differentiating yields

$$g(y) = \begin{cases} \dfrac{f\left(-\sqrt{4-y}\right)}{4\sqrt{4y} - y^{3/2}} - \dfrac{f\left(-\sqrt{4 + \sqrt{y}}\right)}{-4\sqrt{4y} + y^{3/2}} + \dfrac{f\left(\sqrt{4 + \sqrt{y}}\right)}{4\sqrt{4y} + y^{3/2}} - \dfrac{f\left(\sqrt{4 - \sqrt{y}}\right)}{-4\sqrt{4y} - y^{3/2}} & \text{for } 0 \le y \le 16, \\[4mm] \dfrac{f\left(\sqrt{4 + \sqrt{y}}\right)}{4\sqrt{4y} + y^{3/2}} - \dfrac{f\left(-\sqrt{4 + \sqrt{y}}\right)}{-4\sqrt{4y} + y^{3/2}} & \text{for } 16 < y \le 25. \end{cases}$$

$$= \begin{cases} \dfrac{f\left(-\sqrt{4 - \sqrt{y}}\right) + f\left(\sqrt{4 - \sqrt{y}}\right)}{4\sqrt{4y} - y^{3/2}} + \dfrac{f\left(-\sqrt{4 + \sqrt{y}}\right) + f\left(\sqrt{4 + \sqrt{y}}\right)}{4\sqrt{4y} + y^{3/2}} & \text{for } 0 \le y \le 16, \\[4mm] \dfrac{f\left(\sqrt{4 + \sqrt{y}}\right) + f\left(-\sqrt{4 + \sqrt{y}}\right)}{4\sqrt{4y} + y^{3/2}} & \text{for } 16 < y \le 25. \end{cases}$$

$$= \begin{cases} \dfrac{4 - \sqrt{y}}{36\sqrt{4y} - y^{3/2}} + \dfrac{4 + \sqrt{y}}{36\sqrt{4y} + y^{3/2}} & \text{for } 0 \le y \le 16, \\[4mm] \dfrac{4 + \sqrt{y}}{36\sqrt{4y} + y^{3/2}} & \text{for } 16 < y \le 25. \end{cases}$$

$$= \begin{cases} \dfrac{\sqrt{4 - \sqrt{y}} + \sqrt{4 + \sqrt{y}}}{36\sqrt{y}} & \text{for } 0 \le y \le 16, \\[4mm] \dfrac{\sqrt{4 + \sqrt{y}}}{36\sqrt{y}} & \text{for } 16 < y \le 25. \end{cases}$$

The density $g(y)$ is illustrated in Fig. 6.5. This function is continuous and everywhere differentiable except for $y = 16$.

After studying the last two examples, the reader should be able to solve similar problems with a nonmonotone function ψ.

We will not treat the general case here, which would be of minor practical importance. Instead we proceed directly to the discrete case.

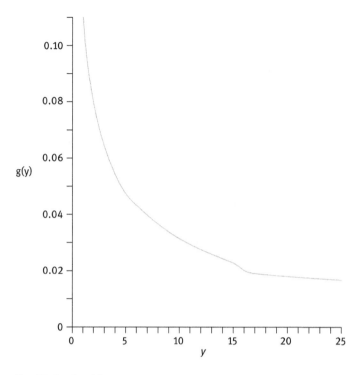

Fig. 6.5: Density of Y.

6.2 Discrete random variables

This topic is much simpler than the continuous case.

Example 6.6: Assume that a participant of a game of chance throws a die and receives the win $Y = \psi(X) = (X - 3)^2 - 3$, where X denotes the result of tossing the die. What is the distribution of Y?

In Tab. 6.1, the corresponding values of X and Y are listed. As in the continuous case, the distribution of Y is obtained by considering equivalent events.

Obviously, both the values -3 and 6 of Y have the probability $1/6$, since the events $Y = -3$ and $Y = 6$ are equivalent to $X = 3$ and $X = 6$, respectively. But, for example, the event $Y = -2$ is equivalent to the event $X \in \{2, 4\}$, where the set $\{2, 4\} = \psi^{-1}(-2)$ is the

Tab. 6.1: Values of X and Y.

X	Y
1	1
2	-2
3	-3
4	-2
5	1
6	6

inverse image of the element −2 with respect to ψ. Thus, $P(Y = -2) = 1/3$. In the same way, we obtain $P(Y = 1) = 1/3$.

Now it should be clear to the reader that the distribution of Y is in general given by

$$P(Y = y) = \sum_{x \in \Psi^{-1}(y)} P(X = x),$$

when X and $Y = \psi(X)$ are discrete random variables.

We still consider a somewhat more complex example which indicates possible applications.

Example 6.7: Assume that a travel agency has sold 20 tourist packages, costing $5,000 per person. To cover the costs for hotels and various touristic events, the company has expanses of 3,500 per client. From previous experience it is known that a client cancels his reservation with probability of 10%. Since the agency guarantees the realization of the journey independently of the number of participants, a high penalty of $500 is charged for each cancelation. For the transport of tourists, the agency may employ one or two buses of the same type, depending on the number of clients appearing at the time of departure. Each bus has seats for 10 passengers and the rent for the entire trip is $6,000 including the driver and a well-trained tour guide.

Solution: The number X of clients appearing at the time of departure is binomially distributed with parameters $n = 20$ and $p = 0.9$, i.e., $P(X = k) = \binom{20}{k} 0.9^k \cdot 0.1^{20-k}$. Disregarding the cost of the buses, the agency wins $1,500 with each customer participating in the trip and $500 with each cancelation. Hence, the profit Y is a random variable given by

$$Y = \psi(X) = \begin{cases} 1{,}500\,X + 500(20 - X) & \text{for } X = 0 \\ 1{,}500\,X + 500(20 - X) - 6{,}000 & \text{for } X = 1, \ldots, 10 \\ 1{,}500\,X + 500(20 - X) - 12{,}000 & \text{for } X = 11, \ldots, 20 \end{cases}$$

Note that for $1 \leq X \leq 10$, one bus is sufficient and for $11 \leq X \leq 20$, two buses are necessary. Simplifying this function yields

$$Y = \psi(X) = \begin{cases} 10,000 & \text{for } X = 0 \\ 1,000X + 4,000 & \text{for } X = 1, \ldots, 10 \\ 1,000X - 2,000 & \text{for } X = 11, \ldots, 20 \end{cases}$$

One obtains easily Tab. 6.2, showing the corresponding values of X and Y.

The possible values of the profit Y are 5,000, 6,000, . . ., 18,000. With the aid of Tab. 6.2, one can easily determine the probabilities of these values. For example, we get

Tab. 6.2: Function of a discrete r.v.

X	Y	X	Y
0	10,000		
1	5,000	11	9,000
2	6,000	12	10,000
3	7,000	13	11,000
4	8,000	14	12,000
5	9,000	15	13,000
6	10,000	16	14,000
7	11,000	17	15,000
8	12,000	18	16,000
9	13,000	19	17,000
10	14,000	20	18,000

$$P(Y = 10,000) = P(X = 0) + P(X = 6) + P(X = 12)$$

$$= 0.1^{20} + \binom{20}{6} 0.9^6 \cdot 0.1^{14} + \binom{20}{12} 0.9^{12} \cdot 0.1^8 \approx 0.0004$$

and in the same way $P(Y = 14,000) = P(X = 10) + P(X = 16) \approx 0.0898$.

Exercises

6.1 Let X be exponentially distributed, e.g., the pdf is $f(x) = ae^{-ax}$ for $x \geq 0$. Determine the pdf of $Y = \sqrt{X}$.

6.2 Let X have the pdf $f(x) = \frac{4}{65}x^3$ for $2 \leq x \leq 3$. Determine the density and the cumulative distribution function of $Y = e^{-X}$. What is the range space of Y?

6.3 If X is uniformly distributed over [0,1], find the pdf of $Y = -a \ln(1 - X)$ for $a > 0$.

6.4 Derive the distribution of $Y = \sin(X)$, where X is uniformly distributed over $[-\pi/2, \pi/2]$.

6.5 The speed of a molecule in a uniform gas at equilibrium is a r.v. V with pdf $f(v) = av^2 e^{-bv^2}$ for $v \geq 0$, where $b = m/2kT$ and k, T, m denote Boltzmann's constant, the absolute temperature and the mass of the molecule, respectively:
(i) Determine the normalization constant a in terms of b.
(ii) Determine the distribution of the kinetic energy $W = mV^2/2$ of the molecule.

6.6 Derive the pdf of $Y = -\ln X$, if X is uniformly distributed over $[2,5]$. What is the range space of Y?

6.7 Express the pdf of $Y = X^2$ in terms of the density $f(x)$ of X.

6.8 Determine the pdf of $Y = [4X(X-1)]^2$ if X has the density $f(x) = \frac{8}{9}x$ over $[0,3/2]$.

6.9 Let X be a continuous random variable with range space $[-1/2,2]$. Determine the cumulative distribution function of $Y = X(X-1)$ in terms of the cdf of X. Illustrate geometrically.

6.10 Let X be a discrete random variable such that $P(X = k) = 1/10$ for $k = 1, \ldots, 10$. Derive the probability function of the variables $Y = (X-5)^2 - 10$ and $Z = |Y|$.

6.11 Tom pretends to celebrate his birthday party with a dinner in a Brazilian restaurant that serves only dishes for two persons. He has invited 10 friends who are nice persons but somewhat unreliable (see Exercise 5.6). Tom assumes that his friends appear independently of each other to his great day and each of them appears with probability 0.7. Every dish for two persons costs $50. If an even number k of friends appears (assuming that they all arrive in time), Tom orders $k/2$ dishes and if an odd number k appears, he orders $(k + 1)/2$ dishes.
What is the distribution of the cost of the dinner (part of the friends)?

6.12 A small factory needs to produce a special tool. Since this tool is rarely produced, the factory has not much experience with the production process. It turns out that the tool is perfect with probability 0.7 and defective with probability 0.3. The production of each tool costs $1,000 and a perfect tool can be sold for $10,000.
Assume that the factory produces tools until the first perfect one is obtained. Denote by X the total number of tools produced and by Y the profit obtained after the sale of the perfect item. Determine the probability functions of X and Y.

6.13 Assume that three dice are thrown having the outcomes X, Y and Z:
(i) Determine the distribution of $X - Y$.
(ii) Determine the distribution of $X - Y + Z$, using the previous item.

7 Bidimensional random variables

In our previous studies we have so far only considered one-dimensional random variables, i.e., we assumed that the outcome of a random experiment could be represented as a single number. However, in many practical situations there are several characteristics associated with the elements of a population or a sample. For example, a physician is interested in several characteristics of a patient, e.g., age, weight, blood pressure, blood sugar values, etc. In evaluating the competitiveness of the countries of a certain community, several characteristics are interesting, as, e.g., an index of unemployment, stock prices, exchange values, etc.

In the following we restrict ourselves to the study of only two characteristics, i.e., we consider bidimensional random variables. We give the following formal definition.

Definition 7.1: Consider a random experiment E with sample space S. Let $X, Y: S \to$ IR be two functions such that the probabilities $P(X \leq c)$ and $P(Y \leq c)$ are defined for any real number c. Then the pair (X, Y) is called a *bidimensional random variable*. The image of the function (X, Y), i.e., the set of pairs $R_{X \times Y} = \{(X(s), Y(s)) \mid s \in S\}$ is called the *range space* of (X, Y).

The range space of a bidimensional r.v. can be interpreted as a subspace of the plane.

If the range space is finite or countably infinite, the r.v. is called *discrete*. In this case the possible values can be represented as (x_i, y_j) for $i = 1, 2, \ldots, j = 1, 2, \ldots$. If (X, Y) can assume all values in some uncountable set of the plane, then it is called a *continuous* r.v. The range set of a continuous bidimensional r.v. can be, for example, a rectangle or an ellipse.

The extension of Definition 7.1 and most other concepts to the n-dimensional case is straight forward.

7.1 Discrete random variables

As a natural generalization of the one-dimensional case we define a (*joint*) *probability function* $p(x, y)$ as a function, satisfying:

$$p(x, y) \geq 0 \text{ for all } x, y, \quad \sum_x \sum_y p(x, y) = 1. \qquad (7.1)$$

Here $p(x, y)$ denotes the probability that X and Y assume the value x and y, respectively; formally we write this as $p(x, y) = P(X = x, Y = y)$. Similar to Section 5.2, the probability of an event B is defined as:

https://doi.org/10.1515/9783111332277-007

$$\sum_{(x,\,y)\in B} p(x, y), \tag{7.2}$$

where the sum runs over all elements (x, y) of the event B.

Example 7.1: Let X, Y denote the number of sons and daughters of a family, randomly chosen from a certain district of a city. The fictive probability distribution is given in the following table.

Tab. 7.1: Bivariate discrete distribution.

X	0	1	2	3	4	$P(Y=i)$
Y						
0	0.02	0.02	0.03	0.08	0.05	0.20
1	0.05	0.10	0.15	0.15	0.05	0.50
2	0.05	0.05	0.10	0.05	0.05	0.30
$P(X=i)$	0.12	0.17	0.28	0.28	0.15	1

It is easily verified that the probabilities satisfy conditions (7.1).

The last column and the last row of Tab. 7.1 contain the sums of the rows and columns, respectively. For example, the first value of the last column is $p(0, 0) + p(1, 0) + \cdots + p(4, 0) = 0.02 + 0.02 + 0.03 + 0.08 + 0.05 = 0.20$. This is the probability that $Y = 0$ and that X has any possible value, i.e., the sum is $P(Y = 0)$. The last column can therefore be interpreted as the distribution of the variable Y. In the same way the last row can be interpreted as the distribution of X. Since the probability distributions of X and Y appear at the margins of the Tab. 7.1, they are called the *marginal distribution* of X and Y, respectively.

We now calculate the probabilities of some events by means of (7.2).

Let be $A = \{X > Y\}$, i.e., A denotes the event that the family has more sons than daughters. We obtain:

$$P(A) = \sum_{x>y} p(x,y) = p(1, 0) + p(2, 0) + p(3, 0) + p(4, 0) + p(2, 1) + p(3,1)$$
$$+ p(4, 1) + p(3, 2) + p(4, 2) + p(4, 3)$$
$$= 0.02 + 0.03 + 0.08 + 0.05 + 0.15 + 0.15 + 0.05 + 0.05 + 0.05 = 0.63.$$

Similarly, the probability of the event $B = \{X = Y\}$ is obtained as $P(B) = p(0, 0) + p(1, 1) + p(2, 2) = 0.02 + 0.1 + 0.1 = 0.22$. Finally, for $C = \{X + Y \geq 4\}$ we get:

$$P(C) = p(2, 2) + p(3, 1) + p(3, 2) + p(4, 0) + p(4, 1) + p(4, 2)$$
$$= 0.1 + 0.15 + 0.05 + 0.05 + 0.05 + 0.05 = 0.45.$$

In order to calculate a probability depending on only one of the variables X, Y, we need only the values of the margins of the table. For example, $P(Y \leq 1) = P(Y = 0) + P(Y = 1) = 0.20 + 0.50 = 0.70$.

One can also calculate conditional probabilities of two events depending on X and Y. For example, the probability that the family has at least three sons if it has no daughter, is:

$$P(X \geq 3 | Y = 0) = \frac{P(X \geq 3, \ Y = 0)}{P(Y = 0)} = \frac{0.08 + 0.05}{0.2} = 0.65.$$

7.2 Continuous random variables

Generalizing Definition 5.4 we obtain the following concept.

Definition 7.2: Let (X, Y) be a continuous bidimensional random variable with range space $R \subset IR^2$. The *joint probability density function* $f(x, y)$ is a function, satisfying the conditions:

$$f(\mathrm{x, \ y}) \geq 0 \text{ for all } (\mathrm{x, \ y}) \in R$$

$$\iint_R f(x, y) \ dy \ dx = 1. \tag{7.3}$$

Without loss of generality the second condition can also be written as

$$\int_{-\infty}^{\infty} \int_{-\infty}^{\infty} f(x, y) \ dy \ dx \tag{7.4}$$

by defining $f(x, y) = 0$ for $(x, y) \notin R$. The probability of an event B is defined as:

$$P(B) = \iint_B f(x, y) \ dy \ dx. \tag{7.5}$$

We illustrate the concept by means of some examples. The reader who needs to refresh knowledge about the integration of bidimensional function is referred to Section 1.4.

Example 7.2: The bidimensional random variable (X, Y) has the joint probability density function $f(x, y) = \frac{1}{9}(2x + 5y)$ over the range space $R = \{(x, y) \,|\, 0 \leq x \leq 2, \ 0 \leq y \leq 1\}$.
(a) Verify that f is a joint pdf.
(b) Illustrate the event $B = \{Y \leq X/3\}$ graphically and evaluate its probability.

Solution:

(a) $\frac{1}{9}\int\limits_{0}^{2}\int\limits_{0}^{1}(2x+5y)\,dy\,dx = \frac{1}{9}\int\limits_{0}^{2}\left(2x+\frac{5}{2}\right)dx = 1.$

(b) $P(B) = \frac{1}{9}\int\limits_{0}^{2}\int\limits_{0}^{x/3}(2x+5y)\,dy\,dx = \frac{1}{9}\int\limits_{0}^{2}\frac{17}{18}x^2\,dx = \frac{68}{243}.$

The event is illustrated in Fig. 7.1.

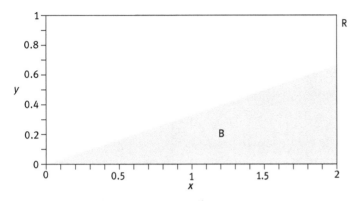

Fig. 7.1: Range space and event B.

Example 7.3: Consider the density $f(x, y) = C(x^2 + 3y)$ over the range space $R = \{(x, y)\,|\,0 \le x \le 2, 0 \le y \le 3\}$.
(a) Determine the normalizing constant C.
(b) Calculate the probability of the event $B = \{Y \le X + 2\}$ and illustrate B geometrically.

Solution:

(a) $\dfrac{1}{C} = \int\limits_{0}^{2}\int\limits_{0}^{3}(x^2+3y)\,dy\,dx = \int\limits_{0}^{2}\left(3x^2 + \dfrac{27}{2}\right)dx = 35$, thus $C = \dfrac{1}{35}$.

Figure 7.2 illustrates that $P(\bar{B})$ can be calculated easier than $P(B)$, since in the second case the upper limit of the inner integral requires consideration of two cases: $y \le 2 + x$ for $x \le 1$ and $y \le 3$ for $x > 1$. Choosing the easier alternative, we obtain:

$$P(\bar{B}) = \frac{1}{35} \cdot \int\limits_{0}^{1}\int\limits_{x+2}^{3}(x^2+3y)\,dy\,dx = \frac{1}{35}\int\limits_{0}^{2}\left(-x^3 - \frac{x^2}{2} - 6x + \frac{15}{2}\right)dx = \frac{7}{60},$$

implying $P(B) = 1 - \dfrac{7}{60} = \dfrac{53}{60}.$

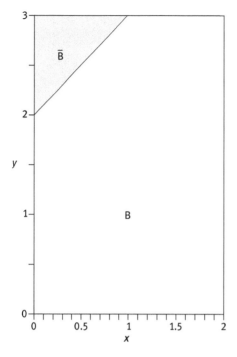

Fig. 7.2: Event B and its complement.

The following example illustrates possible applications.

Example 7.4: Assume that the landing coordinates of a meteorite (in km) represent a bidimensional random variable with density $f(x, y) = C(100 - x^2 - y^2)$ over the range space $R = \{(x, y) \,|\, x^2 + y^2 \leq 100\}$, where the origin of the coordinate systems represent the originally calculated exact point of impact.
(a) Determine the value of C.
(b) What is the probability that the meteorite lands within a radius of 1 km around the point $(0, 0)$?

Solution: (a) By introducing polar coordinates (see Example 1.8) we obtain:

$$\frac{1}{C} = \iint_R (100 - x^2 - y^2)\ dx\ dy = \int_0^{10} \int_0^{2\pi} (100 - r^2 \cos^2(\phi) - r^2 \sin^2(\phi))\ r\ d\phi\ dr$$

$$= \int_0^{10} \int_0^{2\pi} (100 - r^2) r\ d\phi\ dr = \int_0^{10} 2\pi r(100 - r^2)\ dr = 5{,}000\pi. \text{ Hence, } C = \frac{1}{5{,}000\pi}.$$

(b) In the same way, we obtain the required probability as:

$$\frac{1}{5{,}000\pi}\int_0^1\int_0^{2\pi}(100-r^2)r\ d\phi\ dr=\frac{199}{10{,}000}=0.0199.$$

Finally, we consider a very particular case of a joint probability distribution. If the function $f(x, y)$ in Definition 7.2 is a constant $C > 0$, we say that (X, Y) is *uniformly distributed* over the region R. In this case (7.3) implies that:

$$\frac{1}{C}=\iint_R dy\ dx,\tag{7.6}$$

where the integral represents the area of R (assumed to be positive and finite).

7.3 Marginal distributions and independent variables

We will now formalize a concept that has already been used informally in the discrete case (Section 7.1).

Definition 7.3: Let $p(x, y) = P(X = x, Y = y)$ denote the joint probability distribution of a discrete bidimensional r.v. (X, Y). The *marginal probability distribution* of X is defined as

$$g(x)=\sum_y p(x, y).\tag{7.7}$$

Analogously we define the *marginal probability distribution* of Y as

$$h(y)=\sum_x p(x, y).\tag{7.8}$$

The variables X and Y are called *independent*, if

$$p(x, y)=g(x)h(y)\tag{7.9}$$

holds for all possible values x of X and y of Y.

Relation (7.9) can be written as:

$$P(X=x, Y=y)=P(X=x)P(Y=y)$$

for all x and y. Hence, it states that the events $\{X = x\}$ and $\{Y = y\}$ are independent for all x and y.

We can see immediately that, e.g., the variables in Example 7.1 are not independent, since

$$P(X = 0, Y = 0) = 0.02 \neq P(X = 0)P(Y = 0) = 0.12 \cdot 0.2 = 0.024.$$

We consider another example for the discrete case.

Example 7.5: A ball is randomly chosen from an urn, containing five white balls numbered 1 to 5 and five black balls numbered 6 to 10 (see Fig. 7.3).

Fig. 7.3: Urn model.

Let X and Y denote the parity and the color of the ball, respectively, such that

$$X = \begin{cases} 1 \text{ if the number is odd} \\ 2 \text{ if the number is even} \end{cases}, Y = \begin{cases} 1 \text{ if the ball is white} \\ 2 \text{ if the ball is black} \end{cases}.$$

We test whether these variables are independent. The joint probability distribution
and the marginal distributions are shown in Tab. 7.2(a). For example, $P(X = 1, Y = 2)$ is
obviously 2/10, since two of the ten balls in the urn are black and have an odd number
(7 or 9). It can be easily verified that condition (7.9) is not satisfied. In particular,

$$P(X = 0, Y = 0) = 0.3 \neq P(X = 0)P(Y = 0) = 0.5 \cdot 0.5 = 0.25.$$

But if we modify the example, assuming that four balls are white, numbered 1 to 4
and six balls are black, numbered 5 to 10, we obtain the probabilities in Tab. 7.2(b).
Then (7.9) is satisfied and the variables X and Y are independent.

Tab. 7.2: Joint probability distributions.

(a)

X	1	2	$P(Y = y)$
Y			
1	3/10	2/10	5/10
2	2/10	3/10	5/10
$P(X = x)$	5/10	5/10	1

(b)

X	1	2	P(Y = y)
Y			
1	2/10	2/10	4/10
2	3/10	3/10	6/10
P(X = x)	5/10	5/10	1

For the case of continuous variables, we adapt the last definition as follows:

Definition 7.4: Let $f(x, y)$ denote the joint probability density function of a continuous bidimensional r.v. (X, Y) over the range space R. The *marginal probability distributions* of X and Y are defined as

$$g(x) = \int f(x,y)dy \qquad (7.10)$$

and

$$h(y) = \int f(x,y)dx. \qquad (7.11)$$

The variables X and Y are called *independent*, if

$$f(x,y) = g(x)h(y) \qquad (7.12)$$

holds for all possible values x of X and y of Y.

Example 7.6: For the joint probability density function,

$$f(x, y) = \frac{1}{4}(2x + y) \text{ for } 0 \leq x \leq 1, \, 0 \leq y \leq 2 \text{ we obtain}$$

$$g(x) = \int_0^2 \frac{1}{4}(2x + y)dy = x + \frac{1}{2} \text{ for } 0 \leq x \leq 1 \text{ and}$$

$$h(y) = \int_0^1 \frac{1}{4}(2x + y)dx = \frac{y + 1}{4} \text{ for } 0 \leq y \leq 2$$

It can be easily verified that g and h are in fact one-dimensional probability density functions and that $f(x, y) \neq g(x)h(y)$, i.e., X and Y are dependent.

The integration limits in (7.10) and (7.11) are trivial when the integration domain R in Definition 7.4 is rectangular as in the last example (most textbooks are limited to this case). In order to illustrate the general case we consider the following somewhat more complicated situation.

Example 7.7: Consider the density $f(x,y) = \frac{5}{4}(x^2 + y)$ over the range space $R = \{(x, y) \mid -1 \le x \le 1, 0 \le y \le 1 - x^2\}$. We get

$$g(x) = \frac{5}{4} \int_{0}^{1-x^2} (x^2 + y) \, dy = \frac{5}{8}\left(1 - x^4\right) \text{ for } -1 \le x \le 1 \text{ and}$$

$$h(y) = \frac{5}{4} \int_{-\sqrt{1-y}}^{\sqrt{1-y}} (x^2 + y) \, dx = \frac{5}{6}(1 + 2y)\sqrt{1-y} \text{ for } 0 \le y \le 1.$$

In general, the integration domain in calculating $g(x_0)$ is the intersection between R and the vertical line $\{x = x_0\}$. Analogically, the integration domain in calculating $h(y_0)$ is the intersection between R and the horizontal line $\{y = y_0\}$ (see Fig. 7.4). From the definition of the integration domain R it follows that possible outcomes of (X, Y) satisfy $0 \le Y \le 1 - X^2$; thus in the intuitive sense of the concept, the variables cannot be independent. In order to check the formal condition (7.12), we consider a point $(x_0, y_0) \notin R$, lying inside of the rectangle $[-1,1] \times [0,1]$, given by the marginal integration domains (for example, $(x_0, y_0) = (0.5, 0.9)$). Such a point satisfies $0 = f(x_0, y_0) \ne g(x_0)h(y_0) > 0$, i.e., (7.12) is not satisfied.

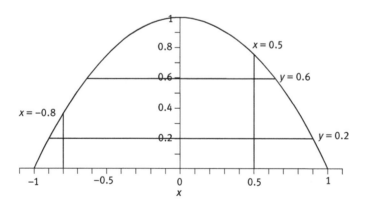

Fig. 7.4: Some integration domains for the marginal densities.

We now consider a case of independent random variables.

Example 7.8: For the density $f(x, y) = 3e^{-x-3y}$ over the range space $R = \{(x, y) \mid x, y \geq 0\}$ we obtain:

$$g(x) = \int_0^\infty 3e^{-x-3y}dy = e^{-x} \text{ for } 0 \leq x < \infty \text{ and}$$

$$h(y) = \int_0^\infty 3e^{-x-3y}dx = 3e^{-3y} \text{ for } 0 \leq x < \infty.$$

Thus $f(x, y) = 3e^{-x-3y} = g(x) h(y) = e^{-x}3e^{-3y}$ and the variables X and Y are independent.

Assume that X and Y are independent r.v. If there are two events A, B such that A depends only on X and B depends only on Y, then A and B are independent events.

We illustrate this fact by means of the last example, defining $A = \{X > 1\}$ and $B = \{Y > 2\}$. Using the marginal distributions, we get:

$$P(A) = \int_1^\infty e^{-x}dx = e^{-1}, \ P(B) = \int_2^\infty 3e^{-3y}dy = e^{-6}, \ P(A \cap B) = \int_1^\infty \int_2^\infty 3e^{-x-3y} \, dy \, dx$$

$$= \int_1^\infty e^{-x-6}dx = e^{-7}.$$

Hence $P(A \cap B) = P(A) P(B)$, i.e., A and B are independent events.

7.4 Conditional distributions, distribution and survival functions

These concepts may be introduced in a very natural way.

Definition 7.5: Let (X, Y) be a discrete bidimensional random variable with probability function p and marginal distributions g and h. Given $Y = y$, the *conditional probability density function* of X is defined as

$$g(x|y) = \frac{p(x, y)}{h(y)} \text{ for } h(y) > 0; \tag{7.13}$$

given $X = x$, the *conditional probability density function* of Y is defined as

$$h(y|x) = \frac{p(x, y)}{g(x)} \text{ for } g(x) > 0. \tag{7.14}$$

For example, for the distribution in Tab. 7.1, we get $g(x|2) = 1/3$ for $x = 2$ and $1/6$ for other x. Moreover, $h(y|1) = 2/17, 10/17, 5/17$ for $y = 0, 1, 2$, respectively.

Definition 7.6: Let (X, Y) be a continuous bidimensional random variable with joint pdf f and marginal densities g and h. Given $Y = y$, the *conditional probability density function* of X is defined as

$$g(x|y) = \frac{f(x, y)}{h(y)} \quad \text{for } h(y) > 0; \tag{7.15}$$

given $X = x$, the *conditional probability density function* of Y is defined as

$$h(y|x) = \frac{f(x, y)}{g(x)} \quad \text{for } g(x) > 0. \tag{7.16}$$

For the continuous distribution in Example 7.7 we obtain, e.g.,

$$g(x|0.2) = \frac{f(x, 0.2)}{h(0.2)} = \frac{1.25 \cdot (x^2 + 0.2)}{\frac{5}{6} \cdot 1.4 \cdot \sqrt{0.8}}$$

$$= \frac{15}{28} \sqrt{5} (x^2 + 0.2) \text{ for } -\sqrt{0.8} \le x \le \sqrt{0.8},$$

$$h(y| - 0.8) = \frac{f(-0.8, y)}{g(-0.8)} = \frac{1.25 \cdot (0.8^2 + y)}{\frac{5}{8} (1 - 0.8^4)}$$

$$= \frac{1,250}{369} (y + 0.64) \text{ for } 0 \le y \le 0.36.$$

By generalizing the one-dimensional cdf (see Definition 5.6) we obtain:

Definition 7.7: For a bidimensional random variable (X, Y) the *cumulative distribution function* is defined as $F(x, y) = P(X \le x, Y \le y)$.

It follows immediately that

$$F(x, y) = \begin{cases} \displaystyle\sum_{s \le x} \sum_{t \le y} p(s, t) & \text{for discrete variables,} \\ \displaystyle\int_{-\infty}^{x} \int_{-\infty}^{y} f(s, t) \, dt \, ds & \text{for continuous variables.} \end{cases}$$

This concept can be extended to functions with n variables. By generalizing Theorem 5.2, it follows that $F(-\infty, -\infty) = 0$, $F(\infty, \infty) = 1$ and $F(x, y)$ is nondecreasing if one of the variables is fixed. The marginal distribution functions of X and Y are given by

$$G(x) = F(x, \infty), \quad H(y) = F(\infty, y). \tag{7.17}$$

In differentiable points of F the density can be obtained as

$$f(x, y) = \frac{\partial^2}{\partial x\, \partial y} F(x, y). \tag{7.18}$$

For the distribution in Example 7.8 we obtain, e.g.,

$$F(x, y) = \int\limits_0^x \int\limits_0^y 3e^{-s-3t} dt\, ds = 1 - e^{-3y} - e^{-x} + e^{-x-3y} \quad \text{for } x, y \ge 0. \tag{7.19}$$

Hence, $F(\infty, \infty) = 1$, $F(x, y) = 0$ for $x \le 0$ or $y \le 0$, $G(x) = F(x, \infty) = 1 - e^{-x}$, $H(y) = F(\infty, y) = 1 - e^{-3y}$, $\frac{\partial^2}{\partial x \partial y}(1 - e^{-3y} - e^{-x} + e^{-x-3y}) = 3e^{-x-3y}$.

For nonrectangular range spaces R the calculation of the distribution function can be complicated, since one must integrate over the region $R \cap]-\infty, x] \times]-\infty, y]$.

Example 7.9: For the density in Example 7.7. we obtain

$$F(x, y) = \int\limits_{-1}^{-\sqrt{1-y}} \int\limits_0^{1-s^2} \frac{5}{4}(s^2 + t)\, dt\, ds + \int\limits_{-\sqrt{1-y}}^{x} \int\limits_0^{y} \frac{5}{4}(s^2 + t)\, dt\, ds$$

$$= \frac{1}{2} + \sqrt{1-y}\left(-\frac{1}{2} + \frac{y}{6} + \frac{y^2}{3}\right) + \frac{5}{12}yx^3 + \frac{5}{8}y^2x \quad \text{for } (x, y) \in R,$$

satisfying $\dfrac{\partial^2}{\partial x \partial y} F(x, y) = \dfrac{\partial}{\partial y}\left(\dfrac{5}{4}yx^2 + \dfrac{5}{8}y^2\right) = \dfrac{5}{4}(x^2 + y) = f(x, y)$.

The integration areas are illustrated in Fig. 7.5.

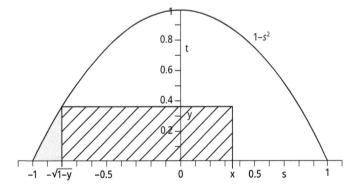

Fig. 7.5: Integration areas of example 7.9.

For $(x, y) \notin R$ the distribution function can be determined similarly. In particular, for $0 \le x \le 1$, $1-x^2 < y < 1$ it holds:

$$F(x, y) = \int_{-1}^{-\sqrt{1-y}} \int_0^{1-s^2} \frac{5}{4}(s^2 + t) \, dt \, ds + \int_{-\sqrt{1-y}}^{\sqrt{1-y}} \int_0^y \frac{5}{4}(s^2 + t) \, dt \, ds + \int_{\sqrt{1-y}}^{x} \int_0^{1-s^2} \frac{5}{4}(s^2 + t) \, dt \, ds$$

$$= \frac{1}{2} + \sqrt{1-y}\left(-1 + \frac{y}{3} + \frac{2y^2}{3}\right) + \frac{5}{8}x - \frac{x^5}{8}.$$

$$(7.20)$$

The reader is advised to illustrate the corresponding integration areas as in Fig. 7.5. The function (7.20) is separable, i.e., is a sum of functions that individually depend on only one of the variables x and y. Hence it is clear that $\dfrac{\partial^2}{\partial x \, \partial y} F(x, y) = 0$.

We finally define the survival functions.

Definition 7.8: For a bidimensional random variable (X, Y) the *survival function* is given by:

$S(x, y) = P(X > x, Y > y)$, i.e.,

$$S(x, y) = \begin{cases} \displaystyle\sum_{s>x}\sum_{t>y} p(s, t) & \text{for discrete variables,} \\[2ex] \displaystyle\int_x^\infty\int_y^\infty f(s, t) \, dt \, ds & \text{for continuous variables.} \end{cases}$$

For the distribution in Example 7.8 we obtain:

$$S(x,y) = \int_x^\infty\int_y^\infty 3e^{-s-3t} \, dt \, ds = e^{-y-3x} \text{ for } x, y \ge 0 \tag{7.21}$$

It is clear that distribution function and survival function are not complementary in the multivariate case, i.e., $R(x,y)$ and $S(x,y)$ do *not* sum up to 1, see Exercise 7.22. For nonrectangular range spaces the calculation of $S(x,y)$ is similar to that of $F(x,y)$ illustrated above. The survival function also satisfies

$$f(x, y) = \frac{\partial^2}{\partial x \, \partial y} S(x, y), \tag{7.22}$$

see (7.18).

7.5 Functions of a random variable

We first consider, very briefly, the discrete case that can be often resolved in a straight forward manner, using again the idea of identifying equivalent events. We restrict ourselves to the functions $M := \max(X, Y)$, $K := \min(X, Y)$ and $S := X + Y$, representing the maximum, the minimum, and the sum of a bidimensional r.v. (X, Y). Assume that the variables take on natural numbers with joint probability function $p_{i,j} = P(X = i, Y = j)$. Since the event $\{M = m\}$ occurs if and only if $\{X, Y \le m\}$ and at least one of the events $\{X = m\}$ and $\{Y = m\}$ occurs, we obtain:

$$P(M = m) = p_{m,1} + \cdots + p_{m,m-1} + p_{m,m} + p_{m-1,m} + \cdots + p_{1,m}. \qquad (7.23)$$

Analogously we get:

$$P(K = k) = p_{k,k} + p_{k,k+1} + p_{k,k+2} + \cdots + p_{k+1,k} + p_{k+2,k} + \cdots \qquad (7.24)$$

and for the sum it obviously holds:

$$P(S = s) = p_{1,s-1} + p_{2,s-2} + \cdots + p_{s-1,1}. \qquad (7.25)$$

In relation (5.4) of Example 5.4, we have already used the intuitive relation (7.25). In the case of independent, identically distributed r.v.s X and Y we get $p_{i,j} = p_i p_j$ for all i,j, where $p_i = P(X = i) = P(Y = i)$ and the above formulas simplify to

$$P(M = m) = p_m^2 + 2p_m(p_1 + \cdots + p_{m-1}), \qquad (7.26)$$

$$P(K = k) = p_k^2 + 2p_k(p_{k+1} + p_{k+2} + \cdots), \qquad (7.27)$$

and

$$P(S = s) = p_1 p_{s-1} + \cdots + p_{s-1} p_1. \qquad (7.28)$$

Example 7.10: Assume that a die is thrown twice and X and Y denote the outcomes. When the die is fair, all probabilities equal 1/6, yielding

$$P(M = m) = \frac{1}{36} + \frac{2m-1}{6 \cdot 6} = \frac{2m-1}{36} \quad \text{for } m = 1, \ldots, 6,$$

$$P(K = k) = \frac{1}{36} + \frac{26-k}{6 \cdot 6} = \frac{13-2k}{36} \quad \text{for } k = 1, \ldots, 6.$$

In the calculation of (7.28) one must be aware that $p_i = 0$ for $i > 6$. In this way one can verify that $S = X + Y$ has the distribution determined in Tab. 5.1(b) of Example 5.1. We can write this as:

$$P(S = s) = \frac{6 - |s - 7|}{36} \quad \text{for } s = 2, \ldots, 12.$$

Now let (X, Y) be a bidimensional continuous r.v. This case is somewhat more complicated. We are interested in the distribution of the r.v. $U = H_1(X, Y)$, assuming that the
function H_1 is continuous; thus U is a continuous variable. The basic idea to determine
its distribution is to define a second r.v. $V = H_2(X, Y)$ (for which one usually makes the
simplest choice) and to determine the joint pdf of U and V, say $g(u, v)$, by means of the
following theorem. The desired pdf of U can be later determined by integration of $g(u,
v)$ with respect to v.

Theorem 7.1: Let (X, Y) be a bidimensional continuous r.v. with joint pdf $f(x, y)$. Consider a *biunique* transformation $(X, Y) \mapsto (U, V)$ between (X, Y) and another bidimensional variable (U, V) such that all functions

$$U = H_1(X, Y), \quad V = H_2(X, Y) \quad \text{and} \quad X = G_1(U, V), \quad Y = G_2(U, V) \tag{7.29}$$

are continuously differentiable. Then the joint pdf of (U, V) is given by

$$g(u, v) = f(G_1(u, v), G_2(u, v))|J| \tag{7.30}$$

where J is the Jacobian of the transformation (7.29), i.e., the determinant

$$J = \begin{vmatrix} \frac{\partial x}{\partial u} & \frac{\partial x}{\partial v} \\ \frac{\partial y}{\partial u} & \frac{\partial y}{\partial v} \end{vmatrix}.$$

The idea of the proof is as follows. The cdf of (U, V) (see Definition 7.1) can be written as

$$G(u, v) = \int_{-\infty}^{u} \int_{-\infty}^{v} g(s, t)\, dt\, ds = \iint_C f(x, y)|J|\, dy\, dx,$$

where C is the event in terms of x and y, equivalent to $\{U \le u, V \le v\}$. The statement of
Theorem 7.1 follows by differentiating the last integral with respect to u and v.

Example 7.11: Suppose that X and Y are independent continuous random variables
with pdfs f_1 and f_2. We are interested in the pdf of the sum $U = X + Y$. We set $U = X + Y$,
$V = X$ and solving the equations for X and Y yields $X = V = G_1(U, V)$, $Y = U - V = G_2(U, V)$.
The Jacobian is therefore $J = \begin{vmatrix} 0 & 1 \\ 1 & -1 \end{vmatrix} = -1$ and due to the independence of the varia-
bles, (7.30) yields $g(u, v) = f(v, u - v)|-1| = f_1(v)f_2(u - v)$. The pdf of U is therefore

$$h(u) = \int_{-\infty}^{\infty} f_1(v)f_2(u - v)\, dv. \tag{7.31}$$

Assume that X and Y present the lifetimes of two light bulbs, which are exponentially distributed, i.e., $f_1(x) = ae^{-a\,x}$ for $x \geq 0$ and 0 for $x < 0$, $f_2(y) = \beta e^{-\beta\,y}$ for $y \geq 0$ and 0 for $y < 0$ ($\alpha > \beta > 0$). Due to (7.31) the total lifetime $X + Y$ has the density

$$h(u) = \int_{-\infty}^{\infty} f_1(v)f_2(u-v)\,dv = \int_0^u ae^{-av}\beta e^{\beta(u-v)}\,dv = \frac{\alpha\beta}{\alpha-\beta}\left(e^{-\beta u} - e^{-au}\right).$$

Definition 7.8: Integral (7.31) is known as the *convolution* of the functions f_1 and f_2.

Example 7.12: Again, let X and Y be independent continuous random variables with pdfs f_1 and f_2. We now wish to determine the pdf of the product $U = XY$ and set $U = XY$, $V = X$. Solving the equations for X and Y yields $X = V = G_1(U, V)$, $Y = U/V = G_2(U, V)$ with Jacobian

$$J = \begin{vmatrix} 0 & 1 \\ 1/v & -1/v^2 \end{vmatrix} = -\frac{1}{v}.$$

The joint pdf of U and V is

$$g(u, v) = \frac{1}{|v|}f\left(v, \frac{u}{v}\right) = \frac{1}{|v|}f_1(v)f_2\left(\frac{u}{v}\right),$$

yielding the pdf of U:

$$h(u) = \int_{-\infty}^{\infty} \frac{1}{|v|}f_1(v)f_2\left(\frac{u}{v}\right)\,dv. \tag{7.32}$$

Assume now that X and Y represent the current and the resistance of an electrical circuit and that both vary randomly according to the pdfs

$$f_1(x) = \tfrac{2}{9}x \quad \text{for } 0 \leq x \leq 3 \text{ and } 0 \text{ for other } x, \tag{7.33}$$

$$f_2(y) = \frac{y^2}{9} \quad \text{for } 0 \leq y \leq 3 \text{ and } 0 \text{ for other } y. \tag{7.34}$$

The values of u and v for which $f_1(v)$ and $f_2(u/v)$ are not equal to zero must satisfy $0 \leq v \leq 3$ and $0 \leq u/v \leq 3$, hence $u/3 \leq v \leq 3$. From (7.32), it follows that the voltage $U = XY$ has the density:

$$h(u) = \int_{u/3}^3 \frac{1}{v}\frac{2}{9}v\frac{u^2}{9v^2}\,dv = \frac{2u^2}{81}\int_{u/3}^3 \frac{1}{v^2}\,dv = \frac{2}{243}u(9-u) \text{ for } 0 \leq u \leq 9. \tag{7.35}$$

Note that the range space of the voltage is $[0,9]$, since X and Y have values between 0 and 3. An easy calculation shows that $h(u)$ is in fact a density.

Exercises

7.1 The discrete bidimensional r.v. (X, Y) has the following joint probability function:

X \ Y	1	2	3
1	1/10	1/6	0
2	0	2/15	1/9
3	2/15	2/15	2/9

Evaluate all marginal and conditional distributions.

7.2 Let X and Y denote the number of items of a certain product manufactured by two production lines. The joint probability distribution is given as follows.

X \ Y	0	1	2	3	4
0	0	0.01	0.05	0.10	0.04
1	0.05	0.08	0.10	0.05	0.02
2	0.08	0.10	0.20	0.08	0.04

Calculate the marginal distributions and the following probabilities:
(i) $P(X \leq Y)$, (ii) $P(X + Y \leq 4)$, (iii) $P(X = Y)$ and (iv) $P(X \leq 2 \mid Y \geq 1)$.

7.3 Consider two independent r.v.s X and Y whose distribution functions are identical with the marginal distributions of the previous exercise. Calculate the probabilities (i) to (iv) for this case.

7.4 Consider an urn, containing 5 white balls, numbered 1, . . ., 5 and 7 black balls, numbered 6, . . ., 12. (i) Assume that two balls are selected randomly without replacement. Let X be the number of white balls in the selection and Y the number of balls with an even number. Determine the joint probability function of (X, Y) and the marginal distributions. Are the variables X and Y independent? (ii) Consider now the variation of item (i), obtained by selecting four balls.

7.5 Given the continuous bidimensional r.v. (X, Y) with fdp $f(x, y) = C(3x^2 + y)$ for $0 \leq x \leq 3$ and $0 \leq y \leq 2$ ($f(x, y) = 0$ for other values of x and y).
(i) Determine C and illustrate geometrically the sample space and the event $B = \{Y \leq X\}$.
(ii) Determine $P(B)$.
(iii) Determine the marginal densities. Are the variables X and Y independent?
(iv) Determine the conditional distributions.
(v) Determine the distribution function of (X, Y).

7.6 Solve the previous exercise for the density
$f(x, y) = C(2x^2 + 3y)$ for $-1 \le x \le 1$ and $0 \le y \le 1 - x$.

7.7 Determine the distribution of the quotient of two independent continuous r.v.s.

7.8 Consider the continuous bidimensional r.v. (X, Y) with pdf $f(x,y) = \frac{1}{6\pi}(x+2)$ $(y+3)$ over the unit circle $R = \{(x, y)\,|\,x^2 + y^2 \le 1\}$.
   (i) Show that this is a pdf.
   (ii) Calculate the probability of the two events $B_1 = \{X^2 + Y^2 \le 1/2\}$, $B_2 = \{X, Y \ge 0\}$.
   (iii) Determine the marginal distributions.
   (iv) Find the conditional distributions.
   (v) Determine the distribution function of (X, Y).

7.9 The life length X of an electronic device has the density $f(x) = 1,000/x^2$ for $x > 1,000$ and $f(x) = 0$ otherwise. If X_1 and X_2 represent the life lengths of two devices of that type, what is the density of X_1/X_2?

7.10 The intensity of light emitted by a candle, received at a point P, is given by $I = C/D^2$, where C is the candlepower and D the distance between P and the candle. Suppose that C is uniformly distributed over [2,3] while D has the pdf $f(d) = e^{-d}$, $d > 0$. Determine the pdf of I, assuming that C and D are independent.

7.11 When a current I flows through a resistance R, the generated power is given by $W = I^2 R$. Assume that I and R are independent r.v.s with the densities $f(i) = \frac{3}{4}i(2-i)$ for $0 \le i \le 2$ and $g(r) = \frac{2}{9}r$ for $0 \le r \le 3$. Determine the pdf of the power and illustrate its graph.

7.12 A bookseller has a small supply of scientific textbooks. Let X and Y denote the number of books sold per day in the fields of mathematics and physics. The joint probability function of (X, Y) is given as follows:

X	0	1	2	3	4
Y					
0	0	0.01	0.03	0.04	0.02
1	0.01	0.05	0.1	0.08	0.06
2	0.02	0.04	0.08	0.07	0.04
3	0.03	0.06	0.07	0.03	0.01
4	0.02	0.05	0.06	0.02	0

   (i) Determine the marginal distributions of X and Y.
   (ii) What is the probability that on a specific day more textbooks on physics than textbooks on mathematics are sold?
   (iii) What is the probability that the sum of textbooks sold on these areas on a specific day is at least 6?

7.13 (i) Show that $f(x, y) = 8x(x + y - 1)$ is a joint pdf over the range space $R = \{(x, y)\,|\,0 \le x, y \le 1, y \ge 1 - x\}$.
   (ii) Determine the cumulative distribution function.

(iii) Find the marginal distribution functions.

(iv) Obtain the marginal densities in two ways: By using (iii) and by integrating the joint pdf with respect to x and y.

(v) Calculate the probabilities

(a) $P(X \geq 3/4)$, (b) $P(Y \leq 1/3)$, (c) $P(X \leq 3/4, Y \leq 1/2)$ and (d) $P(X \leq 1/3, Y \geq 4/5)$ using the results of the previous items.

7.14 Assume that the range space of the continuous bidimensional r.v. is the rectangle $R = \{(x, y) \mid 0 \leq x \leq a, 0 \leq y \leq b\}$. Assume that the cumulative distribution function F (x, y) is known for each point of R. Express $F(x, y)$ for points *outside* of R by means of the known values.

7.15 If (X, Y) has the joint pdf $f(x, y) = \frac{2}{3}(x + y)e^{-y}$ for $0 \leq x \leq 1$ and $y \geq 0$, determine the conditional probabilities

(i) $P(Y \leq 1 \mid Y \geq 2X)$, (ii) $P(X + Y \leq 2 \mid X \leq 1/2)$.

7.16 The body mass index of a person is defined as $B = Y/X^2$, where X is the height in meters and Y the weight in kg. An adult is considered obese, if $B \geq 30$. If an adult is randomly chosen from a population for which (X, Y) is uniformly distributed over the rectangle $R = \{(x, y) \mid 1 \leq x \leq 2, 50 \leq y \leq 120)\}$, what is the probability that this person is obese?

7.17 Illustrate the marginal pdfs of Example 7.7 graphically. Show that two r.v.s can only be independent when the range space is rectangular.

7.18 Verify that the function $h(u)$ obtained in Example 7.11 is in fact a density. Illustrate it geometrically.

7.19 Show that the functions in (7.13)–(7.16) satisfy all requirements for a one-dimensional pdf.

7.20 Given the probability function $p(x, y) = 1/2^{x+y}$ for $x, y \geq 1$ of a discrete bidimensional r.v. (X, Y), determine the cdf and the marginal cdfs.

7.21 Determine the survival function corresponding to the density in Example 7.7 for the cases

(i) $(x, y) \in R$, (ii) $-1 \leq x \leq 0, 1 - x^2 < y \leq 1$.

7.22 Consider the two-dimensional uniform distribution with density $f(x, y) = 1$ for $0 \leq x, y \leq 1$. Illustrate geometrically that $F(x, y) + S(x, y) < 1$ for $1 < x, y < 0$.

8 Characteristics of random variables

In some situations one does not need the full information about the distribution of a random variable. It may be sufficient to characterize the distribution in terms of a few pertinent parameters. For example, for the calculation of the rates of a life insurance it might be sufficient to have some information about the average lifespan and a measure of variation.

We first introduce the concept of the expected value, which is certainly one of the most important characteristics of an r.v.

8.1 The expected value of a random variable

Assume that a manufacturer produces items such that 10% are defective and 90% perfect. If a defective item is produced, the manufacturer loses $1, which corresponds to the production cost, while a perfect item brings a profit of $5. We are interested in the expected profit in the sale of one item. This is a random variable X whose possible values $x_1 = -1$ and $x_2 = 5$ occur with probabilities $p_1 = 0.1$ and $p_2 = 0.9$, respectively.

Assume that a large number of items is produced, say e.g., 1,000 items. (It becomes clear from the following, that this numerical value could be substituted by any other "large" value). Intuitively we expect that (approximately) 10% are defective. Thus we expect that 100 items are defective and 900 are perfect, resulting in a profit of $-1 \cdot 100 + 5 \cdot 900$. Dividing this by 1,000 we obtain the *expected profit per item*, i.e., the expected value of X as $-1 \cdot 0.1 + 5 \cdot 0.9 = 4.4 = x_1 p_1 + x_2 p_2$.

The last formula motivates the following general definition of the expected value.

Definition 8.1: Let X be a discrete random variable with possible values $x_1, x_2, \ldots$. Let $p_i = P(X = x_i)$ for $i = 1, 2, \ldots$. Then the *expected value* of X (also known as the *expectation of X*), denoted by $E(X)$ is defined as

$$E(X) = \sum_{i=1}^{\infty} x_i p_i$$

if the series converges absolutely, i.e., if

$$\sum_{i=1}^{\infty} |x_i| p_i < \infty.$$

This number is also referred to as the *mean value* of X.

Example 8.1: Assume that a die is thrown and X denotes the outcome. From the above definition it follows that $E(X) = 1 \cdot \frac{1}{6} + 2 \cdot \frac{1}{6} + \cdots + 6 \cdot \frac{1}{6} = \frac{7}{2}$.

https://doi.org/10.1515/9783111332277-008

This example makes clear that the expected value of an r.v. need not be a possible outcome.

Theorem 8.1: Let X be a binomially distributed r.v. with parameters n and p, i.e.,

$$P(X = k) = \binom{n}{k} p^k (1-p)^{n-k} \text{ for } k = 0, 1, \ldots, n. \text{ Then } E(X) = np.$$

Proof: From Definition 8.1 we obtain

$$E(X) = \sum_{k=0}^{n} k \frac{n!}{k!(n-k)!} p^k (1-p)^{n-k} = \sum_{k=1}^{n} \frac{n!}{(k-1)!(n-k)!} p^k (1-p)^{n-k}.$$

Using the substitution $s = k - 1$ we get

$$E(X) = \sum_{s=0}^{n-1} \frac{n!}{s!(n-1-s)!} p^{s+1} (1-p)^{n-1-s} = np \sum_{s=0}^{n-1} \binom{n-1}{s} p^s (1-p)^{n-1-s}.$$

The last sum is the sum of binomial probabilities with n replaced by $(n-1)$ and hence equals 1. Thus $E(X) = np$.

The theorem corresponds to our intuition. Suppose that the probability of an event A is, e.g., $p = 0.2$ when an experiment is performed. If we realize this experiment, say $n = 1,000$ times, then we expect A to occur about $np = 1,000 \times 0.2 = 200$ times. This is exactly the statement of Theorem 8.1. We now extend the concept of expected value to the continuous case.

Definition 8.2: Let X be a continuous random variable with pdf $f(x)$. Then the *expected value* of X is defined as

$$E(X) = \int_{-\infty}^{\infty} x f(x) dx.$$

Since this improper integral may not converge we say that $E(X)$ exists if and only if

$$\int_{-\infty}^{\infty} |x| f(x) dx$$

is finite.

Example 8.2: Suppose that X is the lifetime of a certain electronic component (in hours). Let $f(x) = \frac{20}{x^2}$ for $10 \leq x \leq 20$ be the pdf. The expected life time is then

$$E(X) = \int\limits_{10}^{20} x \frac{20}{x^2}\, dx = 20 \int\limits_{10}^{20} \frac{1}{x}\, dx = 20[\ln x]_{10}^{20} \approx 13.9.$$

Theorem 8.2: Let X be uniformly distributed over the interval $[a,b]$. Then $E(X) = \frac{a+b}{2}$.

Proof: The pdf is $\frac{1}{b-a}$ over $[a,b]$. Thus,

$$E(X) = \int\limits_{a}^{b} \frac{x}{b-a}\, dx = \frac{1}{b-a}\frac{x^2}{2}\Big|_{a}^{b} = \frac{a+b}{2}.$$

This is the midpoint of the interval $[a,b]$ as we would expect intuitively.

8.2 Expectation of a function of a random variable

Let the r.v. Y be a function of the r.v. X, i.e., $Y = \Psi(X)$. If the distribution of Y is known (see Chapter 6) we can naturally use the probability function or the pdf of Y and apply one of the definitions in Section 8.1 to determine the expected value of Y. However, as the following theorem shows, $E(Y)$ can be determined without knowing the distribution of Y.

Theorem 8.3: Let X be the r.v. and let $Y = \Psi(X)$.
(a) If X is discrete with values x_i and probability function $p_i = P(X = x_i)$, then it holds:

$$E(Y) = E(\Psi(X)) = \sum_{i=1}^{\infty} \Psi(x_i)p_i. \tag{8.1}$$

(b) If X is continuous with pdf $f(x)$, then it holds

$$E(Y) = E(\Psi(X)) = \int\limits_{-\infty}^{\infty} \Psi(x)f(x)\, dx. \tag{8.2}$$

The discrete case is intuitively clear, since the *not necessarily different* values $\Psi(x_i)$ occur with probabilities p_i, and as in Definition 8.1 we sum over the products of possible values and its probabilities.

Example 8.3: Assume that a player throws a die and wins the amount $Y = 2X - 6$ in \$, where X denotes the outcome. The expected win is therefore $\sum_{i=1}^{6}(2i-6)\frac{1}{6} = \frac{1}{6}(-4-2+0+2+4+6) = 1\$.$

Example 8.4: Assume that the profit Y in selling the electronic component in Example 8.2 is somehow related to the lifespan, say $Y = X^{3/2}$. The expected profit in the sale of these articles is then

$$E\left(X^{3/2}\right) = \int\limits_{10}^{20} x^{3/2}\frac{20}{x^2}\,dx = 20\int\limits_{10}^{20} x^{-1/2}dx = 40\left[x^{1/2}\right]_{10}^{20} = 40\left(\sqrt{20} - \sqrt{10}\right) \approx 52.4.$$

For the purpose of control it is interesting to compare the result with that obtained by applying Definition 8.2 to the density of Y.

By means of Theorem 6.1 we obtain the pdf of Y as

$$g(y) = f\left(y^{2/3}\right)\frac{2}{3}y^{-1/3} = \frac{20}{y^{4/3}}\frac{2}{3}y^{-1/3} = \frac{40}{3y^{5/3}}.$$

The range space is the interval $[10^{3/2}, 20^{3/2}] \approx [31.6, 69.4]$. Hence Definition 8.2 yields

$$E(Y) = \int\limits_{10^{3/2}}^{20^{3/2}} yg(y)\,dy = \int\limits_{10^{3/2}}^{20^{3/2}} \frac{40}{3y^{2/3}}\,dy = 40\left(\sqrt{20} - \sqrt{10}\right),$$

which is identical to the previous result.

The idea of expressing the expectation of $\Psi(X)$ in terms of the distribution of X works also when the original r.v. is of higher dimension. As before, we restrict ourselves to the bidimensional case.

Theorem 8.4: Let (X, Y) be a bidimensional r.v. and let $Z = \Psi(X, Y)$,
(a) If (X, Y) is discrete with values (x_i, y_j) and probability function $p_{i,j} = P(X = x_i, Y = y_j)$, then it holds:

$$E(Z) = \sum_{i=1}^{\infty}\sum_{j=1}^{\infty} \Psi\left(x_i, y_j\right)p(x_i, y_i). \tag{8.3}$$

(b) If (X, Y) is continuous with joint pdf $f(x, y)$, then it holds:

$$E(Z) = \int\limits_{-\infty}^{\infty}\int\limits_{-\infty}^{\infty} \Psi\left(x, y\right)f(x, y)\,dy\,dx. \tag{8.4}$$

Example 8.5: Consider the r.v. $Z = XY$, where X and Y are the outcomes of rolling a die twice. From Theorem 8.4(a) we obtain the expected value of Z as

$$E(Z) = \sum_{i=1}^{6}\sum_{j=1}^{6} i \cdot j \cdot \frac{1}{36} = \frac{1}{36}\sum_{i=1}^{6} i(1 + \cdots + 6) = \frac{21}{36}\sum_{i=1}^{6} i = \frac{49}{4}.$$

Example 8.6: Consider again the r.v. $U = XY$ of Example 7.12, where X and Y are independent continuous r.v.s representing the current and the resistance of an electrical network, having the pdfs (7.33) and (7.34), respectively. From Theorem 8.4(b) we obtain the expected voltage as

$$E(U) = \int_0^3\int_0^3 xyf_1(x)f_2(y)dydx = \int_0^3\int_0^3 xy\frac{2}{9}x\frac{y^2}{9}dydx$$

$$= \int_0^3 \frac{2}{9}x^2 \int_0^3 \frac{y^3}{9}dydx = \int_0^3 \frac{2}{9}x^2\frac{9}{4}dx = \frac{9}{2}.$$

Alternatively, by using pdf (7.35) of U we obtain $E(U) = \int_0^9 uh(u)du = \int_0^9 \frac{2}{243}u^2(9-u)\,du = \frac{9}{2}.$

8.3 Properties of the expected value

We now list some properties of the expected value, which are useful, when we calculate $E(X)$ for a composed r.v. X. We always assume that the expected value to which we refer exists. The proof will be given only for continuous variables. The discrete case is obtained by replacing integrals by summations.

Property 8.1: If $X = C$ where C is a constant, then $E(X) = C$.

Proof:

$$E(X) = \int_{-\infty}^{\infty} Cf(x)dx = C\int_{-\infty}^{\infty} f(x)dx = C.$$

Property 8.2: Suppose that X is an r.v. and C a constant, then:

$$E(CX) = CE(X).$$

Proof:

$$E(CX) = \int_{-\infty}^{\infty} Cxf(x)dx = C\int_{-\infty}^{\infty} xf(x)dx = CE(X).$$

Property 8.3: Let (X, Y) be a bidimensional r.v. with joint pdf $f(x, y)$. For functions $U = H_1(X, Y)$ and $V = H_2(X, Y)$ we obtain

$$E(U + V) = E(U) + E(V).$$

Proof: From Theorem 8.4 we obtain

$$E(U + V) = \int_{-\infty}^{\infty} \int_{-\infty}^{\infty} [H_1(x, y) + H_2(x, y)] f(x, y) \, dy \, dx$$

$$\int_{-\infty}^{\infty} \int_{-\infty}^{\infty} H_1(x, y) f(x, y) dy \, dx + \int_{-\infty}^{\infty} \int_{-\infty}^{\infty} H_2(x, y) f(x, y) dy \, dx$$

$$= E(U) + E(V).$$

As a most important special case of the statement we obtain:

Property 8.4: Let X and Y be any two random variables. Then

$$E(X + Y) = E(X) + E(Y).$$

This follows immediately from Property 8.3 by setting $H_1(X, Y) = X$ and $H_2(X, Y) = Y$.

By combining the previous properties and using mathematical induction we can easily derive the following *linearity condition* of the expected value.

Property 8.5: Let $X_1, \ldots, X_n$ be random variables and $C_1, \ldots, C_n$ constants. Then

$$E\left(\sum_{i=1}^{n} C_i X_i\right) = \sum_{i=1}^{n} C_i E(X_i).$$

Example 8.7: In a gambling hall three types of slot machines are available. The machines of type A, B, and C make an expected profit of $E_A = 20$, $E_B = 50$, and $E_C = 100$ per game (in \$), respectively. If on a certain day machines of type A, B, and C are used 300, 150, and 80 times, respectively, then the overall expected profit for this day is $300 \cdot E_A + 150 \cdot E_B + 80 \cdot E_C = 21{,}500$ (in \$).

Property 8.6: Let X and Y be *independent* random variables with individual densities f_1 and f_2. Then $E(XY) = E(X) \, E(Y)$.

Proof:

$$E(XY) = \int_{-\infty}^{\infty} \int_{-\infty}^{\infty} xyf(x, y)dydx = \int_{-\infty}^{\infty} \int_{-\infty}^{\infty} xf_1(x)yf_2(y)dydx$$

$$= \int_{-\infty}^{\infty} xf_1(x)dx \int_{-\infty}^{\infty} xf_2(x)dx = E(X)E(Y).$$

The attentive reader may have noticed, that the condition $E(XY) = E(X)E(Y)$ was satisfied in Example 8.6.

8.4 Variance, moments, and shape characteristics

The expected value is definitely not sufficient to characterize a distribution. In particular we also want to know the magnitude of the deviations between the outcomes of an r.v. and the expectation. The most important measure for the deviations is the following:

Definition 8.3: Let X be a random variable. The *variance* of X, denoted by $V(X)$ or σ_X^2 is defined as:

$$V(X) = E\left[(X - E(X))^2\right].$$

This is the expectation of the squared deviation $(X - E(X))^2$ between an outcome of an r.v. and its expectation. The *standard deviation* of X is defined as $\sigma_X = \sqrt{V(X)}$.

The variance is expressed in square units of X. If for example X is the height of a randomly chosen person in cm, the unit of $V(X)$ is cm^2. But the standard deviation is always expressed in the same units as X.

For calculations the following formula for $V(X)$ is often more suitable than the above definition.

Theorem 8.5:

$$V(X) = E(X^2) - [E(X)]^2.$$

Proof: By using the above properties of the expectation we obtain

$$V(X) = E[(X - E(X))^2] = E[X^2 - 2XE(X) + (E(X))^2]$$

$$= E(X^2) - 2E(X)E(X) + (E(X))^2 = E(X^2) - (E(X))^2.$$

Example 8.8: If X presents the outcome of a fair die, then $E(X) = \frac{7}{2}$ (see Example 8.1), $E(X^2) = 1^2 \cdot \frac{1}{6} + 2^2 \cdot \frac{1}{6} + \cdots + 6^2 \cdot \frac{1}{6} = \frac{91}{6}$. Hence from Theorem 8.5, it follows that $V(X) = \frac{91}{6} - \left(\frac{7}{2}\right)^2 = \frac{35}{12}$.

Example 8.9: For the lifetime X of the electronic component in Example 8.2 we obtain $E(X^2) = \int_{10}^{20} x^2 \frac{20}{x^2} dx = 200$. Hence Theorem 8.5 yields $V(X) = E(X^2) - [E(X)]^2 \approx 200 - 13.9^2 = 6.79$ (in square hours).

The standard deviation is $\sqrt{6.79} \approx 2.6$ h.

We now study some properties of the variance.

Property 8.7: If C is a constant, then

$$V(X + C) = V(X).$$

Proof:

$$V(X + C) = E\left[((X + C) - E(X + C))^2\right] = E[(X - E(X))^2] = V(X).$$

The result is intuitively clear, since adding a constant to the variable simply "shifts" the values to the right or to the left.

Property 8.8: If C is a constant, then

$$V(CX) = C^2 V(X).$$

Proof:

$$V(CX) = E[(CX)^2] - (E(CX))^2 = C^2 E(X^2) - C^2(E(X))^2$$
$$= C^2\left(E(X^2) - (E(X))^2\right) = C^2 V(X).$$

Property 8.9: If X and Y are *independent* random variables, then

$$V(X + Y) = V(X) + V(Y).$$

Proof:

$$V(X + Y) = E[(X + Y)^2] - (E(X + Y))^2$$
$$= E(X^2 + 2XY + Y^2) - (E(X))^2 - 2E(X)E(Y) - (E(Y))^2$$
$$= E(X^2) + E(Y^2) - (E(X))^2 - (E(Y))^2 \text{ (using Property 8.6)}$$
$$= V(X) + V(Y).$$

The next statement follows immediately by mathematical induction.

Property 8.10: Let $X_1, \ldots, X_n$ be *independent* random variables. Then $V(X_1 + \ldots + X_n) = V(X_1) + \ldots + V(X_n)$.

The above properties have many practical and theoretical applications.

Example 8.10: We determine the variance of a binomially distributed variable (Definition 5.3). The distribution function is $P(X = k) = \binom{n}{k} p^k (1-p)^{n-k}$. One possibility to calculate $V(X)$ is to determine $E(X^2) = \sum_{k=0}^{n} k^2 \binom{n}{k} p^k (1-p)^{n-k}$ similarly to the proof of Theorem 8.1 and to apply Theorem 8.3. But there exists a more elegant way:

We define auxiliary variables Y_i for $i = 1, \ldots, n$ by setting $Y_i = 1$ if the event A occurs in the ith realization of E and $Y_i = 0$ otherwise. Obviously it holds $X = Y_1 + \ldots + Y_n$ and since the event A occurs with probability p we obtain for any i:

$$P(Y_i = 1) = p \text{ and } P(Y_i = 0) = 1 - p, \quad \text{thus } E(Y_i) = 1p + 0(1-p) = p,$$

$$E(Y_i^2) = 1^2 p + 0^2 (1-p) = p \text{ and finally}$$

$$V(Y_i) = E(Y_i^2) - (E(Y_i))^2 = p - p^2 = p(1-p).$$

From Property 8.10 we obtain

$$V(X) = V(Y_1 + \cdots + Y_n) = V(Y_1) + \cdots + V(Y_n) = np(1-p).$$

We still introduce the concepts of moments that generalize expectation and variance. The terms are analogous to the moments in mechanics. Among others, they are useful to describe the shape of a distribution.

Definition 8.4: Let X be a (discrete or continuous) random variable. We define the kth *moment about the origin* as $\mu_k' = E(X^k)$ and the kth *central moment* as $\mu_k = E[(X - E(X))^k]$. The kth *factorial moment* is defined by $\mu_{[k]} = E[X(X-1)\ldots(X-k+1)]$.

It follows immediately that μ_1' is the expectation and μ_2 the variance of X.

Example 8.11: Let X be the result of tossing a die, then

$$\mu_3' = E(X^3) = 1^3 \cdot \frac{1}{6} + 2^3 \cdot \frac{1}{6} + \cdots + 6^3 \cdot \frac{1}{6} = \frac{147}{2}.$$

$$\mu_3 = E[(X - E(X))^3] = \frac{1}{6}[(1 - 3.5)^3 + (2 - 3.5)^3 + \cdots + (6 - 3.5)^3] = 0 \text{ (since} E(X) = 3.5).$$

$$\mu_{[3]} = E[X(X-1)(X-2)] = E[X^3 - 3X^2 + 2X] = E(X^3) - 3E(X^2) + 2E(X)$$

$$= \frac{147}{2} - 3\frac{91}{6} + 2\frac{7}{2} = 35.$$

Example 8.12: Let X be the continuous r.v. with pdf $f(x) = \dfrac{3}{26}x^2$ for $1 \le x \le 3$. For example,

$$\mu_k{'} = \int\limits_1^3 \frac{3}{26}x^{k+2}dx = \frac{3}{26}\frac{3^{k+3}-1}{k+3}$$

$$\mu_4 = \int\limits_1^3 \frac{3}{26}x^2\left(x - \frac{30}{13}\right)^4 dx \approx 0.1645 \quad (\text{since } E(X) = 30/13).$$

We can now introduce the following coefficients due to Pearson, which are related to the shape of a distribution. The *coefficient of skewness* is defined as:

$$\alpha_3 = \frac{\mu_3}{\sigma^3}, \tag{8.5}$$

where σ is the standard deviation, and the *coefficient of kurtosis* is given by:

$$\alpha_4 = \frac{\mu_4}{\sigma^4} = \frac{\mu_4}{\mu_2{}^2}. \tag{8.6}$$

Both measures represent more or less successful attempts to express geometric concepts in terms of a single number. Despite the shortcomings, the concepts are widely used in statistics, and therefore we study them in detail in this book. The definitions are exact, but the geometric interpretations are very difficult and, particularly in the case of kurtosis, one finds contradictory or unclear statements in the statistical literature. The basic geometric ideas are the following:

The coefficient α_3 is a measure of asymmetry of a probability distribution. Fig. 8.1 schematically illustrates three basic forms of the density: (a) *positively skewed*, i.e., the right tail is longer or fatter than the other, (b) *symmetric*, (c) *negatively skewed*, the opposite case to (a). The coefficient α_3 should be positive, zero, or negative in case (a), (b), (c), respectively, and the absolute value of the coefficient should express the difference in the tails.

The coefficient α_4 measures the flatness or "peakedness" and assumes only nonnegative values. In most statistical texts one considers the three different shapes in Fig. 8.2.

The middle line corresponds to the standardized normal distribution with kurtosis 3 (see Exercise 8.27). A distribution close to this line is called mesokurtic. If the peak is higher we speak about a leptokurtic distribution. The platykurtic distribution indicates a shape flatter than the standardized normal distribution. The coefficient α_4 was defined such that values greater than 3, equal to 3, and smaller than 3 correspond to a leptokurtic, mesokurtic, or platykurtic curve, respectively.

We illustrate the measures by means of the gamma distribution.

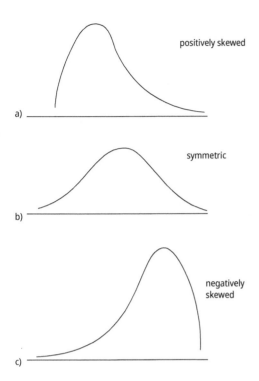

Fig. 8.1: Forms of symmetry.

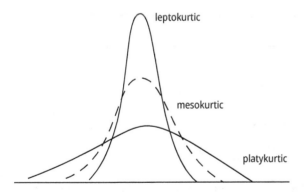

Fig. 8.2: Forms of kurtosis.

Using the table of distributions in the appendix we get:

$$f(x) = \frac{\lambda^r}{\Gamma(r)} x^{r-1} e^{-\lambda x}, \quad \mu_2 = \frac{r}{\lambda^2}, \quad \mu_3 = \int_0^\infty \left(x - \frac{r}{\lambda}\right)^3 f(x) dx = \frac{2r}{\lambda^3},$$

$$\mu_4 = \frac{3r(2+r)}{\lambda^4}, \quad \alpha_3 = \frac{\mu_3}{\mu_2^{3/2}} = \frac{2}{\sqrt{r}}, \quad \alpha_4 = \frac{\mu_4}{\mu_2^2} = 3 + \frac{6}{r}.$$

Both measures are independent from the scaling parameter λ. Fig. 8.3 illustrates some special cases of the gamma distribution for $\lambda=1$ and different values of r.

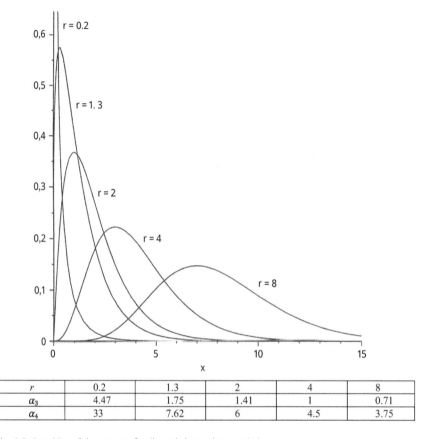

r	0.2	1.3	2	4	8
α_3	4.47	1.75	1.41	1	0.71
α_4	33	7.62	6	4.5	3.75

Fig. 8.3: Densities of the gamma family and shape characteristics.

The corresponding values of the coefficients of skewness and kurtosis are shown in Fig. 8.3. For small values of the parameter r the densities are highly positively skewed and leptokurtic. If r increases both coefficients α_3 and α_4 decrease and converge to 0 and 3, respectively. With the help of Exercises 8.23–8.26 the reader should form his own opinion about the appropriateness of the shape coefficients. In Chapter 17, we will return to the study of shape characteristics in the context of descriptive statistics.

8.5 Probability inequalities

Intuitively we guess that large deviations between the outcomes of an r.v. and its expectation are rare. The following general statement is a mathematical concretization of this idea.

Theorem 8.6: Let X be a random variable and h a nonnegative continuous function. Then for any $K > 0$ it holds

$$P(h(X) \geq K) \leq \frac{E(h(X))}{K}. \tag{8.7}$$

Proof: We will again restrict ourselves to continuous r. v.s. The discrete case is analogous. For $A = \{x \in IR \mid h(x) \geq K\}$ we get:

$$E(h(X)) = \int_{-\infty}^{\infty} h(x)f(x)dx \geq \int_A h(x)f(x)dx \geq \int_A Kf(x)dx = KP(h(X) \geq K).$$

The proof is surprisingly simple in view of the wide applicability of the theorem. We present two special cases, using the standard notations $\mu = E(X)$ and $\sigma^2 = V(X)$. Setting $h(x) = (x - \mu)^2$ and $K = \varepsilon^2$ ($\varepsilon > 0$) results in *Tchebycheff's inequality*:

$$P(|X - \mu| \geq \varepsilon) \leq \frac{\sigma^2}{\varepsilon^2}. \tag{8.8}$$

The probability in (8.8) is schematically illustrated by the shaded area in Fig. 8.4.
By setting $h(x) = |x|^r$ and $K = \varepsilon^r$ ($r, \varepsilon > 0$) we obtain *Markov's inequality*:

$$P(|X| \geq \varepsilon) \leq \frac{E(|X|^r)}{\varepsilon^r}. \tag{8.9}$$

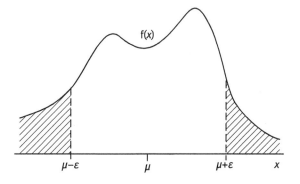

Fig. 8.4: Illustration of Tchebycheff's inequality.

Example 8.13: Let X be uniformly distributed over $[-1,1]$. Then it holds $\mu = 0$, $\sigma^2 = 1/3$ and

$$E\left(|X|^r\right) = \int_{-1}^{1} \frac{1}{2}|x|^r dx = 2\int_{0}^{1}\frac{1}{2}x^r dx = \frac{1}{r+1}. \tag{8.10}$$

Tchebycheff's inequality yields

$$P(|X| \geq \varepsilon) \leq \frac{1}{3 \cdot \varepsilon^2}.$$

Choosing, e.g., $\varepsilon = 3/4$ we obtain

$$P\left(|X| \geq \frac{3}{4}\right) \leq \frac{16}{27} \approx 0.5926. \tag{8.11}$$

Using Markov's inequality and (8.10) we get:

$$P\left(|X| \geq \frac{3}{4}\right) \leq \frac{(4/3)^r}{r+1}. \tag{8.12}$$

The right-hand side in (8.12) assumes the minimum value ≈ 0.5865 for $r \approx 2.476$. Evidently, the exact probability to be estimated is $P(|X| \geq 3/4) = \frac{1}{4}$.

One can observe that the upper limits in (8.11) and (8.12) need not be close to the exact probabilities. This occurs very frequently in applications. However, the inequalities are useful when the exact probability is not easily available.

We now consider a discrete distribution.

Example 8.14: Assume that $n = 1,000$ articles are produced each of which is perfect with probability $p = 0.9$ and defective with probability 0.1. Let X be the number of perfect items obtained. This r.v. is binomially distributed and has therefore the expectation $np = 900$ and the variance $np(1-p) = 90$. From (8.8) we obtain by setting, e.g., $\varepsilon = 15$:

$$P(|X - 900| \geq 15) \leq \frac{\sigma^2}{\varepsilon^2} = \frac{90}{225} = 0.4.$$

The exact probability is now $1 - \sum_{k=886}^{914} \binom{1,000}{k} 0.9^k \cdot 0.1^{1,000-k} \approx 0.1261$. Again we note that the upper limit is not sharp; however ithe calculation of the exact probability is relatively complicated.

We still consider two relations between expected values.

Theorem 8.7 (Jensen's inequality): Let X be a random variable and h a *convex* continuously differentiable function. Then it holds:

$$E(h(X) \geq h(E(X)). \qquad (8.13)$$

Proof: From the convexity it follows that for any x_0 exists a straight line $L(x) = h(x_0)$ $+ h'(x_0) (x - x_0)$ passing through the point $(x_0, h(x_0))$ such that $h(x) \geq L(x)$ for all x (see Fig. 8.5). Hence,

$$E(h(X)) \geq E(L(X)) = E(h(x_0) + \lambda(X - x_0)) = h(x_0) + \lambda(E(X) - x_0)$$

and setting $x_0 = E(X)$ yields relation (8.13).

By selecting specific convex functions in (8.13) one obtains various interesting relations, e.g.,

$$E\left(|X|^p\right) \geq |E(X)|^p \quad \text{for } h(x) = |x|^p \text{ with } p \geq 1,$$

$$E\left(\frac{1}{X}\right) \geq \frac{1}{E(X)} \qquad \text{for } h(x) = 1/x \text{ with } x > 0,$$

$$E\left(e^X\right) \geq e^{E(X)} \qquad \text{for } h(x) = e^x.$$

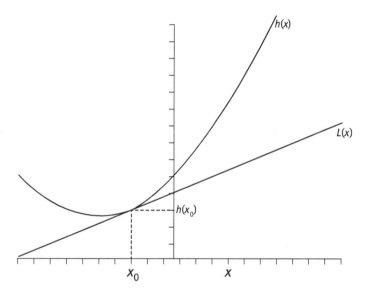

Fig. 8.5: Illustration of Jensen's inequality.

Theorem 8.8 (Cauchy-Schwartz inequality): For random variables X and Y it holds

$$(E(XY))^2 \leq E(X^2)E(Y^2). \qquad (8.14)$$

Proof: Consider the function

$$g(\lambda) = E\left[(X - \lambda Y)^2\right] = E\left(X^2 - 2\lambda XY + \lambda^2 Y^2\right) = E(X^2) - 2\lambda E(XY) + \lambda^2 E(Y^2), \qquad (8.15)$$

which is a quadratic function of the form $g(\lambda) = a - b\lambda + c\lambda^2$ with $c > 0$. The minimum value is

$$\frac{4ac - b^2}{4c} = \frac{E(X^2)E(Y^2) - (E(XY))^2}{E(Y^2)}. \qquad (8.16)$$

Since g is nonnegative, the numerator of the last quotient must also be nonnegative, yielding the statement (8.14).

8.6 Covariance and correlation

In Section 8.4, we studied the variance and some generalizations. All concepts developed there are related to a single random variable. Given a pair of variables (X, Y) it is natural to ask whether there is a parameter that measures in a certain sense the "degree of association" between X and Y. This idea leads us to the following concepts.

Definition 8.5: Let (X, Y) be a bidimensional random variable such that $E(X)$ and $E(Y)$ exist. Then the covariance is defined as

$$\mathrm{Cov}(X, Y) = E[(X - E(X))(Y - E(Y))] = E(XY) - E(X)E(Y). \qquad (8.17)$$

This parameter may assume any real value. If the variances $V(X)$ and $V(Y)$ are positive, we define the *coefficient of (linear) correlation* between X and Y as

$$\rho_{x,y} = \frac{\mathrm{Cov}(X,\ Y)}{\sqrt{V(X)V(Y)}}. \qquad (8.18)$$

This coefficient is dimensionless. If there is no question which r.v.s are involved, we may simply write ρ instead of $\rho_{x,y}$. Let us first examine some properties of the measure.

Theorem 8.9: If X and Y are independent then $\rho = 0$.

Proof: From Property 8.6 it follows immediately that $\mathrm{Cov}(X, Y) = 0$ and thus $\rho = 0$ holds for independent v.a.s X and Y.

However, the converse of Theorem 8.9 is not true.

Example 8.15: Consider the discrete bidimensional r.v. (X, Y) with joint probability function:

$$p(x, y) = \begin{cases} 1/3 & \text{for } x = -1, \ y = 1, \\ 1/3 & \text{for } x = 0, \ y = 0, \\ 1/3 & \text{for } x = 1, \ y = 1, \\ 0 & \text{for other values} \end{cases}$$

and marginal distributions $g(x) = 1/3$ for $x = -1$, 0 and 1, $h(y) = 1/3$ for $y = 0$ and $2/3$ for $y = 1$. The variables X and Y are dependent, since in particular $p(-1, 0) = 0 \neq g(-1) \ h(0) = 1/9$.

On the other hand, it holds that $E(X) = 0$, $E(Y) = 2/3$, $E(XY) = -1 \cdot \frac{1}{3} + 0 \cdot \frac{1}{3} + 1 \cdot \frac{1}{3} = 0$. Hence, $\text{Cov}(X, Y) = 0$ and $\rho = 0$.

Theorem 8.10: It holds that $-1 \leq \rho \leq 1$.

Proof: Consider the r.v.s $W = X - E(X)$, $Z = Y - E(Y)$. We obtain $E(W) = E(Z) = 0$, $V(W) = V(X) = E(W^2)$, and $V(Z) = V(Y) = E(Z^2)$. From (8.17) and (8.18) it follows now that $\rho_{x,y}$ can be expressed as

$$\rho_{x,y} = \frac{\text{Cov}(W, Z)}{\sqrt{V(W)V(Z)}} \tag{8.19}$$

and from the Cauchy–Schwartz inequality (Theorem 8.8) it follows that the square of (8.19) is smaller or equal than 1.

Theorem 8.11: Assume that X and Y are r.v.s, such that $Y = aX + b$, where a and b are constants. Then it holds that $\rho^2 = 1$.

If $a > 0$, then $\rho = 1$ and if $a < 0$, then $\rho = -1$.

Proof: We obtain $E(XY) = E(X(aX + b)) = E(aX^2 + bX) = aE(X^2) + bE(X)$,
$E(Y) = E(aX + b) = aE(X) + b$, $V(Y) = V(aX + b) = a^2 V(X)$.
Thus, $\text{Cov}(X, Y) = aE(X^2) + bE(X) - a(E(X))^2 - bE(X) = aV(X)$ and

$$\rho_{x,y} = \frac{aV(X)}{\sqrt{V(X)a^2 V(X)}} = \frac{a}{|a|},$$

which completes the proof.

Without demonstration we state that the converse of Theorem 8.11 is also true.

The coefficient ρ measures the "degree of linearity" between the variables X and Y. If ρ is "close" to 1 or −1, the relation between X and Y is "approximately linear." If ρ is "close" to 0, there is no linear relation. However, this does not exclude the possibility that another relation between the variables exists. For instance, the correlation between the variables in Example 8.15 is 0, but they satisfy the quadratic relation $Y = X^2$.

We illustrate the meaning of ρ by means of three further examples.

Example 8.16: Let (X, Y) be the r.v. of Example 7.6. By means of the marginal densities calculated there we obtain $E(X) = 7/12$, $E(X^2) = 5/12$, $V(X) = 11/144$, $E(Y) = 7/6$, $E(Y^2) = 5/3$, $V(Y) = 11/36$.

Moreover, $E(XY) = \int_0^1 \int_0^2 \frac{1}{4} xy(2x + y)\, dy\, dx = \frac{2}{3}$

Hence

$$\rho_{x,y} = \frac{\frac{2}{3} - \frac{7}{12} \cdot \frac{7}{6}}{\sqrt{\frac{11}{144} \cdot \frac{11}{36}}} = -\frac{1}{11}.$$

According to our intuition there is no linear correlation between the variables.

Example 8.17: Let (X, Y) be a discrete bidimensional r.v. with the joint probability function given in Tab. 8.1.

Tab. 8.1: Joint probability function.

X	1	2	3	P(Y = i)
Y				
2	0.3	0.02	0.02	0.34
4	0.02	0.3	0.01	0.33
6	0.02	0.01	0.3	0.33
P(X = i)	0.34	0.33	0.33	1

One can observe that with probability 0.9 the linear relation $Y = 2X$ is satisfied. Hence, intuitively one expects a linear correlation coefficient close to 1. This is in fact the case. From the marginal distributions in Tab. 8.1 we obtain $E(X) = 1.99$, $E(X^2) = 4.63$, $V(X) = 0.6699$, $E(Y) = 3.98$, $E(Y^2) = 18.52$, $V(Y) = 2.6796$, and $E(XY) = \sum_{x=1}^{3} \sum_{\substack{y=2 \\ y\,\text{even}}}^{6} xy\, p(x, y) = 9.04$.

Hence,

$$\rho_{x,y} = \frac{9.04 - 1.99 \cdot 3.98}{\sqrt{0.6699 \cdot 2.6796}} \approx 0.9695$$

Example 8.18: Assume that (X, Y) is a continuous bidimensional r.v., uniformly distributed over the triangular range space $R = \{(x, y) \mid 0 \leq x \leq 1,\ 5x \leq y \leq 5\}$ (see Fig. 8.6). Since the area of the triangle R is 5/2, the density has the constant value 2/5, yielding the marginal densities:

$$g(x) = \int_{5x}^{5} \frac{2}{5} dy = 2(1 - x) \quad \text{for } 0 \leq x \leq 1, \quad h(y) = \int_{0}^{y/5} \frac{2}{5} dx = \frac{2y}{25} \quad \text{for } 0 \leq y \leq 5.$$

From these densities one obtains $E(X) = 1/3$, $E(X^2) = 1/6$, $V(X) = 1/18$, $E(Y) = 10/3$, $E(Y^2) = 25/2$, and $V(Y) = 25/18$. The expectation of XY is $E(XY) = \int_0^1 \int_{5x}^5 \frac{2}{5} xy \, dy \, dx = \frac{5}{4}$. Finally,

$$\rho_{x,y} = \frac{\frac{5}{4} - \frac{1}{3} \cdot \frac{10}{3}}{\sqrt{\frac{1}{18} \cdot \frac{25}{18}}} = \frac{1}{2}.$$

This could be interpreted as a very slight degree of linearity imposed by the triangular shape of the range space. Intuitively one might have expected a higher value of $\rho_{x,y}$, since $Y \geq 5X$ must hold due to the definition of R.

Theorem 8.12: Let (X, Y) be a bidimensional r.v. and $W = aX + b$, $Z = cY + d$, where a, b, c, d are constants with $a \neq 0 \neq c$. Then it holds:

$$\rho_{w,z} = \frac{ac}{|ac|} \rho_{x,y}.$$

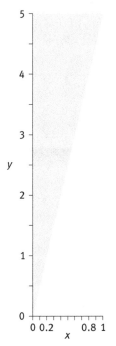

Fig. 8.6: Range space.

The theorem can be easily proved and states that the coefficient of correlation is invariant with respect to linear transformations when both are increasing or both are decreasing.

Exercises

8.1   A die is tossed until the number 6 appears for the first time. What is the expected number of tossings?

8.2   What is the expected number of fixed points of a random permutation? Give a good justification for your answer.

8.3   Determine $E(X)$ and $E(|X|)$ if X has the density $f(x) = \frac{a}{2}e^{-a|x|}$ for $-\infty < x < -\infty$, $a > 0$.

8.4   Given the pdf $f(x) = C/x^2$ for $1 \le x \le 5$. Determine C and $E(X)$ and $V(X)$.

8.5   The percent of lead contained in a metal alloy is an r.v. X with pdf $f(x) = Cx^2(100 - x)$ for $0 \le x \le 100$.

(i)   Determine the normalizing constant C, the expected percent of lead, and the variance.

(ii) Suppose that P, the net profit (per kg) realized in selling this alloy is given by the function $a + bX + cX^2$. What is the expected profit per kg?

8.6   An electronic device has a life length X (in hours). The pdf of the r.v. X is $f(x) = ae^{-ax}$ for $x > 0$ with $a = 0.005$. The cost of manufacturing one such item is \$3. The manufacturer sells the item for \$8 but guarantees a total refund if the device fails after less than 150 h. What is his expected profit per item? What is the variance?

8.7   Show that $f(x, y) = \frac{1}{2\pi}e^{-(x^2+y^2)/2}$ is a joint density over the plane. Are the variables X and Y independent? Determine the expectation of XY.

8.8   Let $P = (X, Y)$ be a randomly chosen point of the square $S := \{(x, y)\,|\,0 \le x,\ y \le 1\}$, i.e., the r.v.s X and Y are independent and uniformly distributed over $[0,1]$.

(i)   Determine the distribution of the distance from the origin $D = \sqrt{X^2 + Y^2}$

(ii) Determine the expected distance from the origin in two ways: using the pdf and applying Theorem 8.4.

8.9   Let $P = (X, Y)$ and $Q = (U, V)$ be two random points as in the previous exercise. Determine the expected distance between P and Q. Evaluate the integral $\int_0^1 \int_0^1 \int_0^1 \int_0^1 \sqrt{(x - u)^2 + (y - v)^2}\ dv\ du\ dy\ dx.$

8.10 Consider a discrete analogy of the previous exercise, i.e., suppose that the coordinates X, Y, U, V assume each of the values $1/n, 2/n, \ldots, n/n$ with probability $1/n$. Calculate the expected distance between P and Q. What is the limit of the expectation for $n \to \infty$?

8.11 Let X and Y be uniformly distributed over $[0, a]$ with $a > 0$.

(i)   Determine the pdf of $X + Y$ and $X - Y$. Use a graphical illustration of the range space of the new variable (U, V) defined as in Theorem 7.1.

(ii) Determine the expected values $E(X + Y)$ and $E(X - Y)$ by using the properties of the expectation and by means of the densities found in (i).

8.12 The population of a hypothetical city is made up of the two ethnic groups A and B. The heights of adults of these groups are random variables with expectations $\mu_A = 170$ cm and $\mu_B = 160$ cm and standard deviations $\sigma_A = 15$ cm and $\sigma_B = 12$ cm, respectively. Let Z denote the height of adults of the city population. Determine $E(Z)$ and $V(Z)$, if the proportion of group A in the city is 30%.

8.13 The weight of an adult of a certain population is an r.v. with expectation μ and variance σ^2. Let Z be the mean weight of a random sample of n adults of this population. Determine $E(Z)$ and $V(Z)$.

8.14 Let X, Y, and Z be independent r.v.s with expectations 5, 10, and 12 and variances 2, 5, and 15, respectively. Determine the expectations and variances of the variables $U = X - Y$, $V = 5X - 3Y + 7Z$, and $W = U - V$.

8.15 Calculate the variance for the uniform distribution.

8.16 Calculate all moments μ'_k, μ_k and $\mu_{[k]}$

   (i)   for the exponential distribution $f(x) = ae^{-a\,x}$ for $x \geq 0$, $a > 0$,

   (ii) for the uniform distribution over $[a, b]$.

8.17 Determine the coefficients of skewness and kurtosis for the density $f(x)$ in exercise 8.5.

8.18 The r.v. X has the density $f(x) = \frac{3}{5 \cdot 10^{10}} x^3 (100 - x)^2$ over the range space $[0,100]$. Use Tchebycheff's inequality to determine an upper limit for the probability $P(|X - \mu| \geq 30)$. Compare with the exact value of this probability.

8.19 Show the relation

$$E(1/X) + E(X^2) + E(\ln(X)) \geq \tfrac{1}{E(X)} + (E(X))^2 + \ln(E(X)),$$

where X assumes only positive values.

8.20 Given the pdf $f(x, y) = 4e^{-(2x+y)2}$ for $x, y \geq 0$, calculate the exact value of $E(XY)$ and the upper limit given by the Cauchy–Schwartz inequality.

8.21 Calculate the covariance and the coefficient of linear correlation for the bidimensional r.v. (X, Y) with pdf $f(x, y) = \tfrac{15}{4}(x^2 + xy)$ over the range space.

8.22 Calculate the covariance and the coefficient of linear correlation for the discrete bidimensional variable with the following joint distribution.

X	0	1	2
Y			
0	3/66	9/66	3/66
1	12/66	18/66	6/66
2	6/66	8/66	1/66

For the following four exercises, software for symbolical calculations should be available. Consult the distribution table in the appendix.

8.23 Determine skewness and kurtosis for the Weibull distribution. Show that both measures are independent of the parameter a and express a_3 and a_4 graphically as a function of the parameter b. Illustrate the Weibull densities for $a = 1$ and $b = 1$, 3.4, 8 and 15. Calculate a_3 and a_4 for these cases.

8.24 Determine a_3 and a_4 for the Rayleigh distribution. Show that the measures do not depend on the parameter s. Illustrate the densities graphically for $s = 1$, 5 and 10. Are the results intuitive?

8.25 (i) Consider the triangular distribution with density:

$$f(x) = \begin{cases} \frac{2}{ab}x & \text{for } 0 \leq x \leq a, \\ & \qquad\qquad 0 < a < b \\ \frac{2}{b-a}\left(1 - \frac{x}{b}\right) & \text{for } a \leq x \leq b. \end{cases}$$

Illustrate $f(x)$ graphically and determine a_3 and a_4.

(ii) Set $a = 1$. Illustrate a_3 graphically as a function of b. Determine $a_3(1)$ and $a_3(\infty)$.

8.26 Determine the kurtosis a_4 of Student's t-distribution. Illustrate the densities for $k = 5$ and $k = 50$. What do you observe?

8.27 Determine the kurtosis of the general univariate normal distribution.

9 Discrete probability models

In Section 5.3, we have already introduced one of the most important discrete models, namely the binomial distribution. In the following, we will get to know some other discrete distributions.

9.1 The Poisson distribution

This model is very useful to describe *rare events* and also occurs as a limiting case of the binomial distribution.

Definition 9.1: Let X be a discrete random variable with possible values 0, 1, 2, . . . If

$$P(X = k) = \frac{e^{-\lambda}\lambda^k}{k!} \quad \text{for } k = 0, 1, 2, \ldots, \tag{9.1}$$

we say that X has a *Poisson distribution* with parameter $\lambda > 0$

To check that this is a legitimate probability distribution, we simply observe that

$$\sum_{k=0}^{\infty} P(X = k) = e^{-\lambda} \sum_{k=0}^{\infty} \frac{\lambda^k}{k!} = e^{-\lambda}e^{\lambda} = 1.$$

Note that the random variable has been defined directly in terms of its range space $R_X = \{0, 1, \ldots\}$ and probability distribution, without reference to any underlying sample space S. Formally, we suppose that S can be identified with R_X and we set $X(s) = s$. Thus, the outcomes of the random experiment are the integers 0, 1, 2, . . ., and their probabilities are given by (9.1). Basically, the same reasoning applies also for the following discrete models of this chapter.

Theorem 9.1: If X has a Poisson distribution with parameter λ, then it holds
(i) $E(X) = \lambda$, (ii) $V(X) = \lambda$.

Proof:

(i) $E(X) = \displaystyle\sum_{k=0}^{\infty} k e^{-\lambda} \frac{\lambda^k}{k!} = \sum_{k=1}^{\infty} e^{-\lambda} \frac{\lambda^k}{(k-1)!}.$

By substituting $s = k - 1$, we get

$$E(X) = \sum_{s=0}^{\infty} e^{-\lambda} \frac{\lambda^{s+1}}{s!} = \lambda \sum_{s=0}^{\infty} e^{-\lambda} \frac{\lambda^s}{s!} = \lambda.$$

(ii) $E(X^2) = \displaystyle\sum_{k=0}^{\infty} k^2 e^{-\lambda} \frac{\lambda^k}{k!} = \sum_{k=1}^{\infty} k \, e^{-\lambda} \frac{\lambda^k}{(k-1)!}.$

https://doi.org/10.1515/9783111332277-009

With the same substitution as above, we obtain

$$E(X^2) = \sum_{s=0}^{\infty}(s+1)e^{-\lambda}\frac{\lambda^{s+1}}{s!} = \lambda\left(\sum_{s=0}^{\infty}se^{-\lambda}\frac{\lambda^s}{s!} + \sum_{s=0}^{\infty}e^{-\lambda}\frac{\lambda^s}{s!}\right)$$

$$= \lambda(E(X)+1) = \lambda^2 + \lambda.$$

Finally, it follows that

$$V(X) = E(X^2) - (E(X))^2 = \lambda.к к$$

The following theorem states that the Poisson distribution approximates the more complicated binomial, whenever n is "large" and p is "small."

Theorem 9.2: Let be $p_k = \binom{n}{k}p^k(1-p)^{n-k}$. Suppose that $n \to \infty$ such that np remains unchanged, i.e., $np = \lambda$ holds for a constant λ, then $\lim_{n\to\infty}p_k = \dfrac{e^{-\lambda}\lambda^k}{k!}$ for $k = 0, 1, 2 \dots$.

Proof: Due to the above assumptions, the binomial probability p_k can be written as

$$p_k = \frac{n\cdot(n-1)\dots(n-k+1)}{k!}\left(\frac{\lambda}{n}\right)^k\left(1-\frac{\lambda}{n}\right)^{n-k}$$

$$= \frac{\lambda^k}{k!n}\frac{n-1}{n}\dots\frac{n-k+1}{n}\left(1-\frac{\lambda}{n}\right)^n\left(1-\frac{\lambda}{n}\right)^{-k}.$$

For $= n \to \infty$, the factors $\dfrac{n-i}{n}$ and $\left(1-\dfrac{\lambda}{n}\right)^{-k}$ converge to 1 and $\left(1-\dfrac{\lambda}{n}\right)^n$ to $e^{-\lambda}$. Thus, p_k converges to $\dfrac{e^{-\lambda}\lambda^k}{k!}$.

The Poisson distribution is applicable whenever the binomial is applicable (see Definition 5.3), and when, in addition, n is "large" and p is "small."

Example 9.1: A health insurance company has $n = 10{,}000$ clients, which can be considered as a random sample of the population. A person suffers from a particular disease with probability $p = 0.002$.
(i) What is the probability that exactly five clients of the company suffer from this disease?
(ii) What is the probability that more than 30 clients suffer from this disease?

We calculate the exact probabilities and the approximations for both cases. Let X denote the number of clients with the considered disease.

(i) Exact probability: $P(X = 5) = \binom{10,000}{5} 0.002^5 0.998^{9995} \approx 5.436 \cdot 10^{-5}$.

Approximation: since $\lambda = np = 20$, we get $P(X = 5) \approx \dfrac{e^{-20} 20^5}{5!} \approx 5.496 \cdot 10^{-5}$.

(ii) By summing up the respective probabilities, we obtain for $P(X > 30) = 1 - P(X \leq 30)$, the exact value 0.013 391 and the approximate value 0.013 475. The relative error made by using the approximation is about 0.6%.

9.2 The Poisson process

In the above, we used the Poisson distribution as an approximation for the binomial. However, its importance goes far beyond this relation. The Poisson distribution represents an adequate model for a large number of observational phenomena. We will only sketch the reasoning to obtain a very important result (see Theorem 9.3).

Assume that we count the observations of a certain type of phenomenon during a time interval. In this context, an observation may be, for example, the registration of an alpha particle by a Geiger counter, the arrival of a call in a switchboard, or the occurrence of an accident at a specific location of a main road. Let X_t denote the number of observations of the considered phenomenon in the time interval $[0, t[$. In principle, we may consider an r.v. X_t for any positive t, so we are dealing with a so-called *stochastic process*. In the general case, the probability $P(X_t = k)$ may or may not depend on the variables X_s, with $s < t$. We will restrict ourselves to the following situation:

(i) the probability of making an observation in a small time interval $[t, t + \Delta t[$ is $\alpha \, \Delta t$ ($\alpha > 0$),

(ii) the probability of making more than one observation in this interval is zero,

(iii) the probability of making an observation in the interval $[t, t + \Delta t[$ does not depend on the observations before the time t.

Theorem 9.3: Let X_t be defined as above and satisfy the conditions (i)–(iii). Then, it holds

$$P(X_t = k) = \frac{e^{-\alpha t} (\alpha t)^k}{k!}, \tag{9.2}$$

i.e., X_t has a Poisson distribution with parameter $\lambda = \alpha t$.

Proof: We first obtain the following relation between probabilities. Since, at most one observation occurs in the interval $[t, \bar{t}[$, with $\bar{t} = t + \Delta t$, the event $X_t = k$ may result in two ways (see Fig. 9.1). In the first case, there have been $k - 1$ observations until the time t and one observation in the interval $[t, \bar{t}[$; in the second case, k observations occurred until t and no observation in the interval $[t, \bar{t}[$.

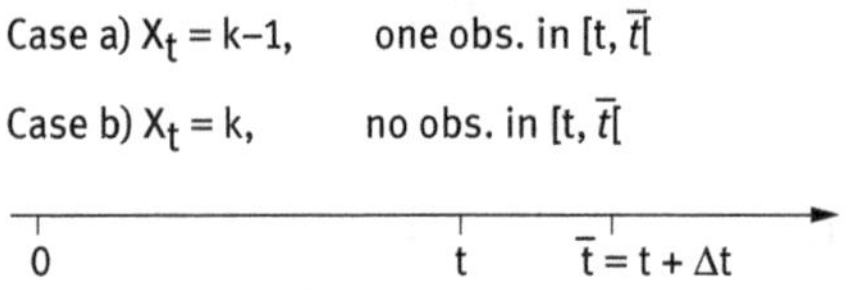

Case a) $X_t = k-1$,   one obs. in $[t, \bar{t}[$

Case b) $X_t = k$,   no obs. in $[t, \bar{t}[$

Fig. 9.1: Poisson process.

Thus, we get

$$P(X_{t+\Delta t} = k) = \alpha \Delta t P(X_t = k-1) + (1 - \alpha \Delta t)P(X_t = k),$$

which can be reformulated as

$$\frac{P(X_{t+\Delta t} = k) - P(X_t = k)}{\Delta t} = \alpha[P(X_t = k-1) - P(X_t = k)].$$

By taking the limit for $\Delta t \to 0$, we obtain the differential equation

$$\frac{\partial}{\partial t}P(X_t = k) = \alpha[P(X_t = k-1) - P(X_t = k)] \qquad (9.3)$$

for $k = 0, 1, 2, \ldots$. It can be verified that the Poisson probabilities in (9.2) satisfy this equation.

Example 9.2: A telephone exchange receives an average of two calls over a period of 5 min. Assuming that we are dealing with a Poisson process, what is the probability that exactly 7 calls are received in a period of 15 min?

The expected number of calls recorded in 15 min is $2 \times 3 = 6$, thus $\lambda = 6$ and from Theorem 9.3, we obtain $P(X_t = 7) = \frac{e^{-6}6^7}{7!} \approx 0.1377$.

A Poisson process need not refer to observations over a time period. When the above conditions (i)–(iii) apply analogously, (9.2) holds also for observations over a distance, over an area, or a volume, etc.

Example 9.3: Suppose that a book of 485 pages contains 41 typographical errors. If the number of errors per page has a Poisson distribution, what is the probability that a section of 12 pages is free of errors?

Let X_t denote the number of errors until page t. The average rate of errors in a selection of 12 pages is $\lambda = \dfrac{41 \cdot 12}{485} \approx 1.0144$. Thus, $P(X_t = k) = \dfrac{e^{-\lambda}\lambda^k}{k!}$ and in particular, $P(X_t = 0) = e^{-\lambda} \approx e^{-1.0144} \approx 0.3626$.

9.3 The geometric distribution

As in the definition of the binomial r.v. (Definition 5.3), we consider an (individual) random experiment E and an event A associated with E such that $P(A) = p$ and $P(\bar{A}) = 1 - p = q$. The experiment E is carried out repeatedly; however, now the number of realizations of E is not predetermined. We perform independent realizations of E until the event A occurs for the first time (first success). The random variable X is defined as the number of realizations of E performed thereby. If, for example, the fifth realization is the first success, i.e., the results of E are $A, A, A, A, \bar{A}$, we have $X = 5$.

Obviously, $X = k$ occurs if and only if the first $(k - 1)$ realizations of E result in $\bar{A}$, while the kth realization results in A. This event happens with probability

$$P(X = k) = (1 - p)^{k-1} p \quad \text{for } k = 1, 2, \ \ldots \tag{9.4}$$

A random variable with probability distribution given by (9.4) is said to have a *geometric distribution* with parameter p. It can be easily verified that this is a legitimate probability distributions, since $P(X = k) \geq 0$ for $k = 1, 2, \ldots$, and from the geometrical series, it is known that $\displaystyle\sum_{k=1}^{\infty} P(X = k) = p\left(1 + q + q^2 + \cdots\right) = p\frac{1}{1-q} = 1.$

Example 9.4: A die is thrown until the number 6 occurs for the first time. What is the probability that the number 6 occurs for the first time in the 10th throw?

Solution: $P(X = 10) = \left(1 - \dfrac{1}{6}\right)^9 \dfrac{1}{6} \approx 0.0323.$

Theorem 9.4: If X has the geometric distribution defined above, then

(i) $E(X) = \dfrac{1}{p}$, (ii) $V(X) = \dfrac{q}{p^2}$.

Proof:

(i) $E(X) = \displaystyle\sum_{k=1}^{\infty} kpq^{k-1} = p\sum_{k=1}^{\infty} \frac{d}{dq} q^k = p\frac{d}{dq}\sum_{k=1}^{\infty} q^k = p\frac{d}{dq}\frac{q}{1-q}$

$\qquad\qquad = p\dfrac{1}{(1-q)^2} = \dfrac{1}{p}.$

The elegant idea of interchanging differentiation and summation is justified by the fact that the series converges for $|q| < 1$. The proof of part (ii) is left to the reader (see Exercise 9.6).

Example 9.5: The execution of a certain experiment in a space capsule costs $10,000. If the experiment fails, an additional cost of $2,000 occurs because certain changes must be performed before the next trial. The probability of a success is 0.15. Assuming independence between the individual trials, what is the expected total cost of all experiments, if they are repeated until the first success?

Solution: If X is the number of trials and C is the cost, we have $C = 10{,}000X + 2{,}000(X - 1)$ $= 12{,}000X - 2{,}000$. Thus, the expected cost is

$$E(C) = 12{,}000E(X) - 2{,}000 = 12{,}000\frac{1}{0.15} - 2{,}000 = \$ 78{,}000.$$

The geometric distribution is characterized by the following interesting property.

Theorem 9.5: Let X be a discrete random variable with possible values $0, 1, 2, \ldots$ The relation

$$P(X > n + m | X > n) = P(X > m) \quad \text{for } n, m = 0, 1, 2, \ldots \tag{9.5}$$

holds if and only if X has a geometric distribution.

Proof:

(i) We first show that the geometric distribution satisfies (9.5). Since $X > k$ holds if and only if the first k realizations of E yield fail, we get $P(X > k) = q^k$ (see also Exercise 9.10). Hence, for the geometric distribution, it holds that

$$P(X > n + m | X > n) = \frac{P(X > n + m, \, X > n)}{P(X > n)} = \frac{P(X > n + m)}{P(X > n)}$$

$$= \frac{q^{n+m}}{q^n} = q^m = P(X > m).$$

(ii) Assume that (9.5) holds, which is equivalent to

$$P(X > n + m) = P(X > n)P(X > m). \tag{9.6}$$

For $n = m = 0$, we obtain $P(X > 0) = P(X > 0)^2$, implying $P(X > 0) = 1$. By setting $p = P(X = 1)$, it follows that $P(X > 1) = 1 - p$ and for $m = 1$ eq. (9.6) yields

$$P(X > n + 1) = P(X > n)(1 - p).$$

From this, we obtain by induction

$$P(X > n) = (1 - p)^n. \tag{9.7}$$

From (9.7), we finally obtain $P(X = n) = P(X > n - 1) - P(X > n) = (1 - p)^{n-1} - (1 - p)^n = p(1 - p)^{n-1}$. Hence, X has a geometric distribution.

The theorem states that the geometric distribution has "no memory." Suppose that no success has occurred during the first n realizations of E. Then, the probability

that there will be no success during the next m realizations is the same as the probability of no success during the first m realizations.

9.4 The Pascal distribution

In the previous section, a random experiment E has been carried out until the first
success was obtained. An obvious generalization arises if we continue performing experiments until the rth success occurs. As before, we assume that the realizations of E
are independent and that the event A (success) occurs with probability p. The probability of a failure is $q = 1 - p$. Our r.v. Y is the number of realizations of E until the
occurrence of the rth success. If, for example, $r = 4$ and the outcomes of E are

$$\bar{A}, \bar{A}, \bar{A}, A, \bar{A}, A, \bar{A}, \bar{A}, \bar{A}, \bar{A}, A, \bar{A}, A, \tag{9.8}$$

we have $Y = 13$, since the fourth A occurs in the 13th trial. The probability that the rth
success occurs in the kth trial is given by

$$P(Y = k) = \binom{k-1}{r-1} p^r q^{k-r} \text{ for } k = r, r+1, \ldots . \tag{9.9}$$

One can justify this as follows. Any sequence of results having the rth success in the
kth trial has $r - 1$ successes and $k - r$ failures on the first $k - 1$ positions. Due to the
independence of the trials, any of these sequences occurs with probability $p^r q^{k-r}$ and
their number is $\binom{k-1}{r-1}$, corresponding to the number of ways to select the positions
of the $r - 1$ successes among the first $k - 1$ positions.

If a random variable has the distribution (9.9), we say that it has a *Pascal distribution*. Of course, the special case for $r = 1$ is the geometric distribution. In order to show
that (9.9) is a legitimate distribution, we make use of the general binomial theorem
(Proposition 1.5), from which we obtain

$$(1-q)^{-r} = \sum_{i=0}^{\infty} \binom{-r}{i} (-q)^i. \tag{9.10}$$

The summand can be rewritten as $\binom{r+i-1}{r-1} q^i$, implying

$$(1-q)^{-r} = \sum_{i=0}^{\infty} \binom{r+i-1}{r-1} q^i$$

and the substitution $k = i + r$ yields

$$(1-q)^{-r} = \sum_{i=r}^{\infty} \binom{k-1}{r-1} q^{k-r}.$$

From this, it follows immediately that the probabilities in (9.9) sum up to 1. Because of the negative exponent $(-r)$ in (9.10), the model (9.9) is also called the *negative binomial distribution*.

Theorem 9.6: If Y has the Pascal distribution (9.9), then (i) $E(Y) = r/p$, (ii) $V(Y) = rq/p^2$.

Proof: Consider the auxiliary random variable Z_i, defined as the number of realizations of E after the $(i-1)$th occurrence of a success, up to and including the ith occurrence of a success, $i = 1, \ldots, r$. (For the example (9.8), we have $Z_1 = 4$, $Z_2 = 2$, $Z_3 = 5$, and $Z_4 = 2$). Then, we have $Y = Z_1 + \cdots + Z_r$, where the Z_i are independent, geometrically distributed r.v.s. From the additivity properties of expectation and variance, we obtain $E(Y) = E(Z_1) + \cdots + E(Z_r) = r(1/p)$ and $V(Y) = V(Z_1) + \cdots + V(Z_r) = r(q/p^2)$

Example 9.6: A pastry shop needs twelve special pies for a birthday party of a prominent person. Each produced pie is suitable for sale with probability 0.7.
(i) What is the probability that more than 15 pies must be produced to obtain 12 that are suitable for sale?
(ii) What is the expected number of pies that must be produced?

Solution:

(i) $P(Y = k) = \binom{k-1}{11} 0.7^{12} 0.3^{k-12}$ for $k = 12, 13, \ldots$

$P(Y > 15) = 1 - P(Y = 12) - \ldots - P(Y = 15)$

$= 1 - 0.7^{12} \left[\binom{11}{11} + \binom{12}{12} \cdot 0.3 + \binom{13}{11} \cdot 0.3^2 + \binom{14}{11} \cdot 0.3^3 \right] \approx 0.7031.$

(ii) $E(Y) = 12/0.7 \approx 17.14$.

We finally cite an interesting relation between the binomial and the Pascal distributions.
Assume that X and Y have a binomial and a Pascal distribution, respectively, i.e.,
$p(X = k) = \binom{n}{k} p^k (1-p)^{n-k}$ for $k = 0, 1, \ldots, n$ and $P(Y = k) = \binom{k-1}{r-1} p^r q^{k-r}$ for $k = r, r + 1, \ldots$, where $r \le n$. Then, it holds

$$P(X \ge r) = P(Y \le n). \tag{9.11}$$

The relation is intuitively clear, since the events $(X \ge r) = \{$there are at least r successes among the first n trials$\}$ and $\{Y \le n\} = \{$it requires at most n trials to obtain r successes$\}$ are equivalent.

As an illustration. we solve Example 9.6(i) alternatively. From (9.11), we get $P(Y \leq 15)$

$$= P(X \geq 12) = \Sigma_{k=12}^{15} \binom{15}{k} 0.7^k 0.3^{15-k} \approx 0.2969. \text{ Thus, } P(Y > 15) \approx 0.7031.$$

9.5 The hypergeometric distribution

Assume that an urn contains m_1 white balls and m_2 black balls ($m = m_1 + m_2$). We chose n balls randomly without replacement ($n \leq m$). Let X denote the number of white balls in the selection. From the studies in Section 3.1, it is clear that

$$P(X = x) = \frac{\binom{m_1}{x} \cdot \binom{m_2}{n-x}}{\binom{m}{n}} \quad \text{for } x = 0, 1, 2, \ldots. \tag{9.12}$$

We may define the above probabilities for all $x = 0, 1, \ldots$ since $\binom{a}{b} = 0$ for integers a, b with $b < 0$ or $b > a$. It is clear from intuition that $P(X = x)$ can only be positive if $x \leq n$, m_1 and $n - x \leq m_2$. If a discrete random variable X has the distribution (9.12), we say that it has a *hypergeometric distribution*.

In order to check whether this is a legitimate distribution, we observe that $\Sigma_{k=0}^{\infty} P(X = k) = 1$ is equivalent to Proposition 1.4.

Theorem 9.7: If X has the distribution (9.12), then

(i) $E(X) = \dfrac{m_1}{m} n,$

(ii) $V(X) = \dfrac{m_1}{m} n \dfrac{m_2}{m} \dfrac{m-n}{m-1}.$

Proof:

(i) $E(X) = \displaystyle\sum_{x=1}^{m_1} x \frac{m_1!}{x!(m_1-x)!} \frac{\binom{m_2}{n-x}}{\binom{m}{n}}$

$\quad = \dfrac{m_1}{\binom{m_1}{n}} \displaystyle\sum_{x=1}^{m_1} \frac{(m_1-1)!}{(x-1)!(m_1-x)!} \binom{m_2}{n-x}.$

By means of the substitution $s = x - 1$, we get

$$E(X) = \frac{m}{\binom{m_1}{n}} \sum_{s=0}^{m_1-1} \binom{m_1-1}{s} \binom{(m-1)-(m_1-1)}{(n-1)-s} = \frac{m_1}{\binom{m_1}{n}} \binom{m-1}{n-1} = \frac{m_1}{m} n.$$

where the sum is evaluated, again applying Proposition 1.4.

The formula for the expectation is very intuitive, stating that the expected number of white balls in the selection is the proportion of white balls in the urn multiplied by the size n of the selection. The proof of part (ii) of the theorem is left to the reader (see exercise 9.16).

Example 9.7: An urn contains 10 white and 15 black balls. If 5 balls are chosen at random without replacement, what is the probability that the number of white balls selected is larger than its expectation.

Solution: The expectation of white balls is $\frac{10}{25} \cdot 5 = 2$. The required probability

is $P(X > 2) = \Sigma_{x=3}^{5} \dfrac{\binom{10}{x}\binom{15}{5-x}}{\binom{25}{5}} \approx 0.3012.$

Before we touch briefly a generalization of the hypergeometric distribution, we illustrate that under certain conditions, (9.12) is approximately given by the binomial distribution.

Assume that the numbers m_1 and m_2 of black and white balls in the urn are large and that only a small number n of balls is selected. In this case, it does not make "much difference" whether the balls are selected with or without replacement. In the latter case, the distribution of the number Y of white balls in the selection is given by

the binomial distribution $P(Y = x) = \binom{n}{x} p^x (1-p)^{n-x}$, where $p = m_1/m$ is the proportion of white balls. Under the above conditions, it holds

$$\frac{\binom{m_1}{x} \cdot \binom{m_2}{n-x}}{\binom{m}{n}} \approx \binom{n}{x} p^x (1-p)^{n-x} \tag{9.13}$$

$$\text{with } p = m_1/m$$

We illustrate this numerically for the case $m = 1{,}000$, $m_1 = 300$, $n = 10$ and $x = 4$. Then, it holds

$$\frac{\binom{300}{4} \cdot \binom{700}{6}}{\binom{1{,}000}{10}} = \frac{300 \cdot \ldots \cdot 297 \cdot 700 \cdot \ldots \cdot 695}{4! \quad 6!} \frac{10!}{1{,}000 \cdot \ldots \cdot 991}$$

$$\approx \frac{10!}{4!\,6!} \frac{300^4 \cdot 700^6}{1{,}000^4 \cdot 1{,}000^6} = \binom{10}{4} 0.3^4 0.7^6.$$

Here, we have $p = m_1/m = 0.3$. The exact value of the hypergeometric probability is 0.200839, and the approximate calculation by the binomial model results in 0.200121.

The experimental situation resulting in the distribution (9.12) has an obvious generalization. We now assume that the urn contains balls of k different colors (types), such that there are m_1 balls of color 1, m_2 balls of color 2, $\cdots$ and m_k balls of color k ($m_1 + \cdots + m_k = m$). We select $n \leq m$ balls at random without replacement. We consider the random vector $X = (X_1, \ldots, X_k)$, where X_i denotes the number of balls with color i in the selection ($i = 1, \ldots, k$). Since n is the number of selected balls, it holds $X_1 + \cdots + X_k = n$. In particular, the variables X_i are not independent, and from the studies in Section 3.1, it follows that

$$P(X_1 = x_1, \ldots, X_k = x_k) = \frac{\binom{m_1}{x_1} \cdot \binom{m_2}{x_2} \cdots \binom{m_k}{x_k}}{\binom{m}{n}} \tag{9.14}$$

for nonnegative integers x_i with $= x_1 + \cdots + x_k = n$.

This distribution is known as the *multivariate hypergeometric distribution*.

As an extension of Theorem 9.7, we obtain:

Theorem 9.8: If X has the distribution (9.14), then it holds

$$E(X_i) = \frac{m_i}{m} n \text{ and } V(X) = \frac{m_i}{n} n \frac{m - m_i}{m} \frac{m - n}{m - 1} \quad \text{for } i = 1, \ldots, k.$$

The proof follows intuitively from the previous theorem, assuming that there are *only two colors*: color i and "the other color," i.e., all colors different from i are identified with each other.

Example 9.8: Assume that there are 500 persons in a Bavarian ballroom: 200 German, 150 Austrian, 80 French and 70 Italian. If 20 persons are chosen at random who will participate in a price draw,
(i) what is the expected number of persons from each country in the selection?
(ii) what is the probability that there are chosen 6 German, 6 Austrian, 5 French and 3 Italian?

Solution:

(i) $E(G) = \dfrac{200}{500} \cdot 20 = 8$, $E(A) = 6$, $E(F) = 3.2$ and $E(I) = 2.8$.

(ii) $\dfrac{\binom{200}{6} \cdot \binom{150}{6} \cdot \binom{80}{5} \cdot \binom{70}{3}}{\binom{500}{20}} \approx 0.005813$.

9.6 The multinomial distribution

This model is obtained as a natural generalization of the binomial distribution (Definition 5.3). Assume that the sample space S of the random experiment E can be partitioned into k mutually exclusive events $A_1, \ldots, A_k$, i.e., when E is performed, exactly one of these events occurs, and the probabilities $p_i = P(A_i)$ are positive, satisfying $p_1 + \cdots + p_k = 1$.

We perform n independent realizations of E during which the probabilities p_i remain constant. Similar to the model (9.14), we consider the random vector $X = (X_1, \ldots, X_k)$, where the X_i denotes the number of occurrences of the event A_i during the n repetitions of E ($i = 1, \ldots, k$). We obtain

$$P(X_1 = x_1, \ldots, X_k = x_k) = \binom{n}{x_1, \ldots, x_k} p_1^{x_1} \cdot \ldots \cdot p_k^{x_k} \tag{9.15}$$

for nonnegative integers x_i with $x_1 + \cdots + x_k = n$, where $\binom{n}{x_1, \ldots, x_k}$ is the multinomial coefficient (Definition 1.6). Formula (9.15) results from the following reasoning. The event $\{X_1 = x_k, \ldots, X_1 = x_k\}$ occurs if and only the series of outcomes in the n realizations of E is

$$A_1, \underset{x_1 \text{times}}{\ldots}, A_1, A_2, \underset{x_2 \text{times}}{\ldots}, A_2, \ldots, A_k, \underset{x_k \text{times}}{\ldots}, A_k, \tag{9.16}$$

or any permutation thereof. Any of these permutations occurs with probability $p_1^{x_1} \cdot \ldots \cdot p_k^{x_k}$ and their number is $\binom{n}{x_1, \ldots, x_k}$

The proof that (9.15) is a legitimate distribution, follows from the multinomial theorem (Proposition 1.6).

Theorem 9.9: If X has the distribution (9.15), then it holds $E(X_i) = np_i$ and $V(X) = np_i(1 - p_i)$ for $i = 1, \ldots, k$.

The theorem follows intuitively from the fact that for the binomial distribution, $E(X) =$ np and $V(X) = np(1 - p)$ hold (see Theorem 8.1 and Example 8.10). We consider the *two* events $A: = A_i$ and $\bar{A}$, which is the union of all events different from A_i.

Example 9.9: During the inspection of produced articles, three categories of items are distinguished: perfect (p), with minor defects (d) and with major defects (D). Assume that an inspected article is classified as (p), (d) and (D) with probability 0.7, 0.2 and 0.1, respectively, and that a random sample of ten articles is selected.
(i)   What is the probability that the sample consists of 6 perfect articles, 2 items with minor defects and 2 articles with major defects?
(ii) What is the probability to encounter more articles with major defects than articles with minor defects?
(iii) Determine the expectation and variance for the number of perfect items.

Solution:
(i)   The required probability is $\begin{pmatrix} 10 \\ 6, 2, 2 \end{pmatrix} 0.7^6 \cdot 0.2^2 \cdot 0.1^2 \approx 0.0593$.
(ii) Denoting the number of items with minor and major defects by i and j, we obtain $\sum_{i=0}^{4}\sum_{j=i+1}^{10-i}\begin{pmatrix} 10 \\ 10-i-j, i, j \end{pmatrix} 0.7^{10-i-j} \cdot 0.2^i \cdot 0.1^j \approx 0.1847$.
(iii) $E(X_1) = 10 \cdot 0.7 = 7$, $V(X_1) = 10 \cdot 0.7 \cdot 0.3 = 2.1$.

We finally mention that the multinomial distribution with probabilities $p_i = m_i/m$ serves as an approximation for the more complex multivariate hypergeometric distribution (9.14) if the number of balls of each color in the urn is "large" and if only a "small" number n of balls is selected. The reasoning is analogous to that in Section 9.5, establishing an approximate relation between the hypergeometric and the binomial distribution (see exercise 9.19).

Exercises

9.1   Assume that the health insurance company of Example 9.1 has n clients and that a person suffers from the considered disease with probability $p = 20/n$. Calculate the probability $P(X = 5)$ for $n = 10^i$, where $i = 2, 3, \ldots, 6$, and verify the convergence to $\dfrac{e^{-20}20^5}{5!}$.

9.2   A radioactive source is observed during 10 time intervals, each of twenty seconds, and the number of particles emitted during each period is counted. Suppose that the number of particles emitted during each period has Poisson distribution with parameter $\lambda = 30$. What is the probability that in all of the 10 time intervals, 25 or more particles are emitted?

9.3 Assume that the number of bacteria per cm^3 in a certain liquid has Poisson distribution with parameter $\lambda = 5$. What is the probability that at least 60 units of bacteria are found in a test tube containing 10 cm^3 of this liquid?

9.4 The number of traffic accidents per month in a certain city has Poisson distribution with $\lambda = 20$. What is the probability that
(i) exactly six accidents occur in a specific week?
(ii) more than two accidents happen on a given day?

9.5 Show that the central moments satisfy the relation

$$\frac{\mu_{r+1}}{\lambda} = r\mu_{r-1} + \frac{d}{d\lambda}\mu_r$$

if X has Poisson distribution with parameter λ.

9.6 Proof part (ii) of Theorem 9.4

9.7 What is the probability that the cost in Example 9.5 is over $100,000?

9.8 A trapeze artist suffers a fatal accident during a show with probability $p = 0.001$.
(i) What is the probability that he dies during the 30th presentation?
(ii) What is the probability that he survives 100 shows?
(iii) For what value of p, the probability to survive 100 shows is 0.95?

9.9 Let k be a fixed positive integer. For what value of p, the probability $P(X = k) = (1-p)^{k-1}p$ assumes its maximum?

9.10 Similar to the generalizations of the binomial distribution mentioned in Section 5.3, one can generalize the geometric distribution, assuming that the probability of a success need not be constant, i.e., we assume that in the ith realization of the experiment E, the event A occurs with probability p_i.
(i) Determine a simple formula for the cumulative distribution function $F(k) = P(X \le k)$, using intuition.
(ii) Determine the probability function, expectation and variance for the specific case with $q_i := \frac{1}{2^i}$ and $p_i := 1 - \frac{1}{2^i}$.

9.11 An archer hits the center of the target with probability 0.35. What is the probability that he hits the center for the third time in the tenth shot?

9.12 The probability that an experiment will succeed is 0.7. If it is repeated until five successful outcomes have occurred, what is the probability that more than ten realizations will be necessary?

9.13 The r.v. Y has a Pascal distribution. The expectation and variance are 50 and 450, respectively. Calculate the probability $P(Y = 10)$.

9.14 Prove Theorem 9.6, evaluating the sums $\sum_{k=r}^{\infty} k^i P(Y = k)$ for $i = 1, 2$, where $P(Y = k)$ is given by (9.9).

9.15 Prove relation (9.11) by induction on n.

9.16 Proof part (ii) of Theorem 9.7.

9.17 Consider again the random selection of Example 9.8. Determine the probabilities of the following events:
(i) the selection does not contain any Italian,

(ii) the selection contains at least 5 French and at least 4 Italian,

(iii) the number of Germans in the selection is greater or equal to the number of Austrian.

9.18 A die is thrown 5 times.

(i) What is the probability to obtain the number 1 once, the number 2 two times and the number 5 two times?

(ii) What is the probability to obtain at least the number 5 two times and the number 6 at least two times?

9.19 Calculate the probabilities of the events in Exercise 9.17 approximately by the multinomial model.

9.20 A marker performs a random walk on a plane, starting at the origin (0,0). In any movement, it makes a step of one unit in any of the four cardinal directions. The marker moves to the right with probability p_1 and to the left with probability p_2. It moves up with probability p_3 and down with probability p_4 ($p_1 + \cdots + p_4 = 1$).

(i) What is the probability that the marker is located at the point (x, y) after n movements?

(ii) Calculate the probability of item (i) for $n = 20$, $x = 5$, $y = 3$ and $p_1 = \cdots = p_4 = 1/4$.

10 Continuous probability models

In this chapter, we will present some continuous distributions, starting with one of the most important ones.

10.1 The normal distribution

This distribution has countless applications in probability and statistics since it has many desirable mathematical properties and serves as an excellent approximation to a large class of distributions of practical importance (see e.g. Section 13.2).

Definition 10.1: The random variable X, assuming all real values, has a *normal (or Gaussian) distribution*, if its pdf is of the form

$$f(x) = \frac{1}{\sqrt{2\pi}\sigma} \exp\left(-\frac{1}{2}\left(\frac{x-\mu}{\sigma}\right)^2\right) \text{ for } -\infty < x < \infty. \tag{10.1}$$

(Due to typographical reasons, we occasionally use the notation $\exp(t)$ instead of e^t). The parameters μ and σ satisfy the conditions $-\infty < \mu < \infty$, $\sigma > 0$. We will see in Theorem 10.1 that they represent the expectation and standard deviation. The use of the symbols μ and σ to denote the parameters is therefore justified. We will write $X \sim N(\mu,\sigma^2)$ if X has the distribution (10.1). We first show that we are dealing with a legitimate distribution. Obviously, it holds $f(x) \geq 0$. In order to show that (10.1) satisfies $\int_{-\infty}^{\infty} f(x)\,dx = 1$, we first apply the substitution $z = (x - \mu)/\sigma$, which transforms the integral in

$$I = \frac{1}{\sqrt{2\pi}} \int_{\infty}^{-\infty} e^{-z^2/2}dz.$$

We have now to prove that $I = 1$. In order to do this, we evaluate the square of the integral:

$$I^2 = \frac{1}{\sqrt{2\pi}} \int_{-\infty}^{\infty} e^{-x^2/2}dx \frac{1}{\sqrt{2\pi}} \int_{-\infty}^{\infty} e^{-y^2/2}dy = \frac{1}{2\pi} \int_{-\infty}^{\infty}\int_{-\infty}^{\infty} e^{-(x^2+y^2)/2}dy\,dx.$$

By introducing polar coordinates $x = r \cos \Phi$ and $y = r \sin \Phi$ (see Example 1.8), we obtain

$$I^2 = \frac{1}{2\pi} \int_0^{2\pi}\int_0^{\infty} e^{-r^2/2}r\,dr\,d\phi = \frac{1}{2\pi} \int_0^{2\pi} d\phi = 1.$$

https://doi.org/10.1515/9783111332277-010

Hence, $I = 1$, as was to be shown.

The graph of density (10.1) has the well-known bell shape, illustrated in Fig. 10.1.

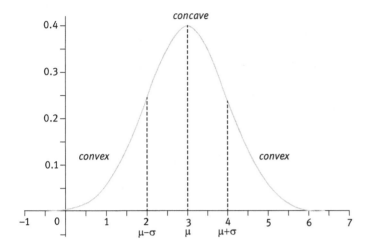

Fig. 10.1: Normal density for $\mu = 3$ and $\sigma = 1$.

To analyze the form of $f(x)$ in more detail, we first observe that the graph is symmetrical with respect to μ. By means of the first three derivatives, which can be written as

$$f'(x) = \frac{\mu - x}{\sigma^2} f(x), \quad f''(x) = \left(\frac{(\mu - x)^2}{\sigma^4} - \frac{1}{\sigma^2} \right) f(x), \quad f'''(x) = \left(\frac{(\mu - x)^3}{\sigma^6} - 3 \frac{\mu - x}{\sigma^4} \right) f(x),$$

(10.2)

we see that $f'(\mu) = 0$ and $f''(\mu) < 0$, i.e., f assumes its maximum in μ. Furthermore, we obtain $f''(\mu - \sigma) = f''(\mu + \sigma) = 0$, $f'''(\mu - \sigma) < 0$ and $f'''(\mu + \sigma) > 0$. Hence, $\mu - \sigma$ and $\mu + \sigma$ are inflection points, i.e., points in which the curvature of the graph changes from being convex to concave (for $\mu - \sigma$) or vice versa (for $\mu + \sigma$). Clearly, $f(x)$ converges to zero for $x \to \pm \infty$. If σ is large, the graph of f tends to be flat and if σ is small, it tends to be "peaked."

Theorem 10.1: For an r.v. X with distribution (10.1), it holds
(i) $E(X) = \mu$,
(ii) $V(X) = \sigma^2$.

Proof:

(i) By definition, we have

$$E(X) = \int_{-\infty}^{\infty} x \frac{1}{\sqrt{2\pi}\sigma} \exp\left(-\frac{1}{2}\left(\frac{x-\mu}{\sigma}\right)^2\right) dx$$

and the substitution $z = (x - \mu)/\sigma$, implying $x = z\sigma + \mu$, $dx = \sigma dz$ yields

$$E(X) = \frac{1}{\sqrt{2\pi}} \int_{-\infty}^{\infty} (\sigma z + \mu) e^{-z^2/2} dz = \frac{1}{\sqrt{2\pi}} \sigma \int_{-\infty}^{\infty} z e^{-z^2/2} dz + \mu \frac{1}{\sqrt{2\pi}} \int_{-\infty}^{\infty} e^{-z^2/2} dz.$$

The first of the above integrals equals zero, since the integrand $g(z) = z \exp(-z^2/2)$ is an odd function, i.e., $g(-z) = -g(z)$. The second integral, including the factor $1/\sqrt{2\pi}$, equals 1, since it is the pdf of the r.v. $X \sim N(0,1)$. Thus, $E(X) = \mu$.

(ii) By applying the same substitution as above, we obtain

$$E(X^2) = \int_{-\infty}^{\infty} x^2 \frac{1}{\sqrt{2\pi}\sigma} \exp\left(-\frac{1}{2}\left(\frac{x-\mu}{\sigma}\right)^2\right) dx$$

$$= \frac{1}{\sqrt{2\pi}} \int_{-\infty}^{\infty} (\sigma z + \mu)^2 e^{-z^2/2} dz$$

$$= \frac{1}{\sqrt{2\pi}} \int_{-\infty}^{\infty} \sigma^2 z^2 e^{-z^2/2} dz + 2\mu\sigma \frac{1}{\sqrt{2\pi}} \int_{-\infty}^{\infty} z e^{-z^2/2} \, dz + \mu^2 \frac{1}{\sqrt{2\pi}} \int_{-\infty}^{\infty} e^{-z^2/2} dz.$$

By means of the same argumentation, as in part (i), we can simplify this as

$$E(X^2) = \sigma^2 \frac{1}{\sqrt{2\pi}} \int_{-\infty}^{\infty} z^2 e^{-z^2/2} dz + \mu^2.$$

Using integration by parts with $h(z) = z$ and $g'(z) = (1/\sqrt{2\pi})z e^{-z^2/2}$, we obtain:

$$\frac{1}{\sqrt{2\pi}} \int_{-\infty}^{\infty} z^2 e^{-z^2/2} dz = \left[\frac{-z e^{-z^2/2}}{\sqrt{2\pi}}\right]_{-\infty}^{\infty} + \frac{1}{\sqrt{2\pi}} \int_{-\infty}^{\infty} e^{-z^2/2} dz = 0 + 1 = 1.$$

Hence, $E(X^2) = \sigma^2 + \mu^2$ and finally $V(X) = \sigma^2$.

The following result states that any linear function of a normally distributed r.v. is also normally distributed.

Theorem 10.2: If X has the distribution $N(\mu,\sigma^2)$, then $Y = aX + b$ (where $a > 0$) has the distribution $N(a\mu + b, a^2\sigma^2)$.

Proof: Using Theorem 6.1, we obtain the density of Y as

$$g(y) = \frac{1}{\sqrt{2\pi}\,\sigma} \exp\left(-\frac{1}{2\sigma^2}\left[\frac{y-b}{a} - \mu\right]^2\right)\frac{1}{|a|}$$

$$= \frac{1}{\sqrt{2\pi}\,\sigma|a|} \exp\left(-\frac{1}{2a^2\sigma^2}[y - (a\mu + b)]^2\right),$$

which is the pdf of an r.v. with distribution $N(a\mu + b, a^2\sigma^2)$.

To simplify formulations, we introduce two definitions. For any random variable X, the *reduced variable* of X is defined as $Y = (X - \mu)/\sigma$, where μ and σ^2 denote the expectation and the variance of X. One can easily show that $E(Y) = 0$ and $V(Y) = 1$ hold. If X has the distribution $N(0,1)$, we say that X has the *standardized normal distribution*. From Theorem 10.2, we obtain:

Corollary: If X has the distribution $N(\mu, \sigma^2)$, then its reduced variable has the distribution $N(0, 1)$.

Consequently, we will denote the pdf and the cdf of the standardized normal distribution by

$$\varphi(x) = \frac{1}{\sqrt{2\pi}}e^{-x^2/2} \text{ and } \phi(x) = \int_{-\infty}^{x} \frac{1}{\sqrt{2\pi}}e^{-s^2/2}ds. \tag{10.4}$$

From the symmetry of $\phi(x)$ (see Fig. 10.2), it follows immediately that $P(X \le -x) = P(X \ge x)$, that is

$$\phi(-x) = 1 - \phi(x). \tag{10.5}$$

It is remarkable that to date, it has not been possible to develop a closed-form representation for the cdf. Therefore, the function $\phi(x)$ has been tabulated (see Appendix).

The table provides the value $\phi(x)$ for $0 \le x \le 3.09$. For $x > 3.09$, it holds $\phi(x) = 1$ with an accuracy of four digits. For negative x, one can calculate $\phi(x)$ by means of (10.5). Formerly, the cdf of the standardized normal distribution could be calculated only by means of such a table, nowadays it is implemented in any statistical standard software.

By means of the cdf in (10.4), one can easily calculate the probability of an event associated to a normally distributed r.v.

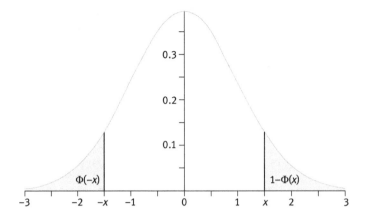

Fig. 10.2: Symmetry relation.

Theorem 10.3: If X has the distribution $N(\mu, \sigma^2)$, then

$$P(a \leq X \leq b) = \phi\left(\frac{b-\mu}{\sigma}\right) - \phi\left(\frac{a-\mu}{\sigma}\right).$$

Proof:

$$P(a \leq X \leq b) = P\left(\frac{a-\mu}{\sigma} \leq Y \leq \frac{b-\mu}{\sigma}\right) = \phi\left(\frac{b-\mu}{\sigma}\right) - \phi\left(\frac{a-\mu}{\sigma}\right),$$

where the last equation follows from the above corollary.

Example 10.1: Assume that the height X of an adult of a certain population is normally distributed with $\mu = 170$ cm and $\sigma = 10$ cm. Determine the following probabilities using the table for the standard normal distribution in the Appendix.
(i) $P(152 \leq X \leq 181)$,
(ii) $P(X \leq 165)$,
(iii) $P(X \geq 190)$.

Solution:
(i)   The required probability is

$$\phi\left(\frac{181-170}{10}\right) - \phi\left(\frac{152-170}{10}\right) = \phi(1.1) - \phi(-1.8) = \phi(1.1) - [1 - \phi(1.8)]$$

$$= 0.8643 - [1 - 0.9641] = 0.8284.$$

(see (10.5) to justify the second equation).

(ii) We want to determine

$$P(-\infty < X \le 165) = \phi\left(\frac{165-170}{10}\right) - \phi(-\infty)$$

$$= \phi(-0.5) - 0 = 1 - \phi(0.5)$$

$$= 1 - 0.6915 = 0.3085.$$

(iii) $P(-190 \le X < \infty) = \phi(\infty) - \phi\left(\frac{190-170}{10}\right)$

$$= 1 - \phi(2)$$

$$= 1 - 0.9772 = 0.0228.$$

At this point, a comment on the practical application of the normal distribution is appropriate. In Definition 10.1, a normally distributed r.v. may assume any real value. However, in many applications of this distribution, the underlying random variable can only assume values in a certain subregion of the real line. For instance, the random variable height in Example 10.1 cannot have negative values, and very large positive values are also practically impossible. Nevertheless, the normal distribution can be applied in such cases, since for adequately chosen parameters μ and σ, the probability that X assumes practically impossible values is almost zero, when calculated according to the model. For instance, the probability that a person in Example 10.1 assumes a negative height is $P(X \le 0) = \phi((0-170)/10) = \phi(-17) = 1 - \phi(17)$, which is practically zero.

As a further illustration, we can subdivide the range space of a normally distributed r.v. by means of points of the form $\mu + i\sigma$, and ask for the probability that X falls in any component of this partition. From Theorem 10.3, we obtain $P(\mu + i\sigma \le X \le \mu + (i+1)\sigma = \phi(i+1) - \phi(i)$. Setting, for example, $i = 0$ and $i = 1$, we get $P(\mu \le X \le \mu + \sigma) = \phi(1) - \phi(0) \approx 0.3413$ and $P(\mu + \sigma \le X \le \mu + 2\sigma) = \phi(2) - \phi(1) \approx 0.1359$, hence $P(X \ge \mu + 2\sigma) \approx 0.0228$.

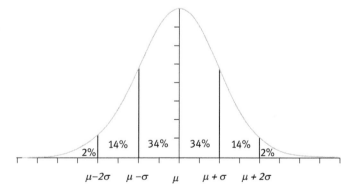

Fig. 10.3: Probability of regions.

The relations are illustrated in Fig. 10.3. Assume, e.g., the monthly net income of a certain population is normally distributed with $\mu = 3{,}000$ and $\sigma = 500$ (in \$), we obtain immediately that about 34% of the persons active on the labor market earn between \$3,000 and \$3,500, about 14% of them earn between \$3,500 and \$4,000, etc.

10.2 The exponential distribution

We now formally introduce a distribution that has already been applied in the course of the text.

Definition 10.2: A continuous random variable X, assuming all values $x \geq 0$, is said to have an *exponential distribution* with parameter $a > 0$, if its pdf is given by

$$f(x) = ae^{-a\,x} \text{ for } x > 0. \tag{10.6}$$

It can be easily shown that this is a legitimate pdf. The model plays an important role in applications, in particular in reliability theory. Expectation, variance and the cdf are given by

$$E(X) = \frac{1}{a}, \tag{10.7}$$

$$V(X) = \frac{1}{a^2}, \tag{10.8}$$

and

$$F(x) = \begin{cases} 1 - e^{-ax} & \text{for } x > 0, \\ 0 & \text{elsewhere} \end{cases} \tag{10.9}$$

(see exercise 10.6). In analogy to the geometric distribution (see Theorem 9.5), the exponential distribution has "no memory." It holds

$$P(X > s + t | X > s) = P(X > t) \quad \text{for} \quad s, t > 0, \tag{10.10}$$

since the left-hand side of (10.10) can be written as

$$\frac{P(X > s + t)}{P(X > s)} = \frac{e^{-a(s+t)}}{e^{-as}} = e^{-a\,t} = P(X > t),$$

and without proof, we state that the exponential distribution is the only continuous distribution satisfying (10.10).

Example 10.2: Assume that the lifetime X of a light bulb is exponentially distributed with $a = 0.003 \frac{1}{h}$. Determine

(i)   the life expectancy and
(ii) the probability that the bulb burns for at least 400 h.

Solution:

(i) $E(X) = \frac{1}{a} \approx 333.3$ h.
(ii) $P(X > 400) = \int_{400}^{\infty} a e^{-a x} dx = e^{-400a} = e^{-1.2} \approx 0.301$.

Example 10.3: Assume that 100 light bulbs of the same type are installed. If the lifetime is distributed as in the above example, what is the probability that at least 30 bulbs are still functioning after 500 h?

Solution: A specific bulbs is still burning after 500 h with probability $e^{-500a} = e^{-1.5}$. By using the binomial distribution, we obtain the requested probability as

$$\sum_{100}^{k=30} \binom{100}{k} e^{-1.5k} \left(1 - e^{-1.5}\right)^{100-k} \approx 0.0456.$$

10.3 The gamma distribution

The gamma distribution has numerous applications in reliability analysis, queuing theory and demography, among others. As the name suggests, this model is based on the gamma function (see (1.3)).

Definition 10.3: A continuous random variable X, assuming all values $x \geq 0$, is said to have a *gamma distribution*, if its pdf is given by

$$f(x) = \frac{a}{\Gamma(r)} (ax)^{r-1} e^{-a x} \quad \text{for } x > 0. \tag{10.11}$$

The distribution depends on the two positive parameters, a and r. The first of them is a scaling parameter and the latter characterizes the shape of the distribution. Using the definition of the gamma function and the substitution $y = ax$, it follows that

$$\int_{0}^{\infty} f(x) dx = \frac{1}{\Gamma(r)} \int_{0}^{\infty} (ax)^{r-1} e^{-ax} a \, dx = \frac{1}{\Gamma(r)} \int_{0}^{\infty} y^{r-1} e^{-y} dy = 1.$$

The pdf is illustrated in Fig. 10.4 for $a = 1$ and some values of r. The gamma distribution reduces to the exponential distribution, setting $r = 1$ in (10.11). One can easily proof the following statement. See Exercise 10.10.

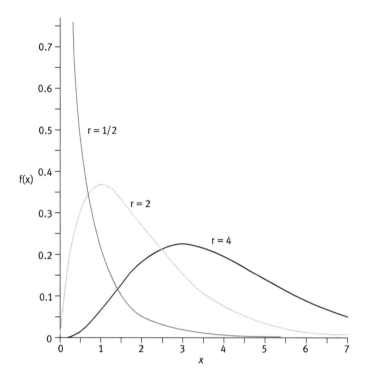

Fig. 10.4: Densities of the gamma family.

Theorem 10.4: If X has the gamma distribution (10.11), then
(i) $E(X) = \frac{r}{a}$
(ii) $V(X) = \frac{r}{a^2}$

Example 10.4: The lifetime X of an electronic component has a gamma distribution with $r = 2$. If the life expectancy is 50 h, what is the probability that the component functions at least 60 h?

Solution: Using Theorem 10.4(i) and (10.11), we obtain $a = 2/50 = 0.04$ and the pdf is $f(x) = a^2 x\, e^{-a\,x}$.

Hence, integration by parts yields

$$P(X \geq 60) = \int\limits_{60}^{\infty} a^2 x e^{-ax} dx \approx 0.3084.$$

Example 10.5:
Assume that the time to failure (in hours) of an antiaircraft missile is a gamma-distributed random variable X.
(i) What is the probability that the system does not fail during a period of 24 h, if the parameters are $\alpha = 0.1$ and $r = 3.2$.
(ii) What is the probability that no failure occurs during 2 days?
(iii) What is the expected time to failure?

Solution:
(i) The required probability is given by

$$P(X > 24) = \int\limits_{24}^{\infty} \frac{0.1}{\Gamma(3.2)} (0.1x)^{2.2} e^{-0.1x} dx \approx 0.6177.$$

(Here, as in some following examples, an appropriate software is necessary for the numerical evaluation of the integral.)
(ii) The analog integral with lower limit 48 gives the probability 0.1176.
(iii) $E(X) = \frac{r}{\alpha} = 32$ h.

10.4 The chi-square distribution

A special, very important case of the gamma distribution is obtained by setting $\alpha = 1/2$ and $r = n/2$ in (10.11), where n is a natural number. Thereby, we obtain the one-parameter density,

$$f(x) = \frac{1}{2^{n/2} \Gamma(n/2)} x^{n/2-1} e^{-x/2} \quad \text{for } x > 0. \tag{10.12}$$

A random variable with this pdf is said to have a *chi-square distribution with n degrees of freedom* and is denoted by χ_n^2
The graph of (10.12) is illustrated in Fig. 10.5 for some values of n. It is monotone decreasing for $n = 1$ and $n = 2$ and for $n > 2$, the distribution is unimodal.
From Theorem 10.4, it follows immediately that $E(X) = n$ and $V(X) = 2n$ holds for an r.v. with chi-square distribution. The distribution χ_n^2 has many applications in statistical inference, some of which will be presented later.
Among others, we will later show (see Theorem 11.7) that $Y = X_1^2 + \cdots + X_n^2$ has distribution χ_n^2 if the X_i are r.v,s with distribution $N(0,1)$.

Example 10.6: Assume that a marker moves randomly on a plane. The position (X,Y) is a bidimensional random variable such that X and Y are independent variables with distribution $N(1,0)$. What is the probability that the distance of the marker from the origin is at most c?

Solution: From the previous remark, it follows that the squared distance $Z = X^2 + Y^2$ has distribution χ_2^2, i.e., the density is $f(z) = \frac{1}{2}e^{-z/2}$. The required probability is therefore easily obtained as $P(Z \le c^2) = \int_0^{c^2} \frac{1}{2}e^{-z/2}dz = 1 - e^{-c^2/2}$.

The chi-square distribution is tabulated (see Appendix) and can also be calculated by any statistical standard software. For large n, one can approximate the chi-square distribution by the normal distribution. Without a proof, we cite the following relation.

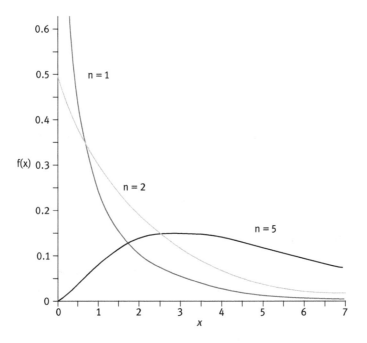

Fig. 10.5: Chi-square distribution.

Theorem 10.5: Assume that X has distribution χ_n^2, where n is sufficiently large. Then, the r.v. $\sqrt{2X}$ has approximately the distribution $N(\sqrt{2n-1}, 1)$.

If X is distributed as in the theorem, the statement yields

$$P(X \le x) = P(\sqrt{2X} \le \sqrt{2x}) \approx \Phi\left(\frac{\sqrt{2x} - \mu}{\sigma}\right) = \Phi\left(\sqrt{2x} - \sqrt{2x-1}\right) \qquad (10.13)$$

(see Theorem 10.3).

10.5 The beta distribution

The following distribution is frequently used in Bayesian analysis, fertility studies and demography, among others.

Definition 10.4: Let X be a continuous random variable that assumes all values over the interval [0, 1]. We say that X has a *beta distribution*, if its pdf is given by

$$f(x) = \frac{1}{B(\alpha, \beta)} x^{\alpha-1}(1-x)^{\beta-1} \quad \text{for } x \in [0, 1].$$ (10.14)

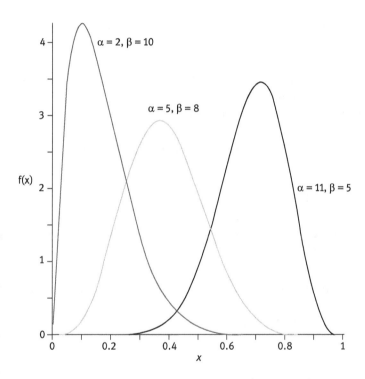

Fig. 10.6: Beta densities.

The expression $B(\alpha, \beta)$ denotes the beta function (see (1.26)), and α and β are two positive shape parameters. Possible forms of the pdf are illustrated in Fig. 10.6. Since the range space is [0, 1], the beta distribution is adequate to model random variables over bounded intervals.

From the definition of the beta function, it follows immediately that (10.14) is a legitimate pdf. It is easy to express the moments about the origin in terms of beta or gamma functions. We obtain

$$E(X^r) = \int_0^1 \frac{x^r}{B(\alpha, \beta)} x^{\alpha-1}(1-x)^{\beta-1} = \frac{1}{B(\alpha, \beta)} \int_0^1 x^{\alpha+r-1}(1-x)^\beta dx$$

$$= \frac{B(\alpha+r, \beta)}{B(\alpha, \beta)} = \frac{\Gamma(\alpha+r)}{\Gamma(\alpha)} \frac{\Gamma(\alpha+\beta)}{\Gamma(\alpha+r+\beta)}. \tag{10.15}$$

When r is an integer, we can make use of the notation of the rising factorial $\alpha^{(r)} = \alpha(\alpha+1) \cdot \cdots \cdot (\alpha+r-1) = \frac{\Gamma(\alpha+r)}{\Gamma(\alpha)}$ (see Section 1.4) and express (10.15) as

$$E(X^r) = \frac{\alpha^{(r)}}{(\alpha+\beta)^{(r)}}. \tag{10.16}$$

In particular, we obtain $E(X) = \frac{\alpha}{\alpha+\beta}$ and $E(X^2) \frac{\alpha(\alpha+1)}{(\alpha+\beta)(\alpha+\beta+1)}$ from which the variance can be easily calculated.

Example 10.7: Suppose that video cameras in a certain delivery are defective according to a beta distribution with parameters $\alpha = 2.3$ and $\beta = 5.4$.
(i)   Compute the probability that the delivery contains between 30 and 50% defective devices.
(ii) Calculate the expected proportion of defective cameras in the delivery.

Solution:
(i)   The required probability is

$$P(0.3 \le X \le 0.5) = \int_{0.3}^{0.5} \frac{1}{B(2.3, 5.4)} x^{1.3}(1-x)^{4.4} dx \approx 0.3394.$$

Note that the probability that a camera is defected is itself a random variable, modeled by the beta distribution.
(ii) The above formula yields the expectation $\frac{\alpha}{\alpha+\beta} = \frac{2.3}{2.3+5.4} \approx 0.30$.

10.6 The multivariate normal distribution

This is a generalization of the univariate normal distribution studied in Section 10.1, which is fundamental for multivariate statistics. Even if one restricts the attention preferentially to the bidimensional case, as we will do, it is convenient to use the vector notation. Let

$$X = (X_1, \ldots, X_n)^T \tag{10.17}$$

denote the column vector of random variables $X_1, \ldots, X_n$. By

$$\mu = E(X) = (\mu_1, \ldots, \mu_n)^T \tag{10.18}$$

we denote the vector of expectations $\mu_i = E(X_i)$, and the matrix of covariances is denoted by

$$\sum = \text{Cov}(X) = \begin{pmatrix} \sigma_{1,1} & \cdots & \sigma_{1,n} \\ \vdots & & \vdots \\ \sigma_{n,1} & \cdots & \sigma_{n,n} \end{pmatrix}, \tag{10.19}$$

where $\sigma_{i,j} = \text{Cov}(X_i, X_j)$ denotes the covariance between X_i and X_j (see Section 8.6). This matrix is symmetric, since $\sigma_{i,j} = \sigma_{j,i}$ for all i, j. In particular, $\sigma_{i,i}$ corresponds to the variance $V(X_i)$. The covariance matrix (10.10) can be written as

$$\text{Cov}(X) = E[(X - E(X)) \cdot \left(X - E(X)^T\right)].$$

Note that the left factor $(E - E(X))$ is a column vector, while $(X - E(X))^T$ is a row vector. Therefore, the multiplication of these vectors results in an $n \times n$-matrix. The pdf of the multivariate normal distribution can now be written as

$$f(x) = \frac{\exp\left(-\frac{1}{2}(x - \mu)^T \Sigma^{-1}(x - \mu)\right)}{\sqrt{\det(\Sigma)}(2\pi)^{n/2}}, \tag{10.20}$$

where $\mu \in IR^n$ and Σ is a positive definite $n \times n$ matrix. These parameter vectors represent the expectations and covariances. By $X \sim N(\mu, \Sigma)$, we denote the fact that X has the multivariate normal distribution (10.20). It can be easily seen that the univariate normal distribution is obtained from the latter by setting $n = 1$. The variables $X_1, \ldots, X_n$ are independent, if and only if (10.19) is a diagonal matrix, i.e., $\sigma_{i,j} = 0$ for $i \neq j$. In this case, (10.20) may be factorized as

$$f(x) = \prod_{i=1}^{n} \frac{1}{\sqrt{2\pi}\sigma_i} \exp\left(\frac{-(x_i - \mu_i)^2}{2\sigma_i^2}\right).$$

In the case $n = 2$, we have $\Sigma = \begin{pmatrix} \sigma_1^2 & \sigma_{1,2} \\ \sigma_{1,2} & \sigma_2^2 \end{pmatrix}$. Thus, $\det(\Sigma) = \sigma_1^2\sigma_2^2 - \sigma_{1,2}^2$. Making use of the linear correlation coefficient $\rho = \sigma_{1,2}/\sigma_1\sigma_2$, we can write the determinant of Σ as

$$\det\left(\sum\right) = \sigma_1^2\sigma_2^2(1 - \rho^2), \tag{10.21}$$

and the inverse of the covariance matrix can be written as

$$\sum^{-1} = \frac{1}{\sigma_1^2 \sigma_2^2 (1 - \rho^2)} \begin{pmatrix} \sigma_2^2 & -\sigma_{1,2} \\ -\sigma_{1,2} & \sigma_1^2 \end{pmatrix}. \tag{10.22}$$

Hence, the exponent of (10.20) is

$$\frac{-1}{2\sigma_1^2 \sigma_2^2 (1 - \rho^2)} (x_1 - \mu_1, x_2 - \mu_2) \begin{pmatrix} \sigma_2^2 & -\sigma_{1,2} \\ -\sigma_{1,2} & \sigma_1^2 \end{pmatrix} \begin{pmatrix} x_1 - \mu_1 \\ x_2 - \mu_2 \end{pmatrix} \tag{10.23}$$

By multiplying out (10.23) and using (10.21), we obtain the pdf as

$$f(x_1, x_2) = \frac{1}{2\pi\sigma_1\sigma_2\sqrt{1-\rho^2}} \exp\left\{ \frac{-1}{2(1-\rho^2)} \left[\left(\frac{x_1 - \mu_1}{\sigma_1}\right)^2 - 2\rho\frac{x_1 - \mu_1}{\sigma_1}\frac{x_2 - \mu_2}{\sigma_2} + \left(\frac{x_2 - \mu_2}{\sigma_2}\right)^2 \right] \right\}. \tag{10.24}$$

The density has the five parameters $\mu_1, \mu_2, \sigma_1, \sigma_2, \rho$, satisfying $\mu_1, \mu_2 \in \mathbb{R}$, $\sigma_1, \sigma_2 > 0$, $-1 \le \rho \le 1$.

Example 10.8: Let X and Y denote the height and the weight of an adult of a certain population (measured in cm and kg, respectively). Suppose that $(X, Y)^T$ has a bivariate normal distribution with $\mu_1 = 170$, $\mu_2 = 80$, $\sigma_1^2 = 20$ and $\sigma_2^2 = 10$. Calculate the probability

$$P(170 \le X \le 180, \ 80 \le Y \le 90) \text{ for } \rho = 0.7 \text{ and } \rho = 0.$$

Solution: By integration of the pdf (10.24), we obtain the probabilities 0.3605 and 0.2433. In the second case, the result can be alternatively computed as $P(170 \le X \le 180) \cdot P(80 \le Y \le 90) \approx 0.4873 \cdot 0.4992 \approx 0.2433$, since X and Y are then independent.

Theorem 10.6: Given the multivariate normal distribution (10.20), for any variable X_i, the marginal distribution is $N(\mu_i, \sigma_i^2)$.

However, the following example shows that the converse of this theorem is not true.

Example 10.9: Consider the bivariate pdfs

$$f(x, y) = \frac{1}{\sqrt{3}\pi} \exp\left(-\frac{2}{3}(x^2 - xy + y^2)\right) \tag{10.25}$$

$$g(x, y) = \frac{3}{4\sqrt{2}\pi} \exp\left(-\frac{9}{16}\left(x^2 - \frac{2}{3}xy + y^2\right)\right) \tag{10.26}$$

and its mixture (convex combination)

$$h(x, y) = af(x, y) + (1 - a)g(x, y). \tag{10.27}$$

For $0 < a < 1$, both marginal distributions of h are $N(0, 1)$, however h is not of the form (10.24), i.e., the mixture is not a bivariate normal distribution (see Exercise 10.22).

The previous theorem states that the marginal distributions of (10.20) are normal as well. Moreover, the normality of a multivariate distribution is maintained in a multiple sense, if one considers diverse "related" distributions. The conditional densities for (10.20) are normal and as the following theorem states, linear functions of normally distributed r.v.'s are normal as well. The latter is often called the *reproductive property*.

Theorem 10.7: Given a random variable $X \sim N(\mu, \Sigma)$ and a linear function $Y = AX + b$, where A and b are a nonsingular matrix and a column vector of appropriate dimensions, then $Y \sim N(A\mu + b, A\Sigma A^T)$.

In particular, if $b = 0$ and A consists of a single row $(a_1, \ldots, a_n)$, it holds $Y = a_1 X_1 + \cdots + a_n X_n$, i.e., Y is a linear combination of the vectors X_i.

Example 10.10: Consider the following rudimentary model to describe the relation between patient characteristics and certain disease risks. Let X and Y denote the weight (in kg) and the systolic blood pressure. Assume that $(X,Y)^T$ has a bivariate normal distribution with parameters $\mu_1 = 80$, $\mu_2 = 120$, $\sigma_1 = 20$ and $\sigma_2 = 30$ and $\rho = 0.6$.

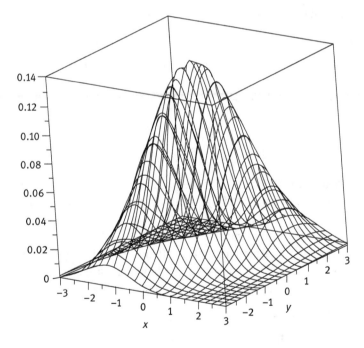

Fig. 10.7: Bivariate normal density.

Furthermore, W and Z denote the probability of developing diabetes and the probability of suffering a heart attack in the next 5 years. It is assumed that $(W,Z)^T$ is a linear function of $(X,Y)^{T,}$ given by

$$\begin{pmatrix} W \\ Z \end{pmatrix} = \begin{pmatrix} 0.004 & 0.002 \\ 0.005 & 0.003 \end{pmatrix} \begin{pmatrix} X \\ Y \end{pmatrix} + \begin{pmatrix} 0.05 \\ -0.1 \end{pmatrix}$$

Determine the variances of W and Z and the covariance between these variables.

Solution: The expectation of $\begin{pmatrix} W \\ Z \end{pmatrix}$ is given as

$$A\mu + b = \begin{pmatrix} 0.004 & 0.002 \\ 0.005 & 0.003 \end{pmatrix} \begin{pmatrix} 80 \\ 120 \end{pmatrix} + \begin{pmatrix} 0.05 \\ -0.1 \end{pmatrix} = \begin{pmatrix} 0.56 \\ 0.76 \end{pmatrix} + \begin{pmatrix} 0.05 \\ -0.1 \end{pmatrix} = \begin{pmatrix} 0.61 \\ 0.66 \end{pmatrix}.$$

Since $\sigma_{1,2} = \rho\sigma_1\sigma_2 = 0.6 \cdot 20 \cdot 30 = 360$, the covariance matrix of $\begin{pmatrix} X \\ Y \end{pmatrix}$ is $\Sigma = \begin{pmatrix} \sigma_1^2 & \sigma_{1,2} \\ \sigma_{1,2} & \sigma_2^2 \end{pmatrix} =$
$\begin{pmatrix} 400 & 360 \\ 360 & 900 \end{pmatrix}$, yielding the following covariance matrix of $\begin{pmatrix} W \\ Z \end{pmatrix}$:

$$A\Sigma A^T = \begin{pmatrix} 0.004 & 0.002 \\ 0.005 & 0.003 \end{pmatrix} \begin{pmatrix} 400 & 360 \\ 360 & 900 \end{pmatrix} \begin{pmatrix} 0.004 & 0.005 \\ 0.002 & 0.003 \end{pmatrix}$$

$$= \begin{pmatrix} 0.004 & 0.002 \\ 0.005 & 0.003 \end{pmatrix} \begin{pmatrix} 2.32 & 3.08 \\ 3.24 & 4.5 \end{pmatrix} = \begin{pmatrix} 0.01576 & 0.02132 \\ 0.02132 & 0.0289 \end{pmatrix},$$

i.e., $V(W) = 0.01576$, $V(Z) = 0.0289$, $\mathrm{Cov}(W,Z) = 0.02132$. The coefficient of linear correlation between W and Z is

$$\frac{0.02132}{\sqrt{0.01576 \cdot 0.0289}} \approx 0.9990.$$

We finally study some geometrical properties of the bivariate normal distribution. Figure 10.7 shows a three-dimensional plot of the bivariate normal density with parameters $\mu_1 = \mu_2 = 0$, $\sigma_1 = 1$ and $\sigma_2 = 2$ and $\rho = 0.6$. The values of the location parameters μ_1 and μ_2 do not influence the shape of the surface.

A good illustration of shape characteristics is also obtained, considering the *level curves* $f(x, y) = c$ of the density. These curves are ellipses, whose axes are given by the eigenvectors of the covariance matrix Σ. For the latter parameters, we get

$$\Sigma = \begin{pmatrix} \sigma_1^2 & \sigma_{1,2} \\ \sigma_{1,2} & \sigma_2^2 \end{pmatrix} = \begin{pmatrix} 1 & 1.2 \\ 1.2 & 2 \end{pmatrix}$$

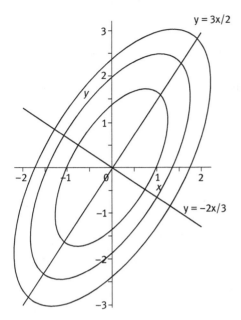

Fig. 10.8: Level curves.

since $\sigma_{1,2} = \rho\sigma_1\sigma_2 = 1 \cdot 2 \cdot 0.6 = 1.2$.

The eigenvalues, i.e., the roots of the characteristic polynomial

$$p(\lambda) = \det(\Sigma - \lambda I) = \begin{vmatrix} 1-\lambda & 1.2 \\ 1.2 & 2-\lambda \end{vmatrix} = (1-\lambda)(2-\lambda) - 1.2^2 = \lambda^2 - 3\lambda + 0.56 \text{ are } \lambda_1 = 2.8$$

and $\lambda_2 = 0.2$. Any eigenvector corresponding to λ_1 is a solution of the system

$$\begin{pmatrix} 1-\lambda_1 & 1.2 \\ 1.2 & 2-\lambda_1 \end{pmatrix} = \begin{pmatrix} 0 \\ 0 \end{pmatrix},$$

i.e., a multiple of $\begin{pmatrix} 2 \\ 3 \end{pmatrix}$. In the same way, we show that any eigenvector corresponding to λ_2 is a multiple of $\begin{pmatrix} -3 \\ 2 \end{pmatrix}$ These vectors give the direction of the axes of the ellipses.

The level curves are illustrated in Fig. 10.8.

If the coefficient of linear correlation ρ is zero, the axes of the ellipses are parallel to the coordinate axes. If furthermore, $\sigma_1 = \sigma_2$ holds, then the ellipses become circles.

10.7 A bivariate exponential distribution

We finally address briefly a bidimensional extension of the exponential distribution. Several extensions of the univariate model have been proposed in the literature. We limit ourselves to the oldest and possibly the simplest version, due to Gumbel. The survival function is given by

$$S(x,y) = \exp(-\alpha x - \beta y - \theta \alpha \beta xy) \text{ for } x, y \geq 0, \tag{10.28}$$

where the parameters satisfy $\alpha, \beta > 0$ and $0 \leq \theta \leq 1$. Deriving this expression with respect to x and y, results in the corresponding density

$$f(x,y) = \left[(1-\theta)\alpha\beta + \theta\alpha^2\beta x + \theta\alpha\beta^2 y + \theta^2\alpha^2\beta^2 xy\right] S(x,y). \tag{10.29}$$

By Definition 7.7, we obtain the distribution function as

$$F(x,y) = 1 - \exp(-\alpha x) - \exp(-\beta y) + \exp(-\alpha x - \beta y - \theta\alpha\beta xy) \text{ for } x, y \geq 0, \tag{10.30}$$

and (7.15) gives the marginal distribution functions of X and Y as

$$G(x) = F(x, \infty) = 1 - e^{-\alpha x}$$

and

$$H(y) = F(\infty, y) = 1 - e^{-\beta y},$$

which are the univariate exponential distributions. It is interesting to observe that these functions do not depend on the parameter θ. Bivariate exponential distributions have applications in reliability theory and telecommunications, among others. If there is, e.g., a two-unit system in which the lifetimes of the units depend upon another in a particular way, the above model may be applicable.

Example 10.11: The lifetimes (X, Y) of two electronic devices (in months) have the density (10.29) with $\alpha = 0.1$, $\beta = 0.15$. What is the probability that the devices survive 11 months and 7 months, respectively. Assume that (i) $\theta = 0$, (ii) $\theta = 0.5$.

Solution: From (10.28), we get immediately $S(11, 7) = 0.1165$ in case (i) and $S(11, 7) = 0.0654$ in case (ii).

The correlation can be expressed as

$$\rho = \frac{1}{\theta}\Gamma\left(0, \frac{1}{\theta}\right) e^{1/\theta}, \tag{10.31}$$

where Γ denotes the incomplete gamma function. Observe that (10.31) is independent of α and β. Finally, it is interesting to know, e.g., which component of a two-unit system will fail first. For this reason, we determine for density (10.29), the probability

$$P(Y \geq X) = \int_{0}^{\infty} \int_{x}^{\infty} f(x,y) \, dydx$$

(10.32)

$$= \frac{1}{2} - \frac{1}{4}\sqrt{\frac{\pi}{\alpha\beta\theta}} \left[erf\left(\frac{\alpha+\beta}{2\sqrt{\alpha\beta\theta}}\right) - 1 \right](\alpha-\beta) \, exp\left(\frac{(\alpha+\beta)^2}{4\alpha\beta\theta}\right).$$

In particular, it follows immediately the intuitive relation $P(Y \geq X) = \frac{1}{2}$ for $\alpha = \beta$.

Exercises

10.1 Assume that the weight X of an adult of a certain population is normally distributed with $\mu = 75$ kg and $\sigma = 10$ kg:
   (i)   Determine the probabilities: $P(52 \leq X \leq 81)$, $P(X \leq 72)$ and $P(X \geq 90)$.
   (ii) For which value of the standard deviation σ, the probability $P(X \leq 72)$ equals 20% ($\mu = 75$)?
   (iii) Which value of μ maximizes the probability $P(52 \leq X \leq 81)$ (for $\sigma = 10$)?

10.2 If X is normally distributed with $\mu = 10$ and $P(X \leq 8) = 0.4$, what is the value of σ?

10.3 Suppose that X has distribution $N(5,4)$. Determine the number c such that $P(X > c)$ $= 3P(X \leq c)$.

10.4 The height of an adult is normally distributed with $\mu = 175$ cm and $\sigma = 15$ cm. What is the probability that among five randomly chosen adults, exactly two persons are taller than 190 cm?

10.5 Assume that the weight X of an adult is normally distributed such that $P(X \leq 80) = 0.4013$ and $P(X > 100) = 0.2266$. Determine μ and σ.

10.6 Prove the relations (10.7)–(10.9).

10.7 Determine $P(X > E(X))$ for the exponentially distributed r.v. X.

10.8 Determine $E(X)$, $V(X)$ and $F(x)$ for the bilateral exponential distribution with pdf $f(x) = \frac{\alpha}{2}e^{-\alpha|x|}$ for $-\infty < x < \infty$. Illustrate the cdf geometrically.

10.9 Suppose that X has an exponential distribution truncated to the left, i.e., the pdf has the form $f(x) = C e^{-\alpha x}$ for $x > t$ and $f(x) = 0$, otherwise. Determine the normalizing constant C and the expectation $E(X)$.

10.10 Prove Theorem 10.4.

10.11 Show that the cdf $F(x)$ of the gamma distribution (10.11) can be expressed in terms of the Poisson distribution as $F(x) = 1 - \sum_{k=0}^{r-1} e^{-\alpha x} \frac{(\alpha x)^k}{k!}$, if r is an integer.

10.12 Determine the mode of the gamma distribution. In which case does it exist?

10.13 Calculate the required probability in Example 10.6, alternatively integrating the joint pdf of (X,Y) over the circle about the origin with radius r.

10.14 Determine the mode of the chi-square density (for $n > 2$).

10.15 Calculate both sides of the approximate relation (10.13) for $n = 100$ and $x = 50, 60,$ . . .,150.

10.16 Consider a one-dimensional random movement, analog to Example 10.6, i.e., a marker moves randomly on the horizontal axis, where the position X has distribution $N(0,1)$. Show that the squared distance from the origin has the chi-square distribution with one degree of freedom.

10.17 Suppose that X has distribution $N(0,36)$. Evaluate $P(4 < X^2 < 9)$.

10.18 In a meteorological model, the probability of precipitation on a certain day in July or August in the Brazilian state Goias has beta distribution with parameters $\alpha = 3.8$ and $\beta = 10.4$.
   (i) What is the probability that it rains at most 30% of the considered period?
   (ii) What is the expected percentage of rainy days?

10.19 Determine the mode of the beta density.

10.20 For which parameter values is the beta distribution symmetric?

10.21 If X has beta distribution with $E(X) = 2/5$ and $V(X) = 1/25$, determine α and β.

10.22 Write the matrix (10.19) for the case $n = 4$.

10.23 (i) Show that the densities f and g in Example 10.9 are of the form (10.24). Use this fact to show that both marginal distributions of h are $N(0,1)$.
   (ii) Determine the coefficient of linear correlation for h.

10.24 Suppose that X has a multivariate normal distribution with $\mu = 0$ and $\Sigma = I$ and $Y = AX$, where A is an orthogonal matrix, i.e., $A^{-1} = A^T$. Show that Y has distribution $N(0,I)$.

10.25 Solve the variants of Example 10.10, obtained by setting $\rho = 0.2$ and $\rho = 0$.

10.26 In a population of couples, X_1 and X_2 represent the height of the man and the woman, respectively. If the joint probability distribution is normal, with $\mu_1 = 180$, $\mu_2 = 165$ and $\Sigma = \begin{pmatrix} 150 & 80 \\ 80 & 120 \end{pmatrix}$, calculate the probability $P(175 \leq X_1 \leq 190,\ 150 \leq X_2 \leq 165)$.

10.27 If $(X,Y)^T$ has a bivariate normal distribution with $E(X) = 1/2$, $E(Y) = 1$, $E(X^2) = E(Y^2) = 2$ and $\rho = \frac{1}{2}$, what is the joint pdf?

10.28 Determine the distribution of $3X + 5Y$ for independent variables $X \sim N(2,16)$ and $Y \sim N(6,25)$.

10.29 Illustrate the level curves of the bivariate normal density geometrically for $\mu_1 = \mu_2 = 0$ and
   (i) $\sigma_1 = 3, \sigma_2 = 5, \rho = 0.7$,
   (ii) $\sigma_1 = 3, \sigma_2 = 5, \rho = 0.95$,
   (iii) $\sigma_1 = 3, \sigma_2 = 5, \rho = 0$,
   (iv) $\sigma_1 = 3, \sigma_2 = 3, \rho = 0$.

10.30 Calculate $P(X \leq 12)$, where X has a chi-square distribution with $n = 20$.
   (i) Determine the exact probability.
   (ii) Calculate the approximate value using (10.13).

10.31 Verify formulas (10.29) and (10.30).

10.32 If you have an appropriate software available, verify relations (10.31) and (10.32). Illustrate (10.31) graphically.

10.33 Consider the electronic devices in Example 10.11. Determine the following probabilities for $\theta=0.5$.
   (i) $P(X \le 12, \ Y \le 8)$,
   (ii) $P(Y \ge X)$,
   (iii) $P(X \le 9, \ Y \ge 6)$.

10.34 Determine the covariance between X and Y for the above bidimensional exponential distribution.

10.35 Determine $P(Y \ge 2X)$, modifying relation (10.32).

10.36 Another bidimensional exponential distribution is given by
   $$f(x,y) = K \exp(-ax - by - cxy) \text{ for } x, y \ge 0,$$
   where a, b and c are positive parameters and K is the normalizing constant:
   (i)   determine K,
   (ii) determine the marginal distributions (which are *not* exponential) and
   (iii) determine the corresponding expectations.

11 Generating functions in probability

In this chapter, we will introduce some very powerful tools that have many applications in probability theory and statistics. In order to motivate what follows, we recall a very elementary property of the logarithm, namely,

$$\ln(xy) = \ln(x) + \ln(y), \tag{11.1}$$

which was formerly used as a computational aid for multiplication, where suitable tables to determine logarithms were used.

Suppose that we want to calculate the product xy of the two positive numbers x and y, where we have no calculator but a table to calculate logarithms available. In order to avoid the multiplication as a relative complicated arithmetic operation, we can proceed as follows:
-   determine the logarithms $\ln(x)$ and $\ln(y)$ by means of the table,
-   add them to determine $\ln(xy)$, see (11.1),
-   determine xy, using the table of logarithms in the "reverse direction."

Thus, the operation of multiplication of x and y is realized, applying a corresponding operation (addition) to the logarithms of x and y.

11.1 The moment-generating function

The above approach works, since the logarithm function assigns *biuniquely* a real number $\ln(x)$ to any positive real number x. In a similar way, an "operation between distributions" can be realized, performing a corresponding operation between the associated moment-generating functions. For example, the distribution of a sum of r.v.s can in principle be obtained, by multiplying the corresponding generating functions.

Definition 11.1: For any random variable X, the *moment-generating function* (mgf) is defined as the expectation of the variable e^{tX}, i.e.,

$$M_X(t) = \sum_{j=1}^{\infty} e^{t\,x_j} p(x_j) \text{ if } X \text{ is discrete and } M_X(t) = \int_{-\infty}^{\infty} e^{tx} f(x) dx \text{ if } X \text{ is continuous.}$$

Without a proof, we note that the mgf is biuniquely assigned to the distribution, i.e., if two mgfs $M_X(t)$ and $M_Y(t)$ are equal for all t, then X and Y have the same distribution. If it is clear to which r.v. the mfg refers, we simply denote it by $M(t)$. This function is defined as an infinite series or (improper) integral, which may not exist for all values of the real variable t. For $t = 0$, the mfg always exists and equals 1.

https://doi.org/10.1515/9783111332277-011

A related concept is the *characteristic function* defined by $E(\exp(itX))$, where $i = \sqrt{-1}$ is the imaginary unit. This function always exists for all values of t, but we will not detail this concept here, in order to avoid calculations with complex numbers.

Before justifying the name "moment-generating function," we will present some discrete and continuous examples.

Example 11.1: Suppose that X has a Poisson distribution., Thus,

$$M_X(t) = \sum_{k=0}^{n} e^{tk} \frac{e^{-\lambda}\lambda^k}{k!} = e^{-\lambda} \sum_{k=0}^{n} \frac{(\lambda e^t)^k}{k!} = e^{-\lambda}\exp(\lambda e^t) = \exp(\lambda(e^t - 1)).$$

Example 11.2: For the binomially distributed variable X, we obtain

$$M_X(t) = \sum_{k=0}^{n} e^{tk} \binom{n}{k} p^k (1-p)^k = \sum_{k=0}^{n} \binom{n}{k} (pe^t)^k (1-p)^{n-k} = [pe^t + (1-p)]^n,$$

where the last equation follows from the binomial theorem.

Example 11.3: Suppose that X is uniformly distributed over the interval $[a, b]$. Then, the mgf is given by

$$M_X(t) = \int_a^b e^{tx} \frac{1}{b-a} dx = \frac{1}{t(b-a)} \left[e^{bt} - e^{at} \right] \text{ for } t \neq 0.$$

Example 11.4: If X has exponential distribution with parameter a, we get

$$M_X(t) = \int_0^\infty e^{tx} a e^{-ax} dx = a \int_0^\infty e^{(t-a)x} dx = \frac{a}{a-t} \text{ for } t < a.$$

The mgf does not exist for $t \geq a$.

Example 11.5: If X has distribution $N(0,1)$, then

$$M_X(t) = \frac{1}{\sqrt{2\pi}} \int_{-\infty}^{\infty} e^{tx} e^{-x^2/2} dx = \frac{1}{\sqrt{2\pi}} \int_{-\infty}^{\infty} e^{t^2/2} e^{-(x-t)^2} dx = e^{t^2/2} \frac{1}{\sqrt{2\pi}} \int_{-\infty}^{\infty} e^{-(x-t)^2/2} dx = e^{t^2/2}.$$

Since the mgf is simply an expected value, we can obtain the mgf of a function $Y = \Psi(X)$ of a r.v. without first obtaining the probability distribution of Y (see Theorem 8.3).

For example, if X has distribution $N(0,1)$, then the mgf of $Y = X^2$ is

$$M_Y(t) = E(\exp(t\,X^2)) = \frac{1}{\sqrt{2\pi}} \int\limits_{-\infty}^{\infty} \exp(tx^2 - x^2/2)\,dx = \frac{1}{\sqrt{1-2t}} \text{ for } t < \frac{1}{2}.$$

Example 11.6: Let X have the Gamma distribution with parameters a and r. Then,

$$M_X(r) = \frac{a}{\Gamma(r)} \int\limits_{0}^{\infty} e^{tx}(a\,x)^{r-1} e^{-ax}\,dx = \frac{a^r}{\Gamma(r)} \int\limits_{0}^{\infty} x^{r-1} e^{-x(a-t)}\,dx$$

where the integral converges for $t < a$. By means of the substitution $y = x(a - t)$, we obtain

$$M_X(t) = \frac{a^r}{(a-t)\Gamma(r)} \int\limits_{0}^{\infty} \left(\frac{y}{a-t}\right)^{r-1} e^{-y}\,dy = \left(\frac{a}{a-t}\right)^r \frac{1}{\Gamma(r)} \int\limits_{0}^{\infty} y^{r-1} e^{-y}\,dy = \left(\frac{a}{a-t}\right)^r. \quad (11.2)$$

By setting $r = 1$ or $a = 1/2$ and $r = n/2$ in (11.2), we obtain the mgf,

$$M_X(t) = \frac{a}{a-t} \quad (11.3)$$

of the exponential distribution and the mgf

$$M_X(t) = (1-2t)^{-n/2} \quad (11.4)$$

of the chi-square distribution.

11.2 Generation of moments

We will now justify why M_X is called a moment-generating function. The exponential function has the series expansion

$$e^x = 1 + x + \frac{x^2}{2!} + \frac{x^3}{3!} + \cdots$$

which converges for all x. Thus,

$$e^{tx} = 1 + tx + \frac{(tx)^2}{2!} + \frac{(tx)^3}{3!} + \cdots,$$

$$M(t) = E(e^{tX}) = E\left(1 + tX + \frac{(tX)^2}{2!} + \frac{(tX)^3}{3!} + \cdots\right),$$

and

$$M(t) = 1 + tE(X) + \frac{t^2 E(X^2)}{2!} + \frac{t^3 E(X^3)}{3!} + \cdots. \tag{11.5}$$

Deriving with respect to t, we obtain

$$M'(t) = E(X) + tE(X^2) + \frac{t^2 E(X^3)}{2!} + \frac{t^3 E(X^4)}{3!} + \cdots. \tag{11.6}$$

(We assume that the conditions are satisfied under which the linearity of expectation and differentiation are also valid for infinite series.) From (11.6), it follows immediately that $M'(0) = E(X)$. Calculating the second derivative, we obtain

$$M''(t) = E(X^2) + tE(X^3) + \frac{t^2 E(X^4)}{2!} + \cdots;$$

thus, $M''(0) = E(X^2)$. Continuing in this manner, we obtain the following statement (assuming that the nth derivative exists).

Theorem 11.1: It holds

$$M^{(n)}(0) = E(X^n),$$

i.e., the nth derivative of $M_X(t)$ evaluated at $t = 0$ yields the nth moment about the origin $E(X^n)$.

Thus, from the knowledge of the mgf, the moments $E(X^n)$ can be "generated." It depends on the respective r.v. X, whether the calculation of moments via Theorem 11.1 is easier than the direct calculation by means of their definition.

Since the general Mclaurin series expansion of a function h is

$$h(t) = h(0) + h'(0)t + \frac{h''(0)t^2}{2!} + \cdots.$$

Theorem 11.1 allows to write the mgf in terms of the moments around the origin (see Definition 8.4),

$$M_X(t) = M_X(0) + M'_X(0)t + \frac{M''_X(0)t^2}{2!} + \cdots$$

$$= 1 + \mu'_1 t + \frac{\mu'_2 t^2}{2!} + \cdots. \tag{11.7}$$

We now apply the theorem to determine the moments of some distributions.

Example 11.7: For the binomial distribution, we obtain (see Example 11.2):

$$M(t) = [pe^t + 1 - p]^n,$$

$$M'(t) = [pe^t + 1 - p]^{n-1} npe^t,$$

$$M''(t) = (n-1)[pe^t + 1 - p]^{n-2}(pe^t)^2 n + [pe^t + 1 - p]^{n-1} npe^t,$$

where the latter moment could be further simplified. Thus, $E(X) = M'(0) = np$ and $E(X^2) = M''(0) = (n-1)p^2 n + np = np(np - p + 1)$. In particular, we get $V(X) = E(X^2) - (E(X))^2 = np(1-p)$, which agrees with our previous results.

Example 11.8: For the standardized normal distribution, it holds (see Example 11.5)

$$M(t) = \exp\left(\frac{t^2}{2}\right), \ M'(t) = t\exp\left(\frac{t^2}{2}\right), \ M''(t) = (1+t^2)\exp\left(\frac{t^2}{2}\right).$$

Thus $E(X) = M'(0) = 0$, $E(X^2) = M''(0) = 1$, $V(X) = 1$.

Example 11.9: Let X have a geometric probability distribution with parameter p, i.e., $P(X = k) = q^{k-1}p$ for $k = 1,2, \ldots 0 < p < 1$ and $q = 1 - p$. The mgf is then

$$M(t) = \sum_{k=1}^{\infty} e^{tk}q^{k-1}p = \frac{p}{q}\sum_{k=1}^{\infty}(qe^t)^k.$$

Assuming that $0 < qe^t < 1$, i.e., $t < \ln(1/q)$, then this is a geometric series and thus

$$M(t) - \frac{p}{q}\frac{qe^t}{1 - qe^t} - \frac{pe^t}{1 - qe^t}.$$

The first two derivatives with respect to t are $M'(t) = \frac{pe^t}{(1 - qe^t)^2}$ and $M''(t) = \frac{pe^t(1+qe^t)}{(1-qe^t)^3}$, yielding $E(X) = M'(0) = 1/p$, $E(X^2) = M''(0) = (1 + q)/p^2$ and $V(X) = q/p^2$.

The following two statements are very important for determining the mgf of the composed random variables.

Theorem 11.2: If M_X is the mgf of a random variable X and $Y = aX + b$. Then, the mgf of Y is given by

$$M_Y(t) = e^{bt}M_X(at).$$

Proof: $M_Y(t) = E(e^{Yt}) = E(e^{(aX+b)t}) = e^{bt}E(e^{atX}) = e^{bt}M_X(at).$

This can be used e.g. to obtain the mgf $N(\mu, \sigma^2)$ from that of $N(0,1)$.

Example 11.10: Assume that Y has a normal distribution $N(\mu, \sigma^2)$.

Then, $X = (Y - \mu)/\sigma$ has the distribution $N(0,1)$ (see Section 10.1). Now, we can express Y as $Y = \sigma X + \mu$, and using Theorem 11.2 and Example 11.5, we obtain $M_Y(t) = e^{\mu t} M_X(\sigma t)$
$= e^{\mu t}\, e^{(\sigma t)^2/2} = e^{\mu t + \sigma^2 t^2/2}$.

Theorem 11.3: Let X and Y be independent random variables and $Z = X + Y$, and let $M_X(t)$, $M_Y(t)$ and $M_Z(t)$ denote the mgfs of X, Y and Z. Then,

$$M_Z(t) = M_X(t)M_Y(t).$$

Proof:

$$M_Z(t) = E\left(e^{Zt}\right) = E\left(e^{(X+Y)t}\right) = E\left(e^{Xt}e^{Yt}\right) = E\left(e^{Xt}\right)E\left(e^{Yt}\right)$$

$$= M_X(t)M_Y(t).$$

Clearly, this theorem can be easily generalized for a sum of n independent r.vs $Z = X_1 + \cdots + X_n$. The corresponding mgfs then satisfy

$$M_Z(t) = M_{X_1}(t) \cdot \ldots \cdot M_{X_n}(t). \tag{11.8}$$

Example 11.11: Assume that Z has a Pascal distribution. Then, $Z = X_1 + \cdots + X_r$, where all X_i have a geometric distribution with the same parameter p (see Section 9.4). From Example 11.9 and eq. (11.8), it follows that Z has the mgf

$$M_Z(t) = \left(\frac{pe^t}{1 - qe^t}\right)^r. \tag{11.9}$$

11.3 Reproductive properties

Theorem 11.3 can also be used to show a certain type of reproductive properties (see Section 10.6). A number of important probabilistic models own the following characteristic. If two or more independent r.v.s have a certain distribution, their sum has a distribution of the same type.

Example 11.12: Suppose that X and Y are independent random variables with distributions $N(\mu_1, \sigma_1^2)$ and $N(\mu_2, \sigma_2^2)$, respectively. Then, $Z = X + Y$ has the mgf

$$M_Z(t) = M_X(t)M_Y(t) = \exp\left(\mu_1 t + \sigma_1^2 t^2/2\right)\exp\left(\mu_2 t + \sigma_2^2 t^2/2\right)$$

$$= \exp\left[(\mu_1 + \mu_2)t + \left(\sigma_1^2 + \sigma_2^2\right)t^2/2\right]$$

(see Example 11.10). Thus, Z has the distribution $N\left(\mu_1 + \mu_2,\ \sigma_1^2 + \sigma_2^2\right)$.

The fact that Z has expectation $\mu_1 + \mu_2$ and variance $\sigma_1^2 + \sigma_2^2$ follows immediately from the additivity properties. But to establish the fact that Z is normally distributed, we have employed the mgf. It is obvious that this result can be extended to the sum Z of n normally distributed variables.

Theorem 11.4: Let $X_1, \ldots, X_n$ be independent random variables with distributions $N\left(\mu_i,\ \sigma_1^2\right)$ for $i = 1, \ldots, n$. Then, $Z = X_1 + \cdots + X_n$ has the distribution $N\left(\Sigma_{i=1}^n \mu_i,\ \Sigma_{i=1}^n \sigma_i^2\right)$.

Example 11.13: The weight in kg of adults of a certain population has distribution $N(70, 100)$. If five adults enter into an elevator that permits a total maximum weight of 400 kg, what is the probability that the sum of their weights exceeds this limit?

Solution: The total weight Y of the passengers has the distribution $N(350, 500)$. The required probability is therefore

$$P(Y \geq 400) = 1 - \phi\left(\frac{400 - 350}{\sqrt{500}}\right) \approx 1 - \phi(2.24) \approx 1 - 0.9874 = 0.0126.$$

The Poisson and chi-square distributions have corresponding properties.

Theorem 11.5: Let $X_1, \ldots, X_n$ be independent random variables such that X_i has Poisson distribution with parameter a_i, $i = 1, \ldots, n$.

Then, $Z = X_1 + \cdots + X_n$ has Poisson distribution with parameter $a = \Sigma_{i=1}^n a_i$.

Proof: We first consider the case $n = 2$. By using Theorem 11.3 and Example 11.1, we obtain the mfg of Z as

$$M_Z(t) = M_{X_1}(t)M_{X_2}(t) = \exp\left(a_1\left(e^t - 1\right)\right)\exp\left(a_2\left(e^t - 1\right)\right) = \exp\left((a_1 + a_2)\left(e^t - 1\right)\right),$$

which is the mfg of a Poisson random variable with parameter

$$a = a_1 + a_2.$$

The general result follows now by mathematical induction.

Example 11.14: Suppose that a telephone exchange receives an average of 20 calls between 8 a.m. and 9 a.m. and an average of 30 calls between 9 a.m. and 10 a.m. (see Example 9.2). If the numbers of calls received in these periods have Poisson distributions, what is the probability that more than 60 calls are received between 8 a.m. and 10 a.m.?

Solution: The number Z of calls received in the respective two-hour period is a Poisson variable with parameter $20 + 30 = 50$. The required probability is therefore

$$P(Z > 60) = 1 - \sum_{k=0}^{60} \frac{e^{-50}50^k}{k!} \approx 0.0722.$$

Theorem 11.6: Let $X_1, \ldots, X_k$ be independent random variables such that X_i has chi-square distribution with n_i degrees of freedom $(i = 1, \ldots, k)$.

Then, $Z = X_1 + \cdots + X_k$ has distribution χ_n^2 with $n = \sum_{i=1}^{k} n_i$.

Proof: From eq (11.4), we get

$$M_{X_i}(t) = (1 - 2t)^{-n_i/2}$$

for $i = 1, \ldots, k$. Thus,

$$M_X(t) = M_{X_1}(t) \cdot \cdots \cdot M_{X_k}(t) = (1 - 2t)^{-(n_1 + \cdots + n_k)/2},$$

which is the mgf of a variable with distribution χ_n^2.

It can be shown that if X has distribution $N(0,1)$, then X^2 has distribution χ_1^2 (see Exercise 10.16). Combining this with Theorem 11.6, we obtain the following statement, which has already been announced in Section 10.4.

Theorem 11.7: Suppose that $X_1, \ldots, X_k$ are independent random variables, each having distribution $N(0,1)$.

Then, $S = X_1^2 + \cdots + X_k^2$ has distribution χ_k^2.

Example 11.15: Let $X_1, \ldots, X_n$ be independent random variables, each with distribution $N(0,1)$. What is the pdf of the random variable, $T = \sqrt{X_1^2 + \cdots + X_n^2}$?

Solution: Theorem 11.7 implies that T^2 has distribution χ_k^2.

Thus, the cdf of T is

$$F(t) = P(T \le t) = P(T^2 \le t^2) = \int_{t^2}^{0} \frac{1}{2^{n/2}\Gamma(n/2)} x^{n/2-1} e^{-x/2} dx,$$

and the Leibniz integral rule yields the pdf

$$f(t) = F'(t) = \frac{2t}{2^{n/2}\Gamma(n/2)} \left(t^2\right)^{n/2-1} e^{-t^2/2} = \frac{t^{n-1}e^{-t^2/2}}{n^{n/2-1}\Gamma(n/2)} \quad \text{for } t \geq 0. \tag{11.10}$$

For $n = 2$, the pdf (11.10) takes the form

$$f(t) = t\, e^{-t^2/2} \quad \text{for } t \geq 0. \tag{11.11}$$

This is known as the *Rayleigh distribution*, and for $n = 3$ we get the *Maxwell's distribution*

$$f(t) = \sqrt{\frac{2}{\pi}} t^2 e^{-t^2/2} \quad \text{for } t \geq 0. \tag{11.12}$$

In the above two specific cases, the r.v. T can be interpreted as the distance of a marker from the origin that performs a random movement in the two- or three-dimensional space (Example 10.6). Accordingly, (11.12) has applications in thermodynamics, since it can be used to model the velocity distribution of a gas molecule.

As a final example for reproductive property, we present the following statement that can be easily proved (see Exercise 11.15)

Theorem 11.8: Let $X_1, \ldots, X_n$ be independent random variables such that X_i has gamma distribution with parameters a and r_i ($i = 1, \ldots, n$).

Then, $Z = X_1 + \cdots + X_n$ has gamma distribution with parameters a and $r = r_1 + \cdots + r_n$.

Setting $r_i = 1$ for $i = 1, \ldots, n$, we obtain the following interesting special case of the theorem.

Theorem 11.9: Let $X_1, \ldots, X_n$ be independent random variables, all of which are exponentially distributed with the same parameter a.

Then, $Z = X_1 + \cdots + X_n$ has gamma distribution with parameters a and n.

Example 11.16: The lifetime of a certain type of light bulb is exponentially distributed with parameter $a = 0.011/\text{h}$.
(i)   Determine the pdf of the average lifetime of a selection of 20 bulbs of this type.
(ii) What is the probability that the average lifetime is at least 90 h?
(iii) What is the probability that a specific light bulb functions at least 90 h?

Solutions:

(i) The sum of lifetimes Z has gamma distribution with parameters $\alpha = 0.01$ and $n = 20$. Its pdf is

$$f(z) = \frac{\alpha}{\Gamma(n)}(\alpha z)^{n-1}e^{-\alpha z} \quad \text{for } z \geq 0.$$

Thus, the average lifetime $W = Z/n$ has the density (see Theorem 6.1),

$$g(w) = f(nw)n = n\frac{\alpha}{\Gamma(n)}(\alpha n w)^{n-1}e^{-\alpha n w} = \frac{0.2}{19!}\left(\frac{w}{5}\right)^{19}e^{-w/5}.$$

(ii) $P(W \geq 90) = \displaystyle\int_{90}^{\infty} \frac{0.2}{19!}\left(\frac{w}{5}\right)^{19}e^{-w/5}dw \approx 0.6509.$

(iii) $P(X \geq 90) = \displaystyle\int_{90}^{\infty} \alpha e^{-\alpha w}dw = e^{-90\,\alpha} = e^{-0.9} \approx 0.4066.$

It might be surprising that the average lifetime exceeds the considered threshold with higher probability than an individual lifetime. This could be made plausible by the fact that only about 37% of the light bulbs exceed the average lifetime (see Exercise 10.7), but according to the shape of the exponential distribution, a few bulbs function a very long time, increasing the average lifetime considerably.

11.4 The probability-generating function

We obtain the following concept related to the mgf.

Definition 11.2: Let X be a discrete random variable, assuming only nonnegative integers. We again use the notation $p_k = P(X = k)$ for $k = 0,1,2, \ldots$. (If some integers are not values of X, we simply attach zero probabilities to them.) The *probability-generating function* (pgf) is defined as

$$G_X(t) = E(t^X) = \sum_{k=0}^{\infty} p_k t^k.$$

Since $G_X(1) = 1$, the series converges absolutely for $|t| \leq 1$.

In particular, $G_X(0) = p_0$.

We will present some examples for the pgf.

If X has a geometric distribution, then $p_k = q^{k-1}p$ for $k = 1,2, \ldots$, where $q = 1 - p$. Thus,

$$G_X(t) = \sum_{k=1}^{\infty} q^{k-1}pt^k = tp\sum_{k=1}^{\infty}(tq)^{k-1} = \frac{tp}{1-tq} \quad \text{for } |t| < \frac{1}{q}. \tag{11.13}$$

For the Poisson distribution, it holds $p_k = \left(e^{-\lambda}\lambda^k\right)/k!$ for $k = 0,1, \ldots$. Thus,

$$G_X(t) = \sum_{k=0}^{\infty} e^{-\lambda}\frac{\lambda^k}{k!}t^k = e^{-\lambda}\sum_{k=0}^{\infty}\frac{(t\lambda)^k}{k!} = e^{-\lambda}e^{t\lambda} = e^{(t-1)\lambda}. \tag{11.14}$$

We get for the binomial distribution $p_k = \binom{n}{k}p^k q^{n-k}$ for $k = 0,1, \ldots, n$, $q = 1 - p$. Thus, we obtain from the binomial theorem,

$$G_X(t) = \sum_{k=0}^{n}\binom{n}{k}(pt)^k q^{n-k} = (pt + q)^n. \tag{11.15}$$

Finally, for a Pascal distribution, we obtain $p_k = \binom{k-1}{r-1}p^k q^{k-r}$ for $k = r, r+1, \ldots$, $q = 1 - p$. Hence,

$$G_X(t) = \sum_{k=r}^{\infty}\binom{k-1}{r-1}p^r q^{k-r}t^k = (p\,t)^r\sum_{k=r}^{\infty}\binom{k-1}{r-1}(q\,t)^{k-r} = (p\,t)^r\frac{1}{(1-qt)^r}$$
$$= \left(\frac{pt}{1-qt}\right)^r. \tag{11.16}$$

The pgf is also biuniquely assigned to the distribution, i.e., if two pgfs $G_X(t)$ and $G_Y(t)$ are equal for all t, then X and Y have the same distribution.

The following two statements are analogous to Theorems 11.2 and 11.3.

Theorem 11.10: If G_X is the pgf of a random variable X and $Y = aX + b$, then the pgf of Y is given by

$$G_Y(t) = t^b G_X(t^a).$$

Proof:

$$G_Y(t) = E(t^Y) = E\left(t^{aX+b}\right) = t^b E\left((t^a)^X\right) = t^b G_X(t^a).$$

Theorem 11.11: Let X and Y be independent discrete random variables with pgfs $G_X(t)$ and $G_Y(t)$, respectively and let $Z = X + Y$. Then,

$$G_Z(t) = G_X(t)G_Y(t).$$

Proof: $G_Z(t) = E(t^{X+Y}) = E(t^X)\,E(t^Y) = G_X(t)\,G_Y(t)$.

Clearly, the last theorem can be easily generalized for a sum of n independent r.vs $Z = X_1 + \cdots + X_n$. The corresponding pgfs then satisfy

$$G_X(t) = G_{X_1}(t) \cdot \cdots \cdot G_{X_n}(t). \tag{11.17}$$

Example 11.17: We use (11.17) to prove (11.15) alternatively. Let X be the *Bernoulli random variable*, i.e., X has the two possible values 1 and 0, which are assumed with probability p and $q = 1 - p$, respectively. We have $p_0 = q$, $p_1 = p$ and $p_k = 0$ for $k = 2, 3, \ldots$. Thus, $E(t^X) = p_0 + p_1 t = q + pt$. A sum of n Bernoulli variables with the same parameter p is binomially distributed with parameters n and p. Therefore, it follows from (11.17) that the pgf of the binomial variable is $(q + pt)^n$.

11.5 Generation of probabilities and factorial moments

Similar to Section 11.2, we can recover the probabilities from the higher derivations of the pgf. Deriving

$$G(t) = p_0 + p_1 t + p_2 t^2 + p_3 t^3 + \cdots \tag{11.18}$$

with respect to t gives

$$G'(t) = p_1 + 2p_2 t + 3p_3 t^2 + 4p_4 t^3 + \cdots,$$
$$G''(t) = 2p_2 + 6p_3\ t + 12p_4 t^2 + \cdots,$$
$$G'''(t) = 6p_3 + 24p_4 t + \cdots.$$

Thus, $G'(0) = p_1$, $G''(0) = 2p_2$, $G'''(0) = 3!\ p_3, \ldots$, yielding the relation

$$p_k = \frac{G^{(k)}(0)}{k!}. \tag{11.19}$$

We illustrate this relation by means of two examples.

Example 11.18: Deriving the pgf of the geometric distribution

$$G(t) = \frac{tp}{1 - tq}$$

repeatedly (see (11.13)) we obtain by induction

$$G^{(k)}(t) = \frac{k!\, pq^{k-1}}{(1 - tq)^k} + \frac{k!\, tpq^k}{(1 - tq)^{k+1}}.$$

Thus,

$$\frac{G^{(k)}(0)}{k!} = p \; q^{k-1}.$$

Example 11.19: For the binomial distribution, we obtain (see (11.15))

$$G(t) = (pt + q)^n,$$

and

$$G^{(k)}(t) = (pt + q)^{n-k} p^k \frac{n!}{(n-k)!}.$$

Hence,

$$\frac{G^{(k)}(0)}{k!} = \binom{n}{k} p^k q^{n-k}.$$

Apart from its use to determine probabilities, the pgf has another interesting application.

Theorem 11.12: Let X be a discrete or continuous random variable. Then, the function $G(t) = E(t^X)$ satisfies

$$G^{(k)}(1) = E[X(X-1) \cdots (X - k + 1)].$$

We illustrate the proof for a continuous variable: Making use of the fact that

$$\frac{\partial^k}{\partial t^k}(t^x) = t^{x-k}[x(x-1) \cdots (x - k + 1)],$$

we obtain

$$G^{(k)}(t) = \frac{\partial^k}{\partial t^k} E(t^X) = \frac{\partial^k}{\partial t^k} \int t^x f(x) dx = \int t^{x-k}[x(x-1) \cdots (x - k + 1)] f(x) dx,$$

implying

$$G^{(k)}(1) = E[X(X-1) \cdots (X - k + 1)],$$

where we assume that $G(t)$ is k times continuously differentiable in the neighborhood of 1.

Due to the property in Theorem 11.12, the pgf is also known as the *factorial moment-generating function*. Using Example 11.19, we obtain immediately for the binomial distribution,

$$E[X(X-1)\dots(X-k+1)] = G^{(k)}(1) = p^k \frac{n!}{(n-k)!}.$$

Example 11.20: For the exponential distribution, we obtain

$$G(t) = \int_0^\infty t^x a e^{-ax} dx = \frac{a}{a - \ln(t)}$$

for $0 < t < 1$. The first two derivatives of G are

$$G'(t) = \frac{a}{t[a - \ln(t)]^2}$$

and

$$G''(t) = \frac{a[2 - a + \ln(t)]}{t^2[a - \ln(t)]^3}.$$

Hence,

$$G'(1) = \frac{1}{a}, \quad G''(1) = \frac{2 - a}{a^2},$$

which can be easily verified by direct computation of the factorial moments.

Example 11.21: In the case of the standard normal distribution, it holds

$$G(t) = \int_{-\infty}^\infty t^x \frac{1}{\sqrt{2\pi}} e^{-x^2/2} dx = \exp\left(\frac{1}{2}\ln(t)^2\right)$$

for $0 < t < 1$, implying

$$G'(t) = G(t)\frac{\ln(t)}{t},$$
$$G''(t) = G(t)\frac{1 - \ln(t) + \ln(t)^2}{t^2},$$

and

$$G'''(t) = G(t)\frac{-3 + 5\ln(t) - 3\ln(t)^2 + \ln(t)^3}{t^3}.$$

Thus,

$$G'(1) = 0, \; G''(1) = 1, \; G'''(1) = -3.$$

In Theorem 11.12, we have assumed that $G^{(k)}(1)$ exists. If the function $G^{(k)}(t)$ is not defined for $t = 1$, one can try to determine a factorial moment as the limit

$$\lim_{t \to 1} G^{(k)}(t).$$

Example 11.22: For the uniform distribution on $[a, b]$, we get

$$G(t) = \int_a^b \frac{t^x}{b-a} dx = \frac{t^b - t^a}{\ln(t)(b-a)}$$

and

$$G'(t) = \frac{\ln(t)\left[at^a - bt^b\right] - t^a + t^b}{t(a-b)\ln(t)^2}.$$

The function $G(t)$ is not defined for $t = 1$, since $\ln(1) = 0$. However

$$\lim_{t \to 1} G'(t) = \frac{a+b}{2}, \qquad (11.20)$$

which is the known expectation of the uniform distribution.

Exercises

11.1 Show that the mgf of the binomial distribution (Example 11.2) converges to that of the Poisson distribution if $n \to \infty$, such that $np = \lambda$ remains constant.

11.2 Determine the kth derivative $M^{(k)}(t)$ of the mgf of the gamma distribution with $r = 3$ and calculate the corresponding moments.

11.3 Determine $M_Y(t)$ in Example 11.5 alternatively, applying Definition 11.1 to the pdf of Y.

11.4 Derive the mgfs of the distributions Rayleigh and Maxwell (see (11.11), (11.12)).

11.5 Determine the mgf $M(t)$ of the discrete distribution
$f(k) = \frac{6}{\pi^2 k^2}$ for $k = 1, 2, \ldots$ Calculate $M^{(k)}(t)$ for $t = 1, 2, 3$.

11.6 Develop the mgf $M(t)$ for the double exponential distribution
$f(x) = \frac{a}{2}e^{-a|x|}$, $-\infty < x < \infty$. Calculate $M^{(k)}(t)$ and the corresponding moments about the origin.

11.7 Determine the mgf and the pgf for the "linear distribution"
$f(x) = ax + 1 - a/2$ for $0 \le x \le 1$, $|a| \le 2$.

11.8   Assume that X is exponentially distributed with parameter a. Determine the mgf of $Y = X^2$.

11.9   Given two densities f and g with their corresponding mgfs

$$M_f(t) = \int_{-\infty}^{\infty} e^{tx} f(x)\ dx \text{ and } M_g(t) = \int_{-\infty}^{\infty} e^{tx} g(x)\ dx.$$

Express the mgf of the mixture density
$h(x) = \lambda f(x) + (1 - \lambda)g(x)$ in terms of $M_f(t)$ and $M_g(t)$.

11.10 Illustrate the proof of Theorem 11.12 for a discrete variable X.

11.11 Determine $G^{(k)}(t)$ for the Pascal distribution and verify (11.19) for $k = 1,2,3$.

11.12 Find the factorial moments of the geometric distribution using the pgf.

11.13 Determine (11.13)–(11.16) alternatively from the corresponding mgfs using the relation $G_X(t) = M_X(\ln(t))$.

11.14 Prove (11.10) in Example 11.22 for the case $a = 0$, $b = 1$ using L'Hôpital's rule.

11.15 Prove Theorem 11.8.

11.16 Calculate the kth derivative of the mgf $M(t)$ of the uniform distribution over $[a,b]$.

12 Construction of new distributions

Application areas of probability theory and statistics have grown enormously in re-
cent decades. Accordingly, more and more situations were encountered for which the
few "classic probability models" (see the table of distributions) are not sufficiently
flexible. This has led to extraordinarily active developments in the theoretical field.
Many new distributions have been constructed that aim to accommodate specific fea-
tures like skewness, bimodality or fat tails, among others. In the following we present
some ideas for constructing new distributions. Two approaches may be distinguished.
One can manipulate and combine known densities or distribution functions to obtain
a useful model. Or one can construct a function $Z = H(X, Y, . . .)$ from one or more random variables $X, Y, . . .$, such that the distribution of Z has the desired characteristics.
The latter can be determined using the theory in Sections 6.1 and 7.5. We start with
the approach of the first kind.

To get the maximum benefit from the chapter, the reader should have software
available that can perform symbolic computation. This is indispensable to verify all
presented results and to solve the exercises.

12.1 A classical construction principle

One of the oldest ideas to construct new distributions from given ones is due to Azzalini (1985). Let U and V be two arbitrary absolutely continuous and independent random variables, symmetric about 0, with densities f and g and distribution functions F
and G, respectively. Then for any real parameter λ the function

$$h(x) = 2f(x)G(\lambda x) \tag{12.1}$$

is a density function. The main feature of this construction is to create skew symmetric distributions, where the skewness can be influenced by choosing the value of λ.
We illustrate this by means of two examples.

Example 12.1: Assume that f is the Student density with 3 degrees of freedom and g
the standardized normal density, i.e.,

$$f(x) = \frac{2}{\sqrt{3}\,\pi\left(1 + \frac{x^2}{3}\right)^2},$$

$$g(x) = \frac{e^{-x^2/2}}{\sqrt{2\pi}}.$$

Substituting these functions in eq. (12.1) results in

https://doi.org/10.1515/9783111332277-012

$$h(x) = \frac{12\sqrt{3}}{\pi\,(3+x^2)^2}\;\Phi(\lambda x) \text{ for } -\infty < x < \infty. \tag{12.2}$$

The density is illustrated in Fig. 12.1. One can derive formulas for the expected value and the second moment of a random variable with density (12.2):

$$E(X) = \frac{3\sqrt{2}\,\lambda\exp\!\left(\frac{3}{2}\lambda^2\right)}{\sqrt{\pi}}\left(1-\operatorname{erf}\left(\lambda\sqrt{3/2}\right)\right), \quad E(X^2) = 3 \text{ for } \lambda > 0,$$

where erf denotes the error function (see Exercise 12.1). The relation between the latter and the distribution function of the standardized normal distribution is given by

$$\Phi(x) = \frac{1}{2}\left(1+\operatorname{erf}\left(\frac{x}{\sqrt{2}}\right)\right).$$

For the special densities in Fig. 12.1, we have compiled the values of mode, expectation and variance in Tab. 12.1.

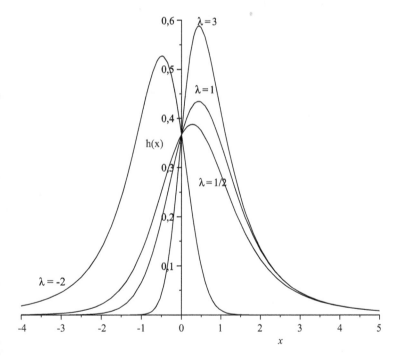

Fig. 12.1: Illustration of the function (12.2) for different parameter values.

Tab. 12.1: Some characteristics of the density in eq. (12.2).

λ	Mode	E(X)	V(X)
-2	-0.4842	-1.0275	1.9443
½	0.2740	0.6730	2.5471
1	0.4335	0.8932	2.2021
3	0.4404	1.0659	1.8638

When the densities f and g are positive over a bounded interval only, the construction may require case distinctions.

Example 12.2: Let be

$$f(x) = \frac{3}{4}\left(1-x^2\right) \quad \text{for } -1 \le x \le 1,$$
$$g(x) = \frac{1}{2} \quad\quad\quad \text{for } -1 \le x \le 1.$$

The latter results in the distribution function

$$G(x) = \begin{cases} 0 & \text{for } x < -1, \\ \frac{x+1}{2} & \text{for } -1 \le x \le 1, \\ 1 & \text{for } x > 1. \end{cases}$$

From eq. (12.1) we obtain the new density

$$h(x) = \frac{3}{2}\left(1-x^2\right) G(\lambda x) \quad \text{for } -1 \le x \le 1. \tag{12.3}$$

In order to specify the function G, the parameter ranges must be taken into account. We restrict ourselves to positive values.

Case 1: $0 \le \lambda \le 1$:

For any $x \in [-1, 1]$, it holds $\lambda x \in [-1, 1]$, i.e.,

$G(\lambda x) = \frac{\lambda x + 1}{2}$ holds for any $x \in [-1, 1]$.

Case 2: $\lambda > 1$:

$$x < -\frac{1}{\lambda} \implies \lambda x < -1 \implies G(\lambda x) = 0,$$

$$-\frac{1}{\lambda} \le x \le \frac{1}{\lambda} \Rightarrow \lambda x \in [-1,\ 1] \Rightarrow G(\lambda x) = \frac{\lambda\, x + 1}{2},$$

$$x > \frac{1}{\lambda} \Rightarrow \lambda x > 1 \Rightarrow G(\lambda x) = 1.$$

The density $h(x)$ is therefore

$0 \le \lambda \le 1$:

$$h(x) = \frac{3}{4}\,(1-x^2)\,(\lambda\,x+1) \qquad \text{for } -1 \le x \le 1. \tag{12.4a}$$

$\lambda > 1$:

$$h(x) = \begin{cases} \frac{3}{4}\,(1-x^2)\,(\lambda\,x+1) & \text{for } -\frac{1}{\lambda} \le x \le \frac{1}{\lambda}, \\ \frac{3}{2}\,(1-x^2) & \text{for } \frac{1}{\lambda} \le x \le 1. \end{cases} \tag{12.4b}$$

Density (12.4) is illustrated in Fig. 12.2.

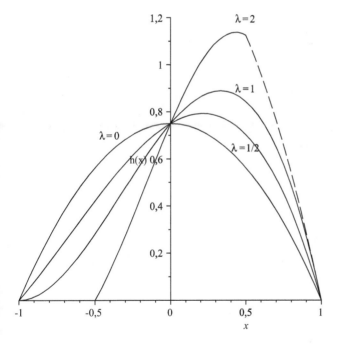

Fig. 12.2: Illustration of density (12.4) for different parameter values.

We determine some characteristics of the density h for the case $0 \le \lambda \le 1$ (see Section 8.4 for the moments):

The mode is $Mo = \left(\sqrt{3\lambda^2 + 1} - 1\right)/3\,\lambda$, and some important moments are

$$E(X) = \frac{\lambda}{5}, \quad V(X) = \frac{1}{5} - \frac{\lambda^2}{25}, \quad \mu_3 = -\frac{6\lambda}{175} + \frac{2\lambda^3}{125},$$

$$\mu_4 = \frac{3}{35} - \frac{18\lambda^2}{875} - \frac{3\lambda^4}{625}.$$

From the last two moments we obtain the coefficients of skewness and kurtosis in dependence of λ:

$$\alpha_3(\lambda) = \frac{\mu_3}{V(X)^{3/2}} = \frac{-\frac{6\lambda}{175} + \frac{2\lambda^3}{125}}{\left(\frac{1}{5} - \frac{\lambda^2}{25}\right)^{3/2}}, \quad \alpha_4(\lambda) = \frac{\mu_4}{V(X)^2} = \frac{\frac{3}{35} - \frac{18\lambda^2}{875} - \frac{3\lambda^4}{625}}{\left(\frac{1}{5} - \frac{\lambda^2}{25}\right)^2}.$$

One can show that $\alpha_3(\lambda)$ is a monotone decreasing and $\alpha_4(\lambda)$ a monotone increasing function of λ with $\alpha_3(0) = 0$, $\alpha_3(1) = -\frac{2}{7}$ and $\alpha_4(0) = \frac{15}{7}$, $\alpha_4(1) = \frac{33}{14}$.

The example shows how an alteration of the parameter influences skewness and kurtosis.

12.2 Exponentiated generalized distribution

Let $F(x)$ be a continuous distribution function. Due to Theorem 5.2, this function is also monotone increasing, such that $F(-\infty) = 0$ and $F(\infty) = 1$. Obviously these three properties are preserved, when F is raised to a positive power, i.e., $G = F^c$ is a new distribution function for any $c > 0$. We speak of the *exponentiated generalized distribution* of F. Let us consider, for example, the distribution function $F(x) = 1 - e^{-ax}$ of the exponential distribution $(a > 0)$. The new distribution function is then $G(x) = (1 - e^{-ax})^c$. Differentiating gives the corresponding density

$$g(x) = cae^{-ax} \left(1 - e^{-ax}\right)^{c-1}. \tag{12.5}$$

Figure 12.3 illustrates some special cases.

The above construction principle can be generalized by defining the distribution function

$$G(x) = \left(1 - (1 - F(x))^a\right)^{\beta}, \tag{12.6}$$

where a, $\beta > 0$. Since the function

$$h(x) = \left(1 - (1 - x)^a\right)^{\beta}$$

is monotone increasing over $[0, 1]$ such that $h(0) = 0$ and $h(1) = 1$, transformation (12.6) preserves the before mentioned characteristics of F. Obviously it reduces to the previ-

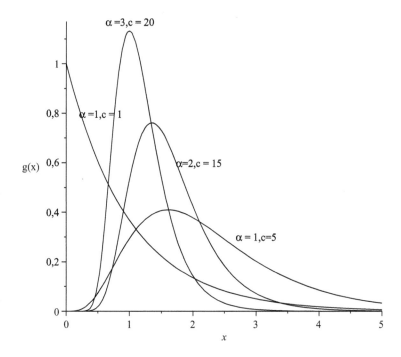

Fig. 12.3: Exponentiated exponential family of densities.

ous construction for $\alpha = 1$. This procedure allows for greater flexibility of the tails in fitting distributions to data and has many applications in engineering and biology.

For illustration we apply (12.6) to the distribution function

$$F(x) = x \text{ for } 0 \le x \le 1$$

of the uniform distribution. The new distribution function is then

$$G(x) = \left(1 - (1-x)^{\alpha}\right)^{\beta}$$

with corresponding density

$$g(x) = \alpha \beta (1-x)^{\alpha-1} \left(1 - (1-x)^{\alpha}\right)^{\beta-1}; \tag{12.7}$$

see Fig. 12.4.

Figures 12.3 and 12.4 show that the densities under consideration can take a wide variety of possible forms.

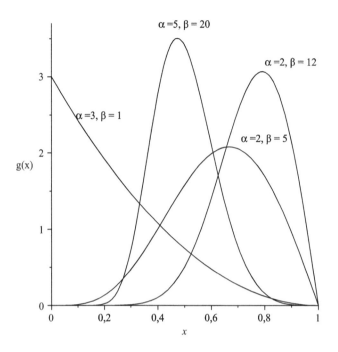

Fig. 12.4: Exponentiated generalized family of densities.

12.3 Mixture distributions

A *probability density function* (or *density* for short) is a function, satisfying the characteristics (i) and (ii) of Definition 5.4. It can be easily shown that

$$f(x) = \sum_{i=1}^{n} \lambda_i f_i \qquad (12.8)$$

is a density, when $f_1, \ldots, f_n$ are densities and $\lambda_1, \ldots, \lambda_n$ are weights, satisfying $\lambda_i > 0$ for $i = 1, \ldots, n$ and $\sum_{i=1}^{n} \lambda_i = 1$. Density (12.8) is then called the *mixture* of the densities f_i and the latter are called the *components* of f. A mixture density arises naturally, when a population is composed of n subpopulations. Let us consider a hypothetical example of a city in the south of Brazil that consists of 50% of Brazilians, 30% of Argentinians and 20% Paraguayans. If the height distributions have densities f_B, f_A and f_P, respectively, the height distribution of the city has the density $f = 0.5 f_B + 0.3 f_A + 0.2 f_P$, where the weights present the proportions of the subpopulations. The mixture distribution has convenient mathematical characteristics since the linearity of eq. (12.8) also holds for the distribution function and the moments. In concrete terms, this means the following:

The distribution function of f in eq. (12.8) is

$$F(x) = \int\limits_{-\infty}^{x} \sum_{i=1}^{n} \lambda_i f_i(x) = \sum_{i=1}^{n} \lambda_i \int\limits_{-\infty}^{x} f_i(x) = \sum_{i=1}^{n} \lambda_i F_i(x), \qquad (12.9)$$

i.e., F is a linear (convex) combination of the distribution functions F_i, corresponding to f_i; the moments about the origin are

$$E(X^j) = \int\limits_{-\infty}^{\infty} X^j \sum_{i=1}^{n} \lambda_i f_i(x) = \sum_{i=1}^{n} \lambda_i \int\limits_{-\infty}^{\infty} X^j f_i(x) = \sum_{i=1}^{n} \lambda_i E_i(X^j), \qquad (12.10)$$

i.e., the jth moment $E(X^j)$ is a linear combination of the moments $E_i(X^j)$ related to the densities f_i. Consider, for example, a mixture of the two normal distributions $N(0,1)$ and $N(3,4)$. The mixture density is then $f(x) = \lambda \dfrac{\exp\left(-\frac{x^2}{2}\right)}{\sqrt{2\pi}} + (1-\lambda) \dfrac{\exp\left(-\frac{(x-3)^2}{2 \cdot 2^2}\right)}{\sqrt{2\pi} \cdot 2}$ for $0 \le \lambda \le 1$ and $-\infty < x < \infty$; see Fig. 12.5.

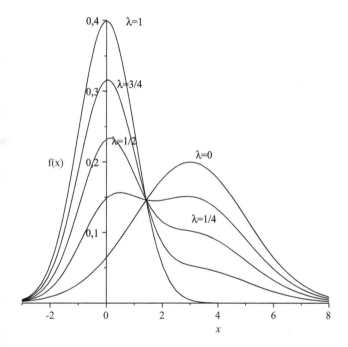

Fig. 12.5: Mixture densities of two normal distributions.

If λ is continuously decreased from 1 to 0, then the graph of $N(0,1)$ gradually merges into the graph of $N(3,4)$.

12.4 An extended distribution class

A specific approach to create new distributions is due to Cortés et al. (2018). A given density $g(x)$ is extended by adding a term $kh(x)g(x)$, where h is a positive continuous function and k a nonnegative constant. In order to ensure that the resulting function is a probability density, it is divided by a normalizing constant C. The new density is therefore

$$f(x) = \frac{g(x) + k\, h(x)g(x)}{C}, \tag{12.11}$$

where

$$C = \int (g(x) + k\, h(x)\, g(x))\, dx = 1 + k \int h(x)g(x)dx = 1 + kK. \tag{12.12}$$

The number $K = \int h(x)g(x)dx$ can be interpreted as the expectation $E(h(X))$, where X is a random variable having distribution with density $g(x)$. Combining the last two expressions yields

$$f(x) = \frac{g(x) + k\, h(x)g(x)}{1 + kK}. \tag{12.13}$$

Obviously this class of distributions contains $g(x)$ as a special case, obtained by setting $k = 0$. Function (12.13) can be rewritten as

$$f(x) = \frac{1}{1 + kK}\, g(x) + \frac{k\, K}{1 + kK}\, \frac{h(x)g(x)}{K}, \tag{12.14}$$

and for $\lambda = \dfrac{1}{1 + kK}$, density (12.14) becomes

$$f(x) = \lambda\, g(x) + (1 - \lambda)\frac{h(x)g(x)}{K}, \tag{12.15}$$

i.e., (12.15) is a mixture distribution, and one can apply (12.9) and (12.10) to determine the distribution function and moments.

Example 12.3: Let g be the standardized normal density and $h(x) = x^2$. Since

$$K = \int\limits_{-\infty}^{\infty} x^2\, \frac{\exp\left(-\frac{x^2}{2}\right)}{\sqrt{2\pi}}\, dx = 1,$$

the extended distribution (12.13) becomes

$$f(x) = \frac{1 + k\, x^2}{1 + k}\, \frac{\exp\left(-\frac{x^2}{2}\right)}{\sqrt{2\pi}}; \tag{12.16}$$

see Fig. 12.6. By increasing the value k from 0 to ∞, the standard normal density is gradually transformed into the so-called *bimodal normal density* $x^2\left(\exp\left(-\left(x^2/2\right)\right)/\sqrt{2\pi}\right)$.

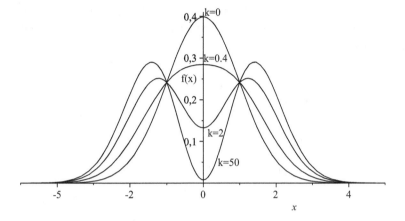

Fig. 12.6: Extended normal distribution.

The symmetric distribution (12.16) has the following characteristics, among others. For $k \leq \frac{1}{2}$ there is a single mode Mo = 0, and for $k > 1/2$ the two modes $\pm\sqrt{\frac{2k-1}{k}}$ exist. The moments about the origin are given by

$$E(X^r) = \begin{cases} \frac{(r-1)(r-3)\cdots\cdots 1\cdot[1+(r+1)\cdot k]}{1+k} & \text{for } r \text{ even,} \\ 0 & \text{for } r \text{ odd,} \end{cases}$$

and the distribution function is

$$F(x) = \frac{1}{2}\left(1 + \operatorname{erf}\left(\frac{x}{\sqrt{2}}\right)\right) - \frac{kx}{\sqrt{2\pi}(1+k)}\, e^{-x^2/2}.$$

There exist two other approaches to construct new distributions which can be considered special cases of the extended distribution introduced above. By setting $k = 1$ and $h(x) = (1-\alpha x)^2$ and $h(x) = \left(1 - \alpha x - \beta x^3\right)^2$ in eq. (12.11), one obtains the *alpha-skew distribution* and the *alpha-beta-skew distribution* of $g(x)$, respectively; see, e.g., Elal-Olivero (2010) and Shah et al. (2023). Both approaches aim to generate an asymmetric distribution that is suitable to fit unimodel and bimodal data sets. We restrict ourselves to an example of the first type.

Example 12.4: As indicated above, the alpha-skew normal distribution is obtained as

$$\frac{1+(1-ax)^2}{2+a^2} \quad \frac{e^{-x^2/2}}{\sqrt{2\pi}},$$

where the denominator $2+a^2$ is the normalizing constant. Figure 12.7 shows this density for different values of α. The densities are bimodal for $\alpha = -3$ and $\alpha = -1.5$ and unimodal for $\alpha = 0.3$ and $\alpha = 1.1$.

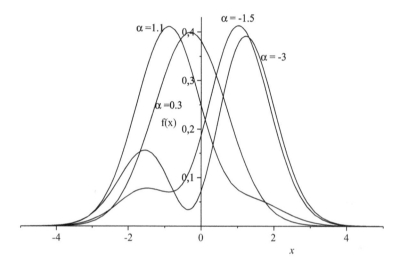

Fig. 12.7: Alpha-skew normal distribution.

We get, e.g., $E(X) = -\dfrac{2\alpha}{2+\alpha^2}$ and $E(X^2) = \dfrac{2+3\alpha^2}{2+\alpha^2}$, and the coefficient of skewness is given by the odd function

$$\alpha_3 = \frac{E[(X-E(X))^3]}{\sigma^3} = \frac{4\alpha^3(3\alpha^2+2)}{(4+4\alpha^2+3\alpha^4)^{3/2}}. \tag{12.17}$$

The skewness assumes its maximum for $\alpha_{max} = 1.76$ and its minimum for $\alpha_{min} = -\alpha_{max}$, where the maximum and minimum skewness are about 0.81 and −0.81.

12.5 Transformations of a random variable

We now construct a new distribution, considering a function of random variables with known distributions. We first restrict ourselves to a function of a single variable. We make use of Theorem 6.1, where ψ is an invertible, differentiable and monotone function over the range space of X. The density of Y is the new distribution we are looking for. According to this theorem, the density of $Y = \psi(X)$ is

$$g(y) = \left| (\psi^{-1})'(y) \right| f(\psi^{-1}(y)),$$

where f is the density of X.

Example 12.5: For $X \sim N(\mu, \sigma^2)$ and $\psi(X) = e^X$ we get $\psi^{-1}(y) = \ln(y)$, $(\psi^{-1})'(y) = 1/y$ and obtain the lognormal distribution with density

$$g(y) = \frac{1}{y} f(\ln(y)) = \frac{1}{\sqrt{2\pi} \, y \, \sigma} \exp\left(\frac{-(\ln(y) - \mu)^2}{2\sigma^2} \right) \text{ for } y > 0. \tag{12.18}$$

This function is illustrated in Fig. 12.8.

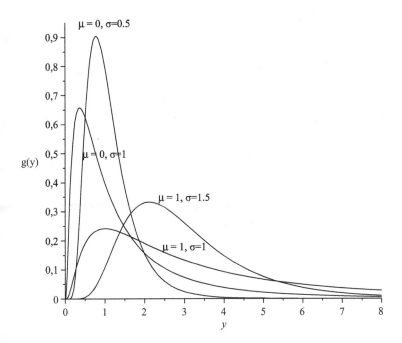

Fig. 12.8: The lognormal distribution.

Some characteristics of the lognormal distribution are $\text{Mo} = \exp(\mu - \sigma^2)$,

$E(Y^r) = \exp\left(\frac{r \cdot (r\sigma^2 + 2\mu)}{2} \right)$ and the distribution function is $G(y) = \frac{1}{2} + \frac{1}{2}\text{erf}\left(\frac{\ln(y) - \mu}{\sqrt{2}\,\sigma} \right)$.

Example 12.6: If $\psi(X) = 1/X$ it follows that $\left(\psi^{-1}\right)(y) = 1/y$, $\left(\psi^{-1}\right)'(y) = (-1)/y^2$. Thus

$$g(y) = \frac{1}{y^2} f\left(\frac{1}{y}\right).$$

This is the *inverse distribution* of the distribution of X. For example, the *inverse exponential distribution* is

$$g(y) = \frac{1}{y^2} a e^{-\frac{a}{y}} \text{ for } y > 0.$$

In particular, we get Mo $= \frac{a}{2}$, $G(y) = e^{-a/y}$.

12.6 Functions of two variables

We now assume that there is a bidimensional continuous random variable (X, Y) with known joint distribution. We consider a biunique function $(X,Y) \mapsto (U,V)$ with the characteristics of Theorem 7.1. The distribution of (U, V) is the new *bivariate* distribution we are looking for. From the literally infinite number of possible constructions we select a few.

Example 12.7: Assume that (X,Y) has the bivariate joint normal distribution with $\mu_1 = \mu_2 = 0$, $\sigma_1 = 1$, $\sigma_2 = 2$, $\rho = \frac{\sqrt{3}}{2}$, see Section 10.6. Note that X and Y are not independent. By means of the biunique mapping $(X,Y) \mapsto (U,V) = (e^X, e^Y)$ we construct a new bivariate distribution over the positive quadrant, which is given by Theorem 7.1. It holds $X = \ln(U)$, $Y = \ln(V)$ and the Jacobian is $J = \begin{vmatrix} 1/u & 0 \\ 0 & 1/v \end{vmatrix} = \frac{1}{uv}$. Since (X, Y) has the joint distribution,

$$f(x,y) = \frac{1}{2\pi} \exp\left(-2x^2 + \sqrt{3}xy - y^2/2\right), \tag{12.19}$$

See eq. (10.24), and the density of (U,V) is

$$g(u, v) = \frac{1}{2\pi\, u\, v} \exp\left(-2(\ln(u))^2 + \sqrt{3}\ln(u)\ln(v) - (\ln(v))^2/2\right) \text{ for } u, v \geq 0. \tag{12.20}$$

If one is interested in a univariate distribution only, one can proceed as in Examples 7.11 and 7.12. In this case one defines a new random variable $U = H_1(X,Y)$, where H_1 is a function $H_1: R^2 \mapsto R$ and defines V as a function of X and Y that is as simple as possible, for example, $V = X$ or $V = Y$. Then one constructs the joint density (7.27) and the desired univariate density is obtained as the marginal distribution of U.

12.7 Slash distributions

In practice one often encounters data sets that cannot be fitted by classical models like the normal distribution since large values occur more frequently than the classical distributions would predict. What is needed is a heavy-tailed distribution. In order to construct one, a random variable X (the so-called parent distribution) is divided by $Y^{1/q}$, where Y is uniformly distributed over [0, 1] and $q > 0$, and X and Y are independent random variables. For $X \sim N(0,1)$ we obtain the density of $U = X/Y^{1/q}$ from Theorem 7.1 as

$$h(u) = \int_0^1 \varphi(uv) \, q \, v^q dv, \tag{12.21}$$

see Exercise 12.19. This is called the *slash distribution*. The case $q = 1$ is of special interest. One can express this density by means of a confluent hypergeometric function:

$$h(u) = \frac{q}{\sqrt{2\pi} \, (q+1)} {}_1F_1\left(\frac{q+1}{2}, \frac{q+1}{2}+1, -\frac{u^2}{2}\right), \tag{12.22}$$

see Zörnig (2019, p. 373, eq. (2)). In particular, for $q = 1$ one obtains

$$h(u) = \int_0^1 \varphi(uv) \, v \, dv = \frac{1 - e^{-u^2/2}}{\sqrt{2\pi} \, u^2}. \tag{12.23}$$

Figure 12.9 illustrates (12.22) for some values of q. The smaller is q, the fatter are the tails, and when q converges to infinity, the parent distribution is obtained as a limiting distribution.

In the relevant literature, slash distributions have already been constructed for various parent distributions, for example, for some variants of the normal distribution, for the elliptical and the logistic distribution, and the construction procedure has also been generalized in several manners; see, e.g., Zörnig (2019).

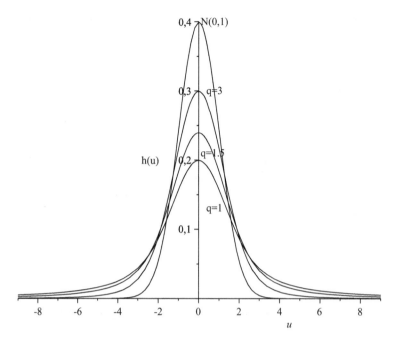

Fig. 12.9: The slash distribution.

Exercises

12.1 Show that the expectation in Example 12.1 is

$$E(X) = \frac{3\sqrt{2}\,\lambda\,\exp\left(\frac{3}{2}\lambda^2\right)}{\sqrt{\pi}}\left(1 + \operatorname{erf}\left(\lambda\sqrt{3/2}\right)\right) \text{ for } \lambda < 0.$$

12.2 Continuing the study in Example 12.2, write a program to calculate the characteristics $E(X)$, $V(X)$, μ_3, μ_4, $a_3(\lambda)$ and $a_4(\lambda)$ for parameter values $\lambda > 1$.
Give explicit formulas for $E(X)$ and $V(X)$ and calculate $a_3(\lambda)$ and $a_4(\lambda)$ for $\lambda = 2, 4, \ldots, 10$.

12.3 Determine h by choosing f and g in (12.1) as the standardized normal density. This results in the *skew normal distribution*. Determine the coefficients $a_3(\lambda)$ and $a_4(\lambda)$ for this density h and illustrate them graphically as functions of λ. Determine $a_3(0)$ and $a_4(0)$ and the limits $a_3(\infty)$ and $a_4(\infty)$.

12.4 Determine the mode of distribution (12.7).

12.5 Construct the squared normal distribution, i.e., determine the density corresponding to $G(x) = \Phi^2(x)$, where Φ is the standardized normal distribution function. Illustrate the density geometrically. Determine mode, skewness a_3 and kurtosis a_4.

12.6 Consider the distribution function $G(x) = F(x)^2$, where F is the Student distribution function with 2 degrees of freedom. Determine the corresponding density, the mode and the expectation.

12.7 It can be observed in Fig. 12.5 that the graphs of all densities intersect in the same point. Do you think that this phenomenon is related to the specific example or is it a general characteristic of the mixture distribution?

12.8 Consider the mixture density

$f(x) = 0.3\, f_1(x) + 0.5 f_2(x) + 0.2 f_3(x)$,

where f_1, f_2, f_3 are the densities of the normal distributions $N(0, \sigma^2)$, $N(2, \sigma^2)$, $N(4, \sigma^2)$, respectively. Determine the cumulative distribution function of f. Illustrate f graphically for $\sigma = 0.4, 0.6, 0.8$ and 1.2 and determine the modes for these cases.

12.9 Construct the mixture density of $f_1(x) = \dfrac{1}{2}\, e^{-|x|}$ (double exponential distribution) and

$$f_2(x) = \frac{\exp\left(-\frac{(x-3)^2}{8}\right)}{2\sqrt{2\pi}}.$$

Determine the cumulative distribution function and the first four moments about the origin. Illustrate the mixture density graphically.

12.10 Determine the moments $E(X^k)$ of the alpha-skew normal distribution. Verify the formula (12.17) for the coefficient of skewness and illustrate α_3 graphically as a function of a.

12.11 Specify the distribution function of the alpha-skew normal distribution.

12.12 Determine the extended exponential distribution $f(x)$ by setting $g(x) = \alpha\, e^{-\alpha x}$ and $h(x) = x^3$ for $x \geq 0$, $\alpha, k > 0$ in eq. (12.13). Illustrate the density $f(x)$ graphically. Determine the moments and the distribution function.

12.13 Create the alpha-skew Weibull distribution for $a = 1$ (see the table of distributions), i.e., study the density

$$f(x) = \frac{1 + (1-ax)^2}{C}\, 2x\, e^{-x^2} \text{ for } x \geq 0,$$

where C is the normalizing constant. Determine C, the moments $E(X^r)$ and the distribution function $F(x)$. Illustrate the density geometrically for $a = -1, 0, 1.5, 2$.

12.14 Determine the density of $Y = X^2$, where X is exponentially distributed. Illustrate the density $g(y)$ geometrically, determine the moments $E(Y^r)$ and the distribution function of Y.

12.15 Construct the inverse Weibull distribution. Determine $g(y)$, $E(Y^r)$, $G(y)$ and the mode. Illustrate $g(y)$ graphically.

12.16 Illustrate the new bivariate density (12.20) graphically. Determine the marginal distributions and calculate the correlation between U and V. Compare with the correlation between X and Y. Calculate the probabilities $P(U \leq 1, V \leq 1)$ and $P(U \geq 1, V \geq 1)$.

12.17 Determine the distribution of $U = X/Y$, where X and Y are independent with $X \sim N(0,1)$ and Y is uniformly distributed over $[-a, a]$, $a > 0$. Obtain the density and the distribution function of U.

12.18 Obtain the distribution of the quotient of two independent exponentially distributed random variables with parameters a and b. Eliminate unnecessary parameters.

12.19 Determine the distribution of $X + Y$, where X and Y are independent with $X \sim N(0,1)$ and Y is exponentially distributed. Illustrate graphically.

12.20 Derive formula (12.21) by means of Theorem 7.1.

12.21 Determine the moments about the origin and the coefficient of kurtosis α_4 for distribution (12.22), if they exist.

12.22 Construct the slash distribution using the exponential as the parent distribution. Illustrate graphically and verify that the slash densities have fatter tails than the exponential.

13 Sums of many random variables

Several random phenomena in the real life arise as sums or average values of many single observations. For example, the total load to be transported by a cargo ship is the sum of the weights of many individual containers, the total duration of a project is the sum of the durations of several individual operations and the temperature of a gas is directly proportional to the average kinetic energy of innumerable molecules.

In particular, the proper concept of probability is based on a large number of repetitions of a random experiment. The relative frequency (see Section 2.3) of an event A can be expressed as $f_A = (1/n)(X_1 + \cdots + X_n)$, where the *indicator variable* X_i is 1 if A results in the ith realization of the random experiment and 0 otherwise ($i = 1, \ldots, n$). We have already noted previously that according to the statistical regularity the relative frequency f_A converges in some sense to the theoretical probability $P(A)$ if the number n of repetitions increases. In the following we want to provide a solid theoretical foundation for these ideas.

13.1 The law of large numbers

With the aid of Tchebycheff's inequality, we now concretize in which sense the relative frequency f_A "converges" to the probability $P(A)$.

Theorem 13.1 (law of large numbers, Bernoulli's form):
Let A be an event associated with a random experiment E. Consider n independent realizations of E, let n_A be the number of times that A occurs among the n realizations of E and let $f_A = n_A/n$ denote the relative frequency of A. Finally, let $P(A) = p$ denote the probability of A, which is assumed to be the same for all experiments.

Then for any number $\varepsilon > 0$ it holds

$$P(|f_A - p| \geq \varepsilon) \leq \frac{p(1-p)}{n\varepsilon^2}.$$

Prove: Obviously, the r.v. n_A is binomially distributed, i.e., $E(n_A) = np$ and $V(n_A) = np(1-p)$. In consequence, $E(f_A) = p$ and $V(f_A) = p(1-p)/n$ and Tchebycheff's inequality (see (8.8)) implies

$$P(|f_A - p| \geq \varepsilon) \leq \frac{p(1-p)}{n\varepsilon^2}. \tag{13.1}$$

The convergence $f_A \to P(A)$ is now clearly defined by the fact that the probability that f_A deviates by more than ε from $P(A)$ is smaller than the right side of (13.1), thereby $\varepsilon > 0$ can be any positive number. However, to obtain a useful statement, ε must be cho-

https://doi.org/10.1515/9783111332277-013

sen such that the upper bound in (13.1) is smaller than 1. Statement (13.1) can be expressed in equivalent forms, e.g., as

$$P(|f_A - p| < \varepsilon) \geq 1 - \frac{p(1-p)}{n\varepsilon^2}. \tag{13.2}$$

In particular, (13.2) implies immediately

$$\lim_{n \to \infty} P(|f_A - p| < \varepsilon) = 1, \tag{13.3}$$

which is equivalent to

$$\lim_{n \to \infty} P(|f_A - p| > \varepsilon) = 0. \tag{13.4}$$

If random variables $X, X_1, X_2, \dots$ satisfy the relation

$$\lim_{n \to \infty} P(|X_n - X| > \varepsilon) = 0, \tag{13.5}$$

we say that the sequence (X_n) *converges in probability* to the r.v. X. Thus, f_A converges in probability to $P(A)$.

Example 13.1: Assume that a certain product is defective with probability $p = 0.1$. Specify a lower limit for the probability that the relative frequency of perfect items in a delivery of 200 items is in the interval [0.85, 0.95]?

By setting $\varepsilon = 0.05$ in (13.2), we obtain the required limit as $1 - \frac{0.9 \cdot 0.1}{200 \cdot 0.05^2} = 0.82$. The exact probability of the event $f_A \in]0.85, 0.95[$ can be calculated by means of the binomial distribution, since this event is equivalent to $170 < n_A < 190$. We obtain the probability

$$\sum_{k=171}^{189} \binom{200}{k} 0.9^k 0.1^{200-k} \approx 0.9756.$$

It must be emphasized that the law of large numbers in no way guarantees that the observed relative frequency f_A lies in the interval $I = [0.85, 0.95]$. There is only a high probability that f_A is located inside of I.

13.2 Normal approximation to the binomial distribution

We have shown above that the law of large numbers is intimately related to the binomial distribution. To understand better, why the relative frequency converges to the exact probability we must study the binomial distribution when n is large.

In this case, one can use Stirling's formula (Section 1.2) to show that the binomial probability function is approximated by a normal density, i.e.,

$$P(X=k) = \binom{n}{k} p^k (1-p)^{n-k} \approx \frac{1}{\sqrt{2\pi\sigma}} \exp\left(-\frac{(k-\mu)^2}{2\sigma^2}\right) = \frac{1}{\sigma}\varphi\left(\frac{k-\mu}{\sigma}\right), \tag{13.6}$$

where $\mu = np$ and $\sigma^2 = np(1-p)$ denote expectation and variance of the binomial distribution and φ is the standard normal density. The derivation of this classical result requires considerable algebraic manipulations and will not be detailed here. Approximation (13.6) is considered good for $\sigma > 3$. By summing up or integrating, respectively, we get

$$P(X \le k) = \int_{-\infty}^{k} \frac{1}{\sigma}\varphi\left(\frac{x-\mu}{\sigma}\right) dx = \phi\left(\frac{x-\mu}{\sigma}\right). \tag{13.7}$$

The result is usually formulated as follows.

Theorem 13.2 (DeMoivre–Laplace approximation to the binomial distribution):
If X has a binomial distribution with parameters n and p and if

$$Y = \frac{X-np}{\sqrt{np(1-p)}}$$

denotes the reduced variable (Section 10.1), then, for large n, Y has approximately a $N(0,1)$ distribution in the sense that

$$\lim_{n \to \infty} P(Y \le y) = \Phi(y) \text{ for all } y.$$

In Fig. 13.1, the values of the binomial distribution with $n = 15$ and $p = 0.4$ (vertical lines) are compared graphically with the normal density in (13.6).

In this section we have approximated the discrete binomial distribution by the continuous normal distribution. Hence, some care must be taken with the end points of the integration domain. The following geometrically motivated *correction for continuity* has proved to be useful:

$$P(X=k) \approx \int_{k-1/2}^{k+1/2} \frac{1}{\sigma}\varphi\left(\frac{x-\mu}{\sigma}\right) dx = \phi\left(\frac{k+1/2-\mu}{\sigma}\right) - \phi\left(\frac{k-1/2-\mu}{\sigma}\right). \tag{13.8}$$

This implies

$$P(k_1 \le X \le k_2) \approx \phi\left(\frac{k_2+1/2-\mu}{\sigma}\right) - \phi\left(\frac{k_1-1/2-\mu}{\sigma}\right) \tag{13.9}$$

for any nonnegative integers $k_1 \le k_2$.

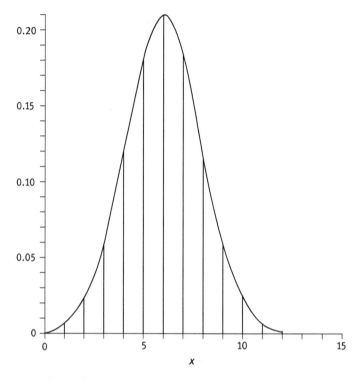

Fig. 13.1: Normal approximation to the binomial distribution.

Example 13.2: Suppose that a system consists of 50 identical components with reliability of 90%, i.e., each component functions properly during a specific time period with probability 0.9. Assuming that the components fail independently and that the whole system functions properly when at least 40 components function, what is the reliability of the system? Determine the exact value of this probability and the approximate value given by (13.9).

Solution: The exact probability is

$$P(X \geq k) = \sum_{40}^{50} \binom{50}{k} 0.9^k \cdot 0.1^{50-k} \approx 0.990645.$$

With $\mu = 50 \cdot 0.9 = 45$ and $\sigma^2 = 50 \cdot 0.9 \cdot 0.1 = 4.5$, the correction for continuity (13.7) yields

$$\phi\left(\frac{50 + 1/2 - 45}{\sqrt{4.5}}\right) - \phi\left(\frac{40 - 1/2 - 45}{\sqrt{4.5}}\right) \approx 0.990478.$$

13.3 The central limit theorem

The above theorem is only a special case of a general statement that we will present now. To realize this, we recall that a binomial variable X can be written as

$$X = X_1 + \cdots + X_n, \tag{13.10}$$

where the X_i are independent Bernoulli variables with common parameter p (see Example 11.17). Theorem 13.2 states that the reduced variable of the specific sum (13.10) is approximately distributed as $N(0,1)$. The following theorem states basically that (almost) any normalized sum of independent r.v.s has the approximate distribution $N(0,1)$.

Theorem 13.3 (central limit theorem, CLT):
Let $X = X_1 + \ldots + X_n$ be a sum of the independent r.v.s X_i, having expectation $E(X_i) = \mu_i$ and variance $V(X_i) = \sigma_i^2$ for $i = 1, \ldots, n$.

Then, under certain conditions (which basically say that the individual components X_i contribute only a negligible amount to the sum X), the reduced variable

$$Z_n = \frac{X - \sum_{i=1}^{n} \mu_i}{\sqrt{\sum_{i=1}^{n} \sigma_i^2}}$$

has approximately a $N(0,1)$ distribution in the sense that

$$\lim_{n \to \infty} G_n(z) = \Phi(z) \text{ for all } z, \tag{13.11}$$

where $G_n(z)$ is the cdf of Z_n.

If (13.11) holds, we say that the variables Z_n *converge in distribution* to the variable $Z \sim N(0,1)$. The basic idea of the proof, which will be omitted here, is to show that the mgfs of the Z_n converge to the mgf $\exp(t^2/2)$ of the standardized normal distribution (see Chapter 11).

The CLT says that the components of the sum $X = X_1 + \ldots + X_n$ may (essentially) have any kind of distribution and yet X is approximately normally distributed. This fact is the basic reason for the importance of the normal distribution in probability theory and statistics. There are innumerous applications of the CLT. In addition to the examples mentioned at the beginning of this chapter, the total consumption of electricity or water of a city during a given time period is the sum of the demands of a large number of individual consumers. The measurement error in a physical experiment is composed of many unobservable small errors that may be considered additive. The CLT generalizes in a certain sense Theorem 11.4, saying that a sum of normally distributed variables is again normally distributed. The CLT states that the summands need not be normally distributed for the sum to be approximated by a normal distribution.

Example 13.3: A queue of 30 cars has been formed before a railway crossing. Suppose that the length L of a car is a random variable with expected value $\mu = 5.3$ m and variance $\sigma^2 = 1.44$ m^2 and that the distance D between two successive vehicles is a r.v. with expected value $\mu_g = 1.1$ m and variance $\sigma_g^2 = 0.09$ m^2. What is the probability that the total length L_{tot} of the queue of cars (consisting of 30 cars and 29 gaps) is between 185 and 195 m?

Solution: The total length of the waiting queue is $L_{tot} = 30\, L + 29\, D$.

Its expectation is therefore $30 \cdot 5.3 + 29 \cdot 1.1 = 190.9$ m, and the variance is $30 \cdot 1.44 + 29 \cdot 0.09 = 45.81$ m^2. Hence, the required probability is

$$\Phi\left(\frac{195-190.9}{\sqrt{45.81}}\right) - \Phi\left(\frac{185-190.9}{\sqrt{45.81}}\right) \approx \Phi\,(0.61) - \Phi\,(-0.87) \approx \Phi\,(0.61) - 1 + \Phi\,(0.87)$$

$$\approx 0.7291 - 1 + 0.8078 \approx 0.5369.$$

An important specific case of the CLT arises when the X_i are identically distributed.

Theorem 13.4: Let $X = X_1 + \ldots + X_n$ be a sum of the independent and identically distributed r.v.s X_i with $E(X_i) = \mu$ and $V(X_i) = \sigma^2$ for $i = 1, \ldots, n$.

Under the conditions of Theorem 13.3, the reduced variable

$$Z_n = \frac{X - n\mu}{\sqrt{n}\sigma}$$

has approximately a $N(0,1)$ distribution in the sense of (13.11).

We now illustrate the practical relevance of Theorem 13.4 by means of some examples.

Example 13.4: Suppose that the weight X_i of an adult has a distribution with expected value $\mu = 75$ kg and variance $V(X_i) = 100$ kg^2. If a tourist bus carries 80 adults, what is the probability that the total weight of the persons exceeds 6.2 tons?

Solution: The total weight W is approximately normally distributed with $\mu_{tot} = 80 \cdot 75$ kg $= 6{,}000$ kg and $\sigma_{tot} = \sqrt{80 \cdot 100}$ kg. Hence, the required probability is

$$P(W > 6200) \approx 1 - \Phi\left(\frac{6200-6000}{\sqrt{80 \cdot 10}}\right) \approx 1 - \Phi\,(2.24) \approx 1 - 0.9874 \approx 0.0126.$$

Example 13.5: Assume that the time of treatment of a patient in a medical practice is a normal random variable with expectation of 20 min and variance of 60 min^2. If there are 10 persons in the waiting room, what is the probability that the physician can treat all of them in less than three hours?

Solution: The total treatment time T is normally distributed with $\mu_{tot} = 10 \cdot 20$ min $= 200$ min and $\sigma_{tot} = \sqrt{600}$ min. Hence, the required probability is $P(T \le 180) \approx \Phi\,((180 - 200)/\sqrt{600}) \approx \Phi\,(-0.82) \approx 1 - \Phi\,(0.82) \approx 1 - 0.7939 \approx 0.2061$.

From the CLT it follows naturally that almost any variable that is a sum of many independent r.v.s can be approximated by the normal distribution. We have discussed this for the binomially distributed r.v. X that can be expressed as a sum of Bernoulli variables (see Section 13.2). Similarly a Poisson variable can be expressed as the sum of occurrences of a certain phenomenon in many small nonoverlapping intervals, or a r.v. with gamma distribution (with parameters a and r) may be represented as a sum of r independent exponentially distributed r.v.s with parameter a. Hence, for large r the gamma distribution can be approximated by the normal distribution.
    We still illustrate this by means of two examples.

Example 13.6: Suppose that calls arrive at a telephone exchange at the rate of 3 per minute. What is the probability that 26 or fewer calls are received during a 10-min period, assuming that the intensity remains the same during this period?

If X is the number of calls received, then the expectation is $E(X) = a = 10 \cdot 3 = 30$, yielding the exact probability

$$P(X \le 26) = e^{-30} \sum_{k=0}^{26} \frac{30^k}{k!} \approx 0.2673.$$

The approximate value due to (13.9) is

$$\Phi\left(\frac{26 + 1/2 - 30}{\sqrt{30}}\right) \approx \Phi\,(-0.64) \approx 1 - 0.7389 \approx 0.2611.$$

Example 13.7: The lifetime T of a certain type of light bulb is exponentially distributed with expectation of 6 months. What is the probability that the total lifetime of 30 bulbs exceeds 18 years? Determine the exact probability and its approximation, using the normal distribution.

Solution: The total life time has a gamma distribution with parameters $a = 1/6$ and $r = 30$; hence, the required exact probability is

$$\int_{216}^{\infty} \frac{1/6}{\Gamma(30)} \left(\frac{x}{6}\right)^{26} e^{-x/6} dx \approx 0.1379.$$

The lower limit of the integral corresponds to 18 years (=216 months). Since the expected total life time and the variance are $E(X) = (r/a) = 180$ and $V(X) = (r/a^2) = 1,080$, we obtain the approximate probability from (13.9) as

$$1 - \Phi\left(\frac{216 + 1/2 - 180}{1,080}\right) \approx 1 - \Phi\,(1.11) \approx 0.1335.$$

Exercises

13.1 How many times a die must be tossed to be at least 99% sure that the relative frequency of obtaining a six is within 0.02 of the theoretical probability 1/6? Make use of the law of large numbers.

13.2 If an item is perfect with probability 0.8, determine a lower limit for the probability that among 150 produced items the number of perfect ones is between 110 and 130.

13.3 Given the discrete random variables X, X_1, X_2, ... with probability functions $f(2)$ = 1 and $f_n(1) = 1/n$, $f_n(2) = 1 - 1/n$ for $i = 1, 2, \ldots$ Show that the sequence (X_n) converges in probability to X.

13.4 Consider a binomial distribution with parameters $n = 30$ and $p = 0.4$. Calculate the exact probabilities $P(X = k)$ for $k = 0,1, \ldots, 30$ and the approximate values, using (13.6) and (13.8).

13.5 Suppose that a fair die is thrown 600 times. What is the probability that the number 6 occurs at least 90 times and at most 105 times? Compare the exact probability with the approximate value obtained by (13.9).

13.6 A teacher has to evaluate 200 written examinations. Suppose that the time of evaluation per exam has expectation of 15 min and standard deviation of 5 min, what is the probability that the total time of evaluation exceeds 52 h?

13.7 A commercial airplane with 300 seats is fully occupied. The passengers are composed of 135 men, 105 women and 60 children. The weights of these groups of persons (in kg) are random variables with expectations 80, 70 and 40 and standard deviations 10, 8 and 5, respectively. Let W denote the total weight of the passengers.
Determine (i) $E(W)$, (ii) $V(W)$, (iii) $P(W < 21,000)$.

13.8 The maximum capacity of a lift is 800 kg. Suppose that the weight of a user is normally distributed with $\mu = 75$ kg and $\sigma = 8.5$ kg.
(i) What is the probability that the capacity is exceeded, if 10 persons use the lift?
(ii) What is this probability if 11 persons use the lift?
(iii) For which value of μ the probability of item (i) equals 10% (for $\sigma = 8.5$ kg unchanged).

13.9 Tom has invited 15 friends to his birthday party who are very fond of beer. Suppose that the beer consumption of a guest is a random variable with expectation $\mu = 3$ L and standard deviation $\sigma = 1.2$ L.
   (i) If Tom has a stock of 50 L on the day of celebration, what is the probability that this quantity is sufficient to quench the thirst of the 15 guests?
   (ii) What must be the quantity Q of the stock of beer such that the demand is satisfied with probability of 99%?

13.10 A chain consists of 50 members. Suppose that the length of each member is a random variable with expectation of 1 cm and variance of 0.04 cm^2. What is the probability that the total length of the chain is between 48 and 53 cm?

13.11 Eight people are waiting in a doctor's office. If the treatment time of any patient (in minutes) is a r.v. with distribution $N(15, 25)$, what is the probability that the total treatment time of the waiting patients is less than 1 h and 50 min?

13.12 A company has to produce 50 specific components of the same type. Each item is perfect with probability 0.7, independent of the results obtained in the previous productions. The manufacturing is continued until the 50th perfect item is obtained. Determine the exact probability that more than 60 components have to be produced and calculate the approximate probability, using the normal distribution. Explain briefly why the CLT applies.

14 Samples and sampling distributions

The main purpose of statistical inference is to make statements about an entire population based on observations in a sample. In many practical situations, there do not
exist alternatives to this kind of reasoning. The entire population is often too large to
investigate all individual elements. If one wants to determine the average income of
the population of a country, it is usually too time consuming and expensive to inter-
view all individuals.

Another problem is that a detailed analysis may destroy the elements. For example, to test the life time of a certain type of light bulbs, one has to switch them on and
wait until they are burned out. So it is evident that this test should not be applied to
the whole population of light bulbs.

To perform a correct statistical inference, we must know how to interpret a sample
observation. To this end we must study the distributions of sample characteristics, e.g.,
the distribution of the average income, the distribution of the longest life time of a light
bulb, etc. A generic term for distributions of sample characteristics is *sampling distributions*. In this chapter, we will study some of the most important sampling distributions.

14.1 Samples and statistics

Let us formalize some notions as follows.

Definition 14.1: Let X be a random variable with a certain probability distribution.
The range of X is considered as a theoretic population. Let $X_1, \ldots, X_n$ denote n independent r.v.s with the same distribution as X. The vector $(X_1, \ldots, X_n)$ is then called a
random sample from the r.v. X.

If $(X_1, \ldots, X_n)$ is a random sample from X, then any function $Y = H(X_1, \ldots, X_n)$ is called
a *statistic*.

Example 14.1: Let X be the height of an adult of a population. If we select a sample of
5 adults, then $(X_1, \ldots, X_5)$ is a random sample where X_i denotes the height of the ith
person of the sample, $i = 1, \ldots, 5$. A possible statistic is the *average height* of the sample, given by $Y = H(X_1, \ldots, X_5) = (X_1 + \cdots + X_5)/5$. This is a random variable with value y
$= (x_1 + \cdots + x_5)/5$ where $x_1, \ldots, x_5$ are the values of $X_1, \ldots, X_5$.

We consider the following important statistics.

Definition 14.2: Let $(X_1, \ldots, X_n)$ be a random sample of a r.v. X.
(i) $\bar{X} = (1/n)\Sigma_{i=1}^{n}X_i$ is called the *sample mean*.
(ii) $S^2 = \left(1/(n-1)\right)\Sigma_{i=1}^{n}(X_i - \bar{X})^2$ is called the *sample variance*.

https://doi.org/10.1515/9783111332277-014

The reason why we divide by $(n-1)$ rather than by n will become clear later (see Theorem 14.2 and the following remark).

(iii) $M = \max(X_1, \ldots, X_n)$ is called the *maximum of the sample*.

(iv) $K = \min(X_1, \ldots, X_n)$ is called the *minimum of the sample*.

(v) $R = M-K$ is called the *sample range*.

(vi) $X_{(j)}$ denotes the *j*th *smallest observation* in the sample.

The r.v.s $X_{(1)}, \ldots, X_{(n)}$ are called the *order statistics* associated with the random sample $(X_1, \ldots, X_n)$. In particular, we have $X_{(1)} = K$ and $X_{(n)} = M$.

We will now state some properties of the above statistics.

Theorem 14.1: Let X denote a r.v. with expectation $E(X) = \mu$ and variance $V(X) = \sigma^2$. Let $\bar{X}$ be the sample mean of a random sample of size n. Then it holds

(i) $E(\bar{X}) = \mu$,

(ii) $V(\bar{X}) = \sigma^2/n$,

(iii) For large n the reduced variable $\dfrac{\bar{X} - \mu}{\sigma/\sqrt{n}}$ has approximately the distribution $N(0, 1)$.

Proof: Parts (i) and (ii) follow immediately from the properties of expectation and variance. Part (iii) follows from the central limit theorem, since $\bar{X} = (1/n) X_1 + \cdots + (1/n) X_n$ is a sum of independent, identically distributed r.v.s.

The theorem states that for a large sample size n, the sample mean is approximately normally distributed with the parameters given in parts (i) and (ii). In generalization of this fact it can be shown that most "well-behaved" functions of $\bar{X}$ have this property.

In particular, the variance of $\bar{X}$ decreases for increasing n, which is intuitively clear (see Exercise 14.1).

Example 14.2: Suppose that the weight of adults of a certain population is a r.v. with $\mu = 80$ kg and $\sigma = 9$ kg. What is the probability that the average weight $\bar{W}$ of 100 adults is between 79 and 81 kg?

Solution: Since $\bar{W}$ has the approximate distribution $N(80, 81/100)$, the required probability is

$$\phi\left(\frac{81 - 80}{0.9}\right) - \phi\left(\frac{79 - 80}{0.9}\right) = \phi\left(\frac{10}{9}\right) - \phi\left(-\frac{10}{9}\right) = 2\phi\left(\frac{10}{9}\right) - 1 \approx 0.733.$$

The next theorem refers to the sample variance.

Theorem 14.2: Suppose that $(X_1, \ldots, X_n)$ is a random sample of a r.v. X with expectation μ and variance σ^2. Then the sample variance (see Definition 14.2(ii)) satisfies

(i) $E(S^2) = \sigma^2$.

(ii) If X is normally distributed, then the statistic $((n-1)S^2)/\sigma^2$ has chi-square distribution with $(n-1)$ degrees of freedom.

Proof:

(i) We rewrite the sum of S^2 as

$$\sum_{i=1}^{n}(X_i - \bar{X})^2 = \sum_{i=1}^{n}(X_i - \mu + \mu - \bar{X})^2 = \sum_{i=1}^{n}(X_i - \mu)^2 + 2(\mu - \bar{X})\sum_{i=1}^{n}(X_i - \mu) + n(\mu - \bar{X})^2$$

$$= \sum_{i=1}^{n}(X_i - \mu)^2 - 2n(\mu - \bar{X})^2 + n(\mu - \bar{X})^2 = \sum_{i=1}^{n}(X_i - \mu)^2 - n(\mu - \bar{X})^2.$$

Hence,

$$E(S^2) = E\left(\frac{1}{n-1}\sum_{i=1}^{n}(X_i - \bar{X})^2\right) = \frac{1}{n-1}\left[n\sigma^2 - n\frac{\sigma^2}{n}\right] = \sigma^2.$$

(ii) We will give the proof only for the case $n = 2$. Then we can write

$$S^2 = (X_1 - \bar{X})^2 + (X_2 - \bar{X})^2 = \frac{(X_1 - X_2)^2}{2}.$$

Since X_i are supposed to be normally and identically distributed, it holds that

$$\frac{X_1 - X_2}{\sqrt{2}\sigma}$$

has distribution $N(0, 1)$. (It can be easily shown that this variable has expectation 0 and variance 1.) Hence,

$$\frac{S^2}{\sigma^2} = \left(\frac{X_1 - X_2}{\sqrt{2}\,\sigma}\right)^2$$

has chi-square distribution with one degree of freedom (see Section 10.4).

Because of part (i), the expectation of S^2 is σ^2, i.e., when the variance of the population is estimated by S^2, the latter "hits the target on the average" (see Chapter 15). This is the reason why we divide by $(n-1)$ in the definition of S^2.

14.2 Order statistics

We now develop results concerning the order statistics in the discrete and in the continuous case. Properties of the statistics M, K and R in Definition 14.2 will arise as consequences. Order statistics may give useful practical information. For example, the maximum temperature and the largest rainfall in a period of 20 years may be useful for planning emergency strategies. The average cost of certain household products can be used to estimate living costs.

Theorem 14.3: Suppose that $(X_1, \ldots, X_n)$ is a random sample of the discrete r.v. X. Let $x_1, x_2, \ldots$ and $p_1, p_2, \ldots$ denote the possible values of X and the corresponding probabilities, respectively $(x_1 < x_2 < \ldots)$. Let F_j and f_j be the cdf and the probability function of the order statistic $X_{(j)}$. Then it holds

$$F_j(x_i) = \sum_{k=j}^{n} \binom{n}{k} q_i^k (1 - q_i)^{n-k} \tag{14.1}$$

and

$$f_j(x_i) = \sum_{k=j}^{n} \binom{n}{k} \left[q_i^k (1 - q_i)^{n-k} - q_{i-1}^k (1 - q_{i-1})^{n-k} \right] \tag{14.2}$$

where $q_0 = 0$ and $q_i = p_1 + \cdots + p_i$ for $i = 1, 2, \ldots$.

Proof: Let Y denote the number of variables $X_1, \ldots, X_n$ having values less or equal to x_i, where i is a fixed index. Then Y is binomially distributed with parameters n and $q_i = P(X_{(j)} \le x_i)$. Now the events $\{X_{(j)} \le x_i\}$ and $\{Y \ge j\}$ are equivalent, both expressing the fact that at least j of the variables $X_1, \ldots, X_n$ have values smaller than or equal to x_i. Hence,

$$F_j(x_i) = P\left(X_{(j)} \le x_i \right) = P(Y \ge j)) = \sum_{k=j}^{n} \binom{n}{k} q_i^k (1 - q_i)^{n-k},$$

i.e., (14.1) holds. By using the relation

$$f_j(x_j) = F_j(x_i) - F_j(x_{i-1}),$$

we obtain (14.2).

For small values of i, one can make use of the binomial theorem to simplify (14.1) and (14.2). In particular, for the sample minimum, we obtain

$$P(K \le x_i) = F_1(x_i) = 1 - (1 - q_i)^n, \tag{14.3}$$

$$P(K = x_i) = P(K \le x_i) - P(K \le x_{i-1}) = (1 - q_{i-1})^n - (1 - q_i)^n. \tag{14.4}$$

Similarly, we get for the sample maximum

$$P(M \le x_i) = F_n(x_i) = q_i^n \tag{14.5}$$

$$P(M = x_i) = f_n(x_i) = q_i^n - q_{i-1}^n. \tag{14.6}$$

Example 14.3: A die is tossed six times. What is the probability that
(i)   the minimum value obtained is 3,
(ii) the maximum value obtained is 5,
(iii) the third-smallest value obtained is 4.

Solution: We have $x_i = i$, $p_i = 1/6$ and $q_i = i/6$ for $i = 1, \ldots, 6$, yielding

(i) $P(K = 3) = (1 - q_2)^6 - (1 - q_3)^6 = \left(\dfrac{4}{6}\right)^6 - \left(\dfrac{3}{6}\right)^6 \approx 0.0722.$

(ii) $P(M = 5) = \tfrac{6}{5} - q_4^6 = \left(\dfrac{5}{6}\right)^6 - \left(\dfrac{4}{6}\right)^6 \approx 02471.$

(iii) $P\left(X_{(3)} = 4\right) = f_3(4) = \sum_{k=3}^{6} \binom{6}{k} \left[\left(\dfrac{4}{6}\right)^k \cdot \left(\dfrac{2}{6}\right)^{6-k} - \left(\dfrac{3}{6}\right)^k \cdot \left(\dfrac{3}{6}\right)^{6-k} \right] \approx 0.2436.$

Since the probabilities in (14.1) satisfy $q_i = F(x_i)$, where F is the cdf of X, we can rewrite this relation as

$$F_j(x) = \sum_{k=j}^{n} \binom{n}{k} F(x)^k (1 - F(x))^{n-k}, \tag{14.7}$$

where the cdf of the order statistic $X_{(i)}$ is expressed in terms of the cdf of X. It is important to emphasize that this relation holds also for continuous random variables, since the reasoning in the proof of Theorem 14.3 is also valid in that case. Moreover, for continuous r.v. X the pdf of $X_{(i)}$ can be obtained by deriving the cdf.

However, we use another approach to determine the pdf of a continuous order statistic. Let f and F denote the pdf and the cdf of the continuous r.v. X. We may assume that with probability one all values of the n order statistics associated with a random sample $(X_1, \ldots, X_n)$ of X are different, i.e., $P(X_{(1)} < \cdots < X_{(n)}) = 1$. Now for a sufficiently small $\Delta > 0$ the pdf of $X_{(k)}$ in a point x is approximately given by

$$f_{(k)}(x) \approx \frac{P\left(X_{(k)} \in [x, x + \Delta]\right)}{\Delta} \tag{14.8}$$

where equality holds for $\Delta \to 0$. But the event $X_{(k)} \in [x, x + \Delta]$ occurs if and only if one sample value lies in $[x, x + \Delta]$, $(k - 1)$ sample values are smaller than x and $(n\text{-}k)$ are greater than $x + \Delta$. Thus from the multinomial model, we get

$$P\left(X_{(k)} \in [x, x + \Delta]\right) = \binom{n}{k - 1,\, 1,\, n - k} P(X < x)^{k-1} \cdot P(X \in [x, x + \Delta]) \cdot P(X > x + \Delta)^{n-k}$$

$$= n \binom{n-1}{k-1} F(x)^{k-1} \cdot P(X \in [x, x + \Delta]) \cdot (1 - F(x + \Delta))^{n-k}.$$

Dividing by Δ and taking the limit $\Delta \to 0$ we obtain from (14.8)

$$f_{(k)}(x) = n \binom{n-1}{k-1} f(x) \cdot F(x)^{k-1} \cdot (1 - F(x))^{n-k}. \tag{14.9}$$

As important specific cases, we obtain the densities of sample minimum and maximum as

$$f_{(1)}(x) = nf(x) \cdot (1 - F(x))^{n-1} \tag{14.10}$$

and

$$f_{(n)}(x) = nf(x) \cdot F(x)^{n-1}. \tag{14.11}$$

The corresponding cumulative distribution functions are

$$P(K \le x) = F_{(1)}(x) = 1 - (1 - F(x))^n \tag{14.12}$$

and

$$P(M \le x) = F_{(n)}(x) = F(x)^n. \tag{14.13}$$

Example 14.4: If the variable X is exponentially distributed, i.e., $f(x) = ae^{-ax}$ and $F(x) = 1 - e^{-ax}$ for $x \ge 0$, one obtains from the above formulas:

$$F_{(1)}(x) = 1 - e^{-anx}, \tag{14.14}$$

$$F_{(n)}(x) = (1 - e^{-ax})^n. \tag{14.15}$$

$$f_{(1)}(x) = ane^{-anx}, \tag{14.16}$$

$$f_{(n)}(x) = ane^{-ax}(1 - e^{-ax})^{n-1}. \tag{14.17}$$

In particular, the example shows the following. If X is exponentially distributed with parameter a, then the minimum K of a random sample of X of size n is exponentially distributed with parameter an.

Example 14.5: Suppose that the life time T of an electronic device is exponentially distributed with $E(X) = 800$ h, i.e., the pdf is $f(x) = ae^{-ax}$ with $a = 1/800$. If 100 such devices are tested, what is the probability
(i)   that the largest observed value exceeds 6,000 h?
(ii) that the smallest observed value is between 10 and 20 h?
(iii) The third-smallest observed value is less than 30 h.

Solution:

(i) From (14.13) it follows that

$$P(M \geq 6{,}000) = 1 - P(M \leq 6{,}000) = 1 - F(6{,}000)^{100}, \text{ where}$$

$$F(6{,}000) = \int_0^{6{,}000} ae^{-a}dx = 1 - e^{-6{,}000a} = 1 - e^{-7.5}.$$

Hence, the required probability is $P(M \geq 6{,}000) = 1 - (1 - e^{-7.5})^{100} = 0.0538$.

(ii) Similarly, we get $F(10) = 1 - e^{-10a}$ and $F(20) = 1 - e^{-20a}$. The required probability is therefore $e^{-10\,a} - e^{-20a} = e^{-1/80} - e^{-1/40} \approx 0.0123$.

(iii) From (14.9), we obtain the pdf of the third-order statistic as

$$f_3(x) = 100 \binom{99}{2} f(x)F(x)^2 (1 - F(x))^{97} = 485{,}100 ae^{-ax}(1 - e^{-ax})^2 e^{-97ax},$$

yielding the required probability $\int_0^{30} f_3(x) \; dx \approx 0.7166$.

By generalizing the approach that led us to pdf (14.9), we can derive the joint densities for order statistics. Without proof we mention that the joint pdf of the two statistics $X_{(i)}$ and $X_{(j)}$ with $i < j$ is given by

$$h(s, t) = \binom{n}{i-1, j-i-1, n-j} f(s)f(t)F(s)^{i-1}[F(t) - F(s)]^{j-i-1}[1 - F(t)]^{n-j}, \qquad (14.18)$$

where f and F denote again the pdf and the cdf of X, respectively. In particular for $i = 1$ and $j = n$, we obtain

$$h(k, m) = n(n - 1)f(k)f(m)[F(m) - F(k)]^{n-2}. \qquad (14.19)$$

With the methods discussed in Section 7.5, we obtain from this the density of the sample range $R = M - K$ as

$$f_R(r) = \int_{-\infty}^{\infty} h(s, s+r)ds = n(n-1) \int_{-\infty}^{\infty} f(s)f(s+r)[F(s+r) - F(s)]^{n-2}ds. \qquad (14.20)$$

By derivation with respect to r, one can easily verify that the cdf of R is

$$F_R(r) = n \int_{-\infty}^{\infty} f(s)[F(s+r) - F(s)]^{n-1}ds. \qquad (14.21)$$

Let us again employ an exponentially distributed r.v. to illustrate the last three formulas. The joint distribution of the minimum and the maximum of a sample is

$$h(k, m) = n(n-1)\alpha^2 e^{-\alpha(k+m)}\left[e^{-\alpha\,k} - e^{-\alpha\,m}\right]^{n-2} \text{ for } 0 \le k \le m. \tag{14.22}$$

The pdf and the cdf of the range are given by

$$f_R(r) = n(n-1)\alpha^2 \int_{-\infty}^{\infty} e^{-\alpha(2s+r)}\left[e^{-\alpha\,s} - e^{-\alpha(s+r)}\right]^{n-2} ds \tag{14.23}$$

$$= \alpha(n-1)(e^{\alpha\,r} - 1)^{n-2} e^{\alpha\,r(1-n)}$$

and

$$F_R(r) = n\alpha \int_{-\infty}^{\infty} e^{-\alpha\,s}\left[e^{-\alpha\,s} - e^{-\alpha(s+r)}\right]^{n-1} ds = \frac{(1 - e^{-\alpha\,r})^n e^{\alpha\,r}}{e^{\alpha\,r} - 1}. \tag{14.24}$$

Example 14.6: Let us continue the study of the electronic devices of the previous example, where the pdf of the life time T is $f(x) = \alpha e^{-\alpha\,x}$ with $\alpha = 1/800$. If 100 such devices are tested, what is the probability
(i)   that all observed life times are between 10 and 6,000 h?
(ii) that the smallest life time is below 10 h while the largest exceeds 5,000?
(iii) that the sample range is smaller than 5,000 h?

Solutions:
(i)   From (14.22) we get

$$P(10 \le K \le M \le 6{,}000) = \int_{10}^{6{,}000} \int_{10}^{6{,}000} 9{,}000 \cdot \alpha^2 e^{-\alpha(k+m)}\left[e^{-\alpha\,k} - e^{-\alpha\,m}\right]^{98} dk\,dm \approx 0.2709.$$

(ii) In analogy to the previous item, we get

$$P(K \le 10, 5000 \le 6{,}000) = \int_{5{,}000}^{\infty} \int_{0}^{10} 9{,}000 \cdot \alpha^2 e^{-\alpha(k+m)}\left[e^{-\alpha\,k} - e^{-\alpha\,m}\right]^{98} dk\,dm \approx 0.1248.$$

(iii) From (14.24) we obtain $P(R \le 5{,}000) = F_R(5{,}000) \approx 0.8259$.

14.3 Sampling in simulation and deterministic applications

As shown above, a sample is useful to obtain some information about unknown parameters of the underlying distribution. However, it can be used for other purposes as well. For example, we can use sample values to approximate certain probabilities that could calculated by other means only with difficulty or not at all. But to this end, one must know how to generate a sample of a given distribution. The following theorem indicates how to do this.

Theorem 14.4: Let X be a random variable with range $[a, b]$. Assume that the pdf f satisfies $f(x) > 0$ for $a < x < b$ and let F denote the cdf.

Then the function $Y = F(X)$ is uniformly distributed over $[a, b]$.
(The r.v. Y is called the *integral transform* of X.)

Proof: Since F is a monotone increasing function, the inverse F^{-1} exists and the cdf of Y satisfies

$$G(y) = P(Y \leq y) = P(F(X) \leq y) = P(X \leq F^{-1}(y)) = F(F^{-1}(y)) = y \text{ for } a \leq y \leq b$$

Hence, the pdf of Y is $g(y) = 1$.

It should be noted that the theorem also holds for discrete r.v.s, but we will confine ourselves to the continuous case.

Now let X be a r.v. from which a sample is required, and let F denote the respective cdf. At first we determine a realization of the uniform distribution over $[0, 1]$. If no random generator is available, one can obtain y_1 from a table of random numbers. By setting $x_1 = F^{-1}(y_1)$ we now generate a first value of X, since $F(X)$ is uniformly distributed over $[0, 1]$. Repeating this procedure, we obtain a sample of X of a desired size.

Example 14.7: We now assume that we want to obtain a sample of size six from a random variable with distribution $N(5, 4)$. Suppose that we get the following values from a table of random numbers: $y_1 = 0.485$, $y_2 = 0.652$, $y_3 = 0.733$, $y_4 = 0.198$, $y_5 = 0.345$, $y_6 = 0.865$. The required sample consists of the six values x_i, satisfying $y_i = \Phi((x_i - 5)/2)$ for $i = 1$, ..., 6. The numerical values are obtained as $(x_1, \ldots, x_6) = (4.92, 5.78, 6.24, 3.30, 4.20, 7.21)$.

In the above manner we can generate a sample of an arbitrary distribution. Nowadays any statistical standard software provides routines to generate samples for the most common distributions.

Example 14.8: Suppose that X has distribution $N(5, 4)$ and we want to study the r.v. $Y = e^{-X} X^3 \sin^2(X)$ (see Fig. 14.1). In particular, we want to evaluate the probability $P(Y \geq 0.1)$. The exact approach consists in expressing the event $B = \{Y \geq 0.1\}$ as a union of intervals and adding the respective probabilities. It turns out that $B = [x_1, y_1] \cup [x_2, y_2] \cup [x_3, y_3]$, where $(x_1, y_1, x_2, y_2, x_3, y_3) \approx (0.765, 2.865, 3.422, 5.853, 6.857, 8.385)$. Hence the required exact probability is

$$\sum_{i=1}^{3} \int_{x_i}^{y_i} \frac{1}{\sqrt{2\pi} \cdot 2} \exp\left(-\frac{(x-5)^2}{8}\right) dx \approx 0.7071.$$

By means of a random generator that constructs samples of a normally distributed r.v., we construct a sample $(X_1, \ldots, X_n)$ of the distribution $N(5, 4)$. For each X_i we define the r.v. $Y_i = e^{X_i} X_i^3 \sin^2(X_i)$ and then evaluate the relative frequency n_A/n, where n_A is the number of Y_i values satisfying $y_i \geq 0.1$. For large n the relative frequency will be "close" to the probability $P(Y \geq 0.1)$ in the sense of the law of large numbers. The reader is advised to verify this by means of some practical calculations.

By combining sample values of various populations, it is possible to simulate outcomes from complicated composed random variables.

Example 14.9: Suppose that the points of impact of two missiles are bidimensional gamma distributed random variables (X,Y) and (W,Z), measured in km. It is assumed that X,Y,W,Z are mutually independent with parameters as given in Tab. 14.1.

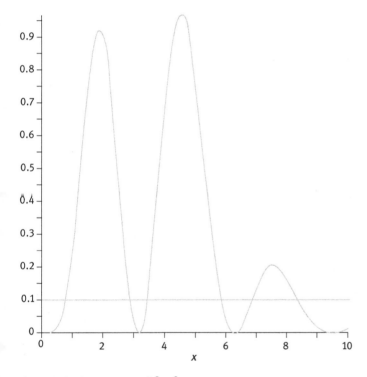

Fig. 14.1: The function $\Psi(x) = e^x x^3 \sin^2(x)$.

Tab. 14.1: Parameter values of coordinates.

	Variable	a	r
Coordinates missile 1	X	1/5	3
	Y	1/4	4
Coordinates missile 2	W	1/2	5
	Z	1/3	4

Let $D = \sqrt{(X-W)^2 + (Y-Z)^2}$ denote the distance between the impact points. Determine
(i) the expected distance $E(D)$ by simulating the outcomes of D and
(ii) the probability that the distance between the impact points is smaller than 10 km.

Solution:
(i) For a predetermined size n we determine samples $(x_1, \ldots, x_n)$, $(y_1, \ldots, y_n)$, $(w_1, \ldots, w_n)$ and $(z_1, \ldots, z_n)$ of the random variables X, Y, W and Z, respectively. From this we generate a sample $(d_1, \ldots, d_n)$ of the distance D by setting

$d_i = \sqrt{(x_i - w_i)^2 + (y_i - z_i)^2}$ for $i = 1, \ldots, n$. (Some software can generate samples of composed r.v.s directly.)

The estimated value of $E(D)$ is then given by the sample mean $(1/n)\sum_{i=1}^{n} d_i$.

(ii) Let n_A denote the number of d_i satisfying $d_i < 10$. The required probability is estimated by the relative frequency n_A/n.

By means of test runs with MAPLE for $n = 10^6$ the numerical values $E(D) \approx 13.08$ and $n_A/n \approx 0.4143$ have been obtained.

The above examples illustrate some basic ideas of a broad class of computational algorithms known as "Monte Carlo methods." Two other application areas for repeated sampling are stochastic optimization and numerical integration. We still illustrate some basic techniques of the latter type.

Suppose that we want to calculate an integral of the form $\int_a^b f(x)dx$, where f is a continuous nonnegative function over the interval $[a, b]$ with maximum value $M = \max\{f(x)|x \in [a, b]\}$. Then we select randomly n points from a rectangle $R = \{(x, y)^T | x \in [a, b], y \in [0, c]\}$ with $c \geq M$ and count the number k of points lying under the graph of f. For sufficiently large n, the proportion k/n is approximately equal to the ratio $\frac{1}{c(b-a)}\int_a^b f(x)dx$ of the integral and the area of the rectangle, i.e., the estimated value of the integral is $c(b-a)k/n$. In this procedure it must be guaranteed that any point of the rectangle R is selected with the same probability. This can be achieved by generating two samples $(x_1, \ldots, x_n)$ and $(y_1, \ldots, y_n)$ of the uniform distributions over $[a,b]$ and over $[0,c]$, respectively, and choosing the ith point as (x_i, y_i) $(i = 1, \ldots, n)$. For obvious reasons this procedure is known as *hit-or-miss Monte Carlo integration*.

Example 14.10: Consider the function $f(x) = 1 + 6x - 3x^2$ over the interval $[0, 2]$. It can be easily shown that $M = 4$ and $\int_0^2 f(x)dx = 6$. Hence we select $c = 4$ and the area of the rectangle is $c(b - a) = 8$. For different values of n, the above approach yielded the estimators as shown in Tab. 14.2.

Tab. 14.2: Integration via simulation.

Sample size n	Estimator $c(b-a)k/n = 8k/n$
1,000	6.032
10,000	6.0168
100,000	6.01184
1,000,000	5.99736

One can verify that the estimators converge to the exact value 6 of the integral. Of course, the practical benefit of the stochastic method is evident when the exact integration cannot be easily performed as in the following example.

Example 14.11: Consider the function $f(x) = e^{-x} x^3 \sin^2(x)$ over the interval $[0, 3\pi]$, illustrated in Fig. 14.1 (see also Example 14.8). The maximum value of f is about 0.97, assumed at the global maximum point $x_0 \approx 4.54$. Thus we choose $c = 1$. The exact value of the required integral is $\int_0^{3\pi} f(x)dx \approx 2.9918$ and the area of the rectangle is 3π. For test runs with increasing n, we have obtained the estimated values for the integral given in Tab. 14.3.

Tab. 14.3: Integration via simulation .

Sample size n	Estimator $3\pi k/n$
1,000	3.0913
10,000	2.9763
100,000	2.9865
1,000,000	2.9918

The above method can be generalized to integrals of functions with various variables. We consider only the two-dimensional case, where the integral $\iint_S f(x, y)dydx$ with a continuous nonnegative function f and a compact region S in the plane is to be calculated. Let $M = \max\{f(x) \mid x \in S\}$ denote the maximum value of f. We now select randomly n points from a cuboid $C = \{(x, y, z)^T \mid x \in [a, b], y \in [s, t], z \in [0, c]\}$ where a, b, s, t and c are chosen such that $c \geq M$ and the rectangle $R = \{(x, y)^T \mid x \in [a, b], y \in [s, t]\}$ contains the domain S. We now count the number k of points lying under the graph of f (i.e., the points $(x, y, z)^T$ satisfying $(x, y) \in S$ and $0 \le z \le c$). For sufficiently large n, the proportion k/n is now approximately equal to the ratio

$$\frac{1}{c(b-a)(t-s)} \iint_S f(x,y)dydx$$

of the integral and the cuboid volume, i.e., the estimated value of the bidimensional integral is $c(b-a)(t-s)k/n$. The points from the cuboid are generated as $\{x_i, y_i, z_i\}$, where $(x_1, \ldots, x_n)$, $(y_1, \ldots, y_n)$ and $(z_1, \ldots, z_n)$ are samples of the uniform distributions over $[a, b]$, $[s, t]$ and $[0, c]$, respectively $(i = 1, \ldots, n)$.

Example 14.12: Consider the function $f(x,y) = 3 + x^2 - 3y^2$ over the circle $S = \{(x, y)^T \mid x^2 + y^2 \le 1\}$, illustrated in Fig. 14.2.

The maximum value of f is 4. Thus we can choose $c = 4$, $a = s = -1$ and $b = t = 1$. The exact value of the required integral is

$$\int_{-1}^{1} \int_{-\sqrt{1-x^2}}^{\sqrt{1-x^2}} \left(3 + x^2 - 3y^2\right) dy \, dx = \frac{5}{2}\pi \approx 7.853983$$

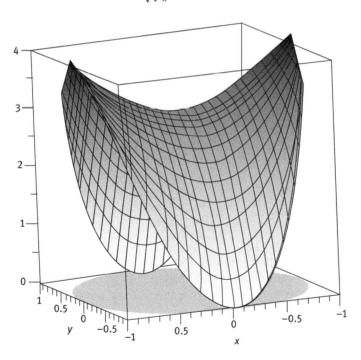

Fig. 14.2: Integral of Example 14.12.

and the volume of the cuboid is 16. The generation of cuboid points may be analogous to the preceding example. For simulations with increasing n we have obtained the estimated values for the integral given in Tab. 14.4.

Tab. 14.4: Bidimensional integration.

Sample size n	Estimator 16k/n
1,000	7.76
10,000	7.7888
100,000	7.82554
1,000,000	7.858256

As an alternative to the above sampling technique, we can directly generate points of the cylinder $\{(x, y, z)^T \mid x^2 + y^2 \le 1, \ 0 \le z \le 4\}$ and determine the proportion of points below the surface $f(x) = 3 + x^2 - 3y^2$.

Exercises

14.1 Suppose that the height X of adults of a population has expectation $\mu = 170$ kg and standard deviation $\sigma = 15$ kg.
   (a) Determine the probability that the average height of a random sample of 150 adults is between 165 and 172 kg.
   (b) If X is normally distributed with the above parameters, what is the probability that the sample variance S^2 is between 200 and 250 kg?
14.2 (a) Determine the cdf for the order statistics $X_{(j)}$ of the geometric distribution.
   (b) Formulate explicitly the distributions of sample minimum and maximum. Of what type is the distribution of the minimum?
14.3 Determine the probability functions for the previous exercise.
14.4 A person of the population suffers from a particular disease with probability 0.002 (see Example 9.1). Suppose that there exist 10 health insurance companies each of which has 10,000 clients.
   (a) What is the probability that the maximum number of disease incidents (of the considered type) among the 10 companies is at most 25?
   (b) What is the probability that the second-largest number of disease incidents among these companies is at most 25?
14.5 Let $(X_1, \ldots, X_n)$ be a random sample of a discrete variable X, assuming the values $1, \ldots, N$.
   (a) Show that the relation
   $$h(i,j) = P(B_{i,j}) - P(B_{i,j-1} \cup B_{i+1,j})$$

holds for $1 \le i < j \le N$, where $h(i, j)$ is the joint probability function of (K, M) (see Definition 14.2) and $B_{i,j}$ denotes the event that all sample values are contained in the set $\{i, i+1, \ldots, j\}$ $(i < j)$.

(b) Derive a formula for $h(k, m)$, when X is the discrete uniform distribution, i.e., $P(X = i) = 1/N$ for $i = 1, \ldots, N$.

(c) Determine for the previous distribution the probability function $f(r)$ of the range R.

(d) Assume that a die is thrown n times and that $(X_1, \ldots, X_n)$ are the results.
   (i)   Determine $h(i, j)$ and $f(r)$.
   (ii) Calculate the conditional probabilities

$$P(M = 6|R = 2), P(K = 1|M = 6) \quad \text{and} \quad P(R = 2|M = 4) \text{ for } n = 10.$$

14.6 Consider a random sample from a continuous random variable X with Rayleigh distribution, i.e.,

$$f(x) = \frac{x}{s^2} \exp\left(\frac{-x^2}{2s^2}\right) \quad \text{for } x \ge 0.$$

Determine the pdf of the largest and the second-largest order statistic.

14.7 Suppose that the height X of an adult (in cm) has distribution $N(170, 100)$ and that a random sample of 20 persons has been determined.
   (a) What is the expected height of the tallest person?
   (b) Illustrate the density of the sample maximum graphically.
   (c) What is the probability that all persons are smaller than 190 cm?
   (d) What is the probability that the second largest person is less than 190 cm?

14.8 The wind speed X (measured in km/h) in a certain region follows the Rayleigh distribution with parameter $s = 10$ (see Exercise 14.6).
   (a) Determine the joint pdf of the minimum K and the maximum M of a sample.
   (b) Assume that 20 measurements of X have been performed. Determine the probabilities $P(M \ge 10\,K)$ and $P(M \ge K^2)$.

14.9 Plot the pdf and the cdf of the range of an exponential sample for $n = 20$ and $\alpha = 5$ (see (14.23) and (14.24)).

14.10 Given a sample of size $n = 13$ of the exponential distribution with parameter $\alpha = 5$. Determine the expectations of the sample minimum, the sample maximum and the range. Calculate the arising integrals and check the results by means of simulation.

14.11 Two markers are moving randomly in the plane, taking on the positions (X_1, X_2) and (Y_1, Y_2). Suppose that these four variables are independent and X_1 and X_2 have distribution $N(0, 4)$ while Y_1 and Y_2 have distribution $N(5, 9)$. Determine by means of simulation
   (i)   the expected distance between the markers,
   (ii) the probability that the distance between the markers is at most 2.

14.12 Assume that the weight of an adult (in kg) is randomly distributed as $N(75, 100)$. Determine by means of simulation the expectation of the heaviest and of the second-heaviest weight encountered in a random sample of 50 persons.

14.13 Suppose that a random sample of size two of a variable X with distribution $N(\mu, \sigma^2)$ is chosen. What is the distribution of the range?

14.14 Let $K_r(x, y)$ denote the circle with center $(x, y)^T \in IR^2$ and radius r and let $Q = \{(x, y)^T | 0 \le x, y \le 1\}$ be a square. Determine by means of simulation the area of the region $R = Q \setminus \{K_1(-0.4, 1.4) \cup K_1(1.2, 1.5) \cup K_1(-0.2, -0.4) \cup K_1(1.2, -0.4)\}$. Illustrate the region geometrically.

14.15 Let $B_r(x, y, z)$ denote the ball with center $(x, y, z)^T \in IR^3$ and radius r. Determine by means of simulation the volume of the intersection of balls $B_{1.5}(0, 2, 0) \cap B_2 (0, 0, 0) \cap B_4(4, 0, 0)$.

14.16 For a continuous distribution the n-quantiles are real numbers $Q_1, Q_2, \ldots, Q_{n-1}$ that divide the range space in n parts of equal probability, i.e.,

$$P(X \le Q_1) = P(Q_1 \le X \le Q_2) = \cdots = P(Q_{n-2} \le X \le Q_{n-1}) = P(X \ge Q_{n-1}) = \frac{1}{n}.$$

In particular for $n = 4$, $n = 10$ and $n = 100$ the quantiles are called quartiles, deciles and percentiles, respectively. It is natural to assume that the expectations of order statistics are "in most cases" close to the respective quantiles. In particular, for $n = 4$ one could guess that $E(X_{(i)})$ is "close" to Q_i for $i = 1, 2, 3$ and for, e.g., $n = 43$ the expectations $E(X_{(11)})$, $E(X_{(22)})$ and $E(X_{(33)})$ should be "close" to Q_1, Q_2 and Q_3, respectively.

(a) Let X have the normal distribution $N(75, 100)$. Determine the following expectations of order statistics by means of simulation:
  (i) $E(X_{(1)})$, $E(X_{(2)})$ and $E(X_{(3)})$ for $n = 3$,
  (ii) $E(X_{(11)})$, $E(X_{(22)})$ and $E(X_{(33)})$ for $n = 43$,
  (iii) $E(X_{(38)})$, $E(X_{(76)})$ and $E(X_{(114)})$ for $n = 151$.
  Compare with the corresponding quartiles!

(b) Repeat item (a) for Maxwells distribution with parameter 2.5, for the exponential distribution with $\alpha = 2.5$, and for the uniform distribution over the interval $[0, 1]$.

15 Estimation of parameters

There are numerous situations in science and daily life where one needs to estimate one or more parameters of a distribution. May be that the health ministry is interested in estimating the average weight of the school children of a certain age. A producer of electronic equipment may be interested in the durability of his/her devices to fix adequate prices and prepare warranty claims. Finally, the voters can hardly wait to get predictions of the election result even if they do not really believe that their preferred candidate would change the politics essentially. In any of these cases, the parameters must be estimated based on observations in a sample, as has been suggested in the previous chapter. In some cases, the choice of the estimator seems to be intuitively clear. For example, one expects that the sample mean is an adequate estimator for the population mean. In particular the average weight of a random sample of school children of a certain age group is expected to be close to the average weight of all the children of this age group in the population. However, the estimation of the population variance is already more involved, as we have seen in Section 14.1. In any way one cannot expect that an estimator provides the exact value of the respective parameter, since the estimate depends on chance. No matter how cleverly the sample information might be used, it is not possible to eliminate any kind of uncertainty. Hence, criteria for the quality of an estimator are required.

15.1 Criteria for estimators

We will first formalize some notions as follows.

Definition 15.1: Let X be a random variable with some probability distribution depending on an unknown parameter θ. Let $(X_1, \ldots, X_n)$ be a sample of X. If $\Theta = g(X_1, \ldots, X_n)$ is a function of the sample to be used for estimating θ, then Θ is called an *estimator* of θ. The value of Θ, i.e., $\hat{\theta} = g(x_1, \ldots, x_n)$, where x_i is the value of X_i for $i = 1, \ldots, n$, is called an *estimate* of θ.

Example 15.1: If the weight X of school children has distribution $N(\mu, \sigma^2)$ and $(X_1, \ldots, X_n)$ is a random sample, then $\Theta_1 = (1/n)\Sigma_{i=1}^{n}X_i$ and $\Theta_2 = \left(1/(n-1)\right)\Sigma_{i=1}^{n}(X_i - \bar{X})^2$ are possible estimators for μ and σ^2. If the values of the weights of a sample with $n = 10$ are (50, 52, 53, 58, 62, 55, 50, 63, 48, 53) (in kg), then the corresponding estimates are $\hat{\mu} = 54.4$ and $\hat{\sigma}^2 = 26.044$.

Definition 15.2: Given a r.v. X depending on a parameter θ. An estimator Θ of θ is called *unbiased*, if $E(\Theta) = \theta$.

https://doi.org/10.1515/9783111332277-015

Example 15.2: The estimators of the previous example are both unbiased, since $E(\Theta_1) = \mu$ and $E(\Theta_2) = \sigma^2$ (see Theorems 14.1(i) and 14.2(i)).

"Unbiasedness" means essentially that the average value of the estimate is close to the true parameter value, when it is used repeatedly. Obviously this is a desirable characteristic of an estimator. Another feature that one would like to encounter is a small variance, i.e., the deviations in repeated estimations should be as low as possible.

Definition 15.3: Let Θ be an unbiased estimator of θ. If all unbiased estimators Θ^* of θ satisfy $V(\Theta) \le V(\Theta^*)$, we say that Θ is an *unbiased minimum variance estimator* of θ.

In the ideal case, an estimator is unbiased and has the smallest possible variance. However, if one has to choose between an estimator with large variance and an unbiased one with small variance, one might prefer the latter, when its expectation is "close enough" to the true parameter value. As Fig. 15.1 illustrates, the advantage of an unbiased estimator might be that large estimation errors are avoided.

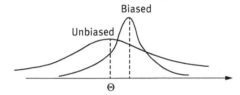

Fig. 15.1: pdf's of a biased and an unbiased estimator.

We also expect that a good estimator works better when the available information, i.e., the sample size increases. This leads us to the following concept.

Definition 15.4: Let Θ be an estimator of θ with value $\hat\theta$, based on a sample $(X_1, \ldots, X_n)$ of size n. The estimator is called *consistent*, if $\hat\theta$ converges in probability to θ for $n \to \infty$, i.e., if

$$\lim_{n\to\infty} P\left(\left|\hat\theta - \theta\right| > \varepsilon\right) = 0 \quad \text{for all } \varepsilon > 0$$

(see (13.5)).

Without a proof we mention the following sufficient condition for the consistency.

Theorem 15.1: Given an estimator Θ as in Definition 15.4. If $\lim_{n\to\infty} E(\Theta) = \theta$ and $\lim_{n\to\infty} V(\Theta) = 0$, then Θ is a consistent estimator for θ.

Example 15.3: A small business owner has launched a new article on the market. To estimate the probability p that a defective article is produced, a sample $(X_1, \ldots, X_n)$ of n articles has been inspected, where $X_i = 1$, if the i-th item is defective and $X_i = 0$ otherwise. If k is the number of defective articles of the sample, the most suggestive estimate for p is the proportion $\hat{p} = k/n$. Let us apply the above criteria to evaluate this approach. Since the number of defective items obeys the binomial distribution, it holds

$$E(\hat{p}) = \frac{1}{n}np = p, \ V(\hat{p}) = \frac{1}{n^2}np(1-p) = \frac{p(1-p)}{n}, \tag{15.1}$$

thus the estimator is unbiased and consistent, since $\lim_{n \to \infty} V(\hat{p}) = 0$.

The following consideration shows that an estimator with these characteristics need not be a "good" one. Theoretically one could define an estimator p^* by setting $p^* = \frac{k^*}{n^*}$, where $n^* < n$ and k^* denotes the number of defective items of the "subsample" $(X_1, \ldots, X_{n*})$. Analogous to (15.1) we obtain

$$E(p^*) = p, \ V(p^*) = \frac{p(1-p)}{n^*}. \tag{15.2}$$

Hence, this estimator is also unbiased and consistent (if one guarantees that n^* increases indefinitely), but obviously the variance of p^* is larger than that of $\hat{p}$ and the former is not a good estimator in the sense of Definition 15.3.

As a final criterion to judge about estimators we introduce the following concept.

Definition 15.5: An estimator of a parameter θ, having the form $\Theta = \sum_{i=1}^{n} a_i X_i$, is called a *linear estimator*. An unbiased linear estimator Θ^* (of θ) is called a *best linear unbiased estimator*, if $V(\Theta^*) \leq V(\Theta)$ holds for any unbiased linear estimator Θ. An estimator that is "good" with respect to all of the above criteria is given by the sample mean. This is summarized in the following statement.

Theorem 15.2: Consider a random variable X with finite expectation μ and variance σ^2. Let $\bar{X} = (1/n)(X_1 + \cdots + X_n)$ be the sample mean based on a random sample of size n. Then $\bar{X}$ is an unbiased, consistent and a best linear unbiased estimator of μ.

Proof: From Theorems 14.1 and 15.1 we get immediately that $\bar{X}$ is unbiased and consistent. We still need to prove the third characteristic:

A linear estimator $\sum_{i=1}^{n} a_1 X_i$ is unbiased, if

$$E\left(\sum_{i=1}^{n} a_i X_i \right) = \sum_{i=1}^{n} a_i E(X_i) = \mu \sum_{i=1}^{n} a_i = \mu, \quad \text{i.e.,} \quad \sum_{i=1}^{n} a_i = 1.$$

Since the variance of such an estimator is

$$V\left(\sum_{i=1}^{n} a_i X_i \right) = \sum_{i=1}^{n} a_i^2 V(X_i) = \sigma^2 \sum_{i=1}^{n} a_i^2,$$

we have to minimize $\sum_{i=1}^{n} a_i^2$ subject to

$$\sum_{i=1}^{n} a_i = 1. \tag{15.3}$$

However, we can write

$$\sum_{i=1}^{n} a_i^2 = \sum_{i=1}^{n} \left(a_i - \frac{1}{n} \right)^2 + \frac{2}{n} \sum_{n=1}^{n} a_i - n \left(\frac{1}{n} \right)^2,$$

and due to restriction (15.3) we get

$$\sum_{i=1}^{n} a_i^2 = \sum_{i=1}^{n} \left(a_i - \frac{1}{n} \right)^2 + \frac{1}{n},$$

which is minimized if $a_i = 1/n$ for $i = 1, \ldots, n$.

It should be emphasized that a sample proportion $\hat{p}$ is a specific case of a sample mean $\bar{X}$. For instance, the proportion of defective items in Example 15.3 can be written as a sample mean $\hat{p} = \bar{X} = (1/n)\sum_{i=1}^{n} X_i$.

Hence, the desirable properties of Theorem 15.2 are in particular valid, if the proportion of a population is estimated by the corresponding statistic $\hat{p}$.

At this point we mention that the estimator Θ_2 for the variance (Example 15.2) also units good properties. Besides being unbiased it is also consistent, which we mention here without proof.

Example 15.4: Suppose that the time to failure, T, of an electronic component is exponentially distributed with density $f(t) = ae^{-at}$ for $t \geq 0$. Since $E(T) = 1/a$, it is reasonable to estimate $1/a$ by the mean value of a sample, i.e., if $(T_1, \ldots, T_n)$ are the times to failure of n tested components we use $\bar{T} = (1/n)\sum_{i=1}^{n} T_i$ as an estimator for the parameter $\beta = 1/a$. We have already seen that this estimator is unbiased and consistent. However, another possibility is to use $K = \min(T_1, \ldots, T_n)$ as an estimator for β. According to Example 14.4, K has the density $g(k) = a n e^{-a n k}$ and $E(K) = 1/a n$. Thus, the estimator nK is also an unbiased estimator for $\beta = 1/a$. But $V(nK) = n^2 V(K) = n^2 \left(1/(a n)^2 \right) = 1/a^2$ is larger than $V(\bar{T}) = V(T)/n^2 = (1/a^2 n^2)$; hence, the former estimator $\bar{T}$ should be preferred in view of Definition 15.3.

However, one could introduce further decision criteria in relation to which the second estimator has advantages. First, when applying the estimator nK, the test may be terminated as soon as the first component of the sample has failed. In case of the estimator $\bar{T}$, one must wait until all components have failed, which can require considerably more time. Hence, when the available time for testing is limited, the estimator nK might be an interesting option. Another advantage of using this estimator might be that only one component of the sample must be destroyed, while the entire sample must be sacrificed when using the estimator $\bar{T}$.

Example 15.5: In Tab. 15.1 the emission of alpha particles of a radioactive source is recorded, where k denotes the number of particles observed in a time unit and n_k the number of intervals in which k particles have been observed. Assuming that the number of particles X, emitted during a time unit obeys a Poisson distribution, it follows that $P(X = k) = e^{-\lambda}\frac{\lambda^k}{k!}$ for $k = 0, 1, 2, \ldots$ and $E(X) = \lambda$. By setting the expectation equal to the sample mean, we obtain the estimate

$$\hat{\lambda} = \bar{x} = \frac{\sum_{k=10}^{10} k n_k}{\sum_{k=0}^{10} n_k} = \frac{10{,}015}{2{,}600} \approx 3.85.$$

Tab. 15.1: Emission of radiation.

k	0	1	2	3	4	5	6	7	8	9	10
n_k	50	200	400	500	550	385	300	150	50	10	5

After the above considerations it should be clear to the reader that the theoretical proof of estimator properties may be a complicated task. An interesting option in this case is to study the expectation, variance and other characteristics by means of simulation (see Exercise 15.11).

15.2 Maximum likelihood estimation

Above we have considered criteria by which we may judge an estimator. However, we do not have yet a general procedure available to find an estimate. There exist several approaches for this purpose. We start with one of the most popular methods.

Example 15.6: Let us reconsider Example 15.3, in which we want to estimate the probability that any of the produced articles is defective. We will use again the sample $(X_1, \ldots, X_n)$ of size n in which k items are defective. However, we want to approach the problem in another way: It is known that a specific sequence of k defective and $n-k$ perfect articles has been produced. If p denotes the *unknown* probability that an article is defective, then the probability that the observed sequence occurs is given by the *likelihood function*:

$$L(p) = P(K = 0)^k P(X = 1)^{n-k} = p^k (1-p)^{n-k}. \tag{15.4}$$

Since the considered event has in fact occurred, there is reason to believe that probability (15.4) is "high." Therefore, one simply estimates the parameter p by the value $\hat{p}$ that maximizes $L(p)$. To this end, we set the derivative of L equal to zero:

$$L'(p) = kp^{k-1}(1-p)^{n-k} - p^k(n-k)(1-p)^{n-k-1} = 0.$$

Dividing the last equation by $p^{k-1}(1-p)^{n-k-1}$ simplifies the equation to

$$k(1-p) - p(n-k) = 0 \Leftrightarrow k - pn = 0 \Leftrightarrow \hat{p} = \frac{k}{n}.$$

In this case, where we sample from the Bernoulli variable X, the estimate is the same as obtained previously.

In general, a random variable X and sample values $(x_1, \ldots, x_n)$ are given.

The variable X *depends on a parameter* θ (which may be a scalar or a vector). We define the likelihood function by

$$L(\theta) = P(X = x_1, \theta) \cdot \cdots \cdot P(X = x_n, \theta)$$

when X is discrete and by

$$L(\theta) = f(x_1, \theta) \cdot \cdots \cdot f(x_n, \theta)$$

when X is continuous, where f denotes the pdf of X. All factors as well as the function $L(\theta)$ depend on the sample *and* on the parameter. But occasionally one simply writes $P(X = x_i)$ instead of $P(X = x_i, \theta)$ and $f(x_i)$ instead of $f(x_i, \theta)$. The estimate $\hat{\theta}$ of θ is given by that value that maximizes the function $L(\theta)$. Rather than maximizing the likelihood function itself, one can also maximize the log likelihood function $l(\theta) = \ln(L(\theta))$, because the logarithm is a monotone increasing function that ensures that the relations

$$L\left(\hat{\theta}\right) \leq L(\theta) \text{ for all } \theta,$$

and

$$\ln(L\left(\hat{\theta}\right)) \leq \ln(L\left(\hat{\theta}\right)) \text{ for all } \theta$$

are equivalent. The above procedure, called *maximum likelihood method* (*ML method*), leads in many cases to reasonable estimates, though sometimes the maximization may be computationally expensive or the estimators may not possess all desirable properties. We illustrate the method by means of some examples.

Example 15.7: Suppose that X has a Poisson distribution, i.e., $P(X = k) = e^{-\lambda}\left(\lambda^k / k!\right)$, for $k = 0, 1, 2, \ldots$ (see Example 15.5). The likelihood function is therefore

$$L(\lambda) = \frac{e^{-\lambda}\lambda^{x_1}}{x_1!} \cdot \cdots \cdot \frac{e^{-\lambda}\lambda^{x_n}}{x_n!}.$$

Hence,

$$l(\lambda) = \ln(L(\lambda)) = [-\lambda + x_1 \ln \lambda - \ln(x_1!)] + \cdots + [-\lambda + x_n \ln \lambda - \ln(x_n!)],$$

and the derivative is

$$l'(\lambda) = \left[-1 + \frac{x_1}{\lambda}\right] + \cdots + \left[-1 + \frac{x_n}{\lambda}\right] = -n + \frac{1}{\lambda}(x_1 + \cdots + x_n) = -n + \frac{n\bar{x}}{\lambda},$$

where $\bar{x} = (1/n)\Sigma_{i=1}^{n}x_i$ denotes the sample mean. Setting this equal to zero yields $(n\,\bar{x}/\lambda) = n \Rightarrow \hat{\lambda} = \bar{x}$. Thus, the estimate is the same as in Example 15.5.

Example 15.8: Assume that X is exponentially distributed. The pdf is then $f(x) = ae^{-ax}$ for $x \geq 0$ and the likelihood function is

$$L(a) = ae^{-a\,x_1} \cdot \cdots \cdot ae^{-a\,x_n} = a^n e^{-a\,n\,\bar{x}}$$

Thus,

$$l(a) = \ln L(a) = n \ln(a) - a\,n\,\bar{x} \Rightarrow l'(a) = \frac{n}{a} - n\,\bar{x}$$

Setting the latter equal to zero yields $\bar{a} = 1/\bar{x}$, which is the same estimator as in Example 15.4.

In this context, it is appropriate to mention that the *ML* estimates possess a very important characteristic, known as *invariance property*: If $\hat{\theta}$ is the *ML* estimate of θ, then $h(\hat{\theta})$ is the *ML* estimate of $h(\theta)$, where h is a biunique function.

In particular, if $\bar{x}$ is the *ML* estimate of $\beta = 1/a$, then $1/\bar{x}$ is the *ML* estimate of a. Moreover, *ML* estimates are consistent under rather general conditions. But they may be biased as we will see e.g. in Example 15.11.

Example 15.9: Assume that X has a gamma distribution with $a = 1$ and unknown parameter r, i.e., the pdf is $f(x) = \frac{1}{\Gamma(r)} x^{r-1} e^{-x}$ for $x \geq 0$.

We obtain

$$L(r) = \frac{1}{\Gamma(r)^n} (x_1 \cdot \cdots \cdot x_n)^{r-1} e^{-(x_1 + \ldots + x_n)} \Rightarrow$$

$$l(r) = \ln(L(r)) = -n\ln\Gamma(r) + (r-1)\ln(x_1 \cdot \cdots \cdot x_n) - (x_1 + \cdots + x_n) \Rightarrow$$

$$l'(r) = -n\frac{\Gamma'(r)}{\Gamma(r)} + \ln(x_1 \cdot \cdots \cdot x_r).$$

Setting the latter equal to zero yields

$$\frac{\Gamma'(r)}{\Gamma(r)} = \frac{\ln(x_1 \cdot \,\cdots\, \cdot x_n)}{n} \tag{15.5}$$

where the left-hand side is the Psi function $\Psi(x)$ (see (1.17)). Since Ψ is monotone decreasing, (15.5) provides a unique estimate $\hat{r}$.

If, for instance, the values of a sample of size $n = 10$ are (5, 6, 5, 7, 5.5, 6, 7, 7.5, 6.5, 6), we get $\ln(x_1 \,\cdots\, x_{10})/10 \approx 1.8077$; hence,

$$\hat{r} = \Psi^{-1}(1.8077) \approx 6.59.$$

In the following example, the maximum value of L cannot be found by simply differentiating.

Example 15.10: We now suppose that X is uniformly distributed over the interval $[0, b]$, where b is an unknown parameter. The pdf is given by $f(x) = 1/b$ for $0 \le x \le b$ and $f(x) = 0$ elsewhere. For sample values $(x_1, \ldots, x_n)$ the likelihood function is given as

$$L(b) = \begin{cases} \frac{1}{b^n} & \text{if } 0 \le x_i \le b \quad \text{for all } i, \\ 0 & \text{otherwise.} \end{cases}$$

In the first case that is equivalent to $b \ge \max(x_1, \ldots, x_n)$, the function $L(b)$ is monotone decreasing. Therefore, the maximum value is obtained for $\hat{b} = \max(x_1, \ldots, x_n)$. We will study some properties of this estimate. According to (14.11) the pdf of this estimator is

$$g(\hat{b}) = n f(\hat{b}) F(\hat{b})^{n-1} = n \frac{1}{b} \left(\frac{\hat{b}}{b}\right)^{n-1} = n \frac{\hat{b}^{n-1}}{b^n} \quad \text{for } 0 \le \hat{b} \le b,$$

resulting in the expectation

$$E(B) = \frac{n}{b^n} \int_0^b \hat{b}^{n-1} d\hat{b} = \frac{n}{b^n} \frac{b^{n+1}}{n+1} = \frac{n}{n+1} b \ne b.$$

Hence, the estimator is not unbiased, but it holds

$$\lim_{n \to \infty} E(B) = b. \tag{15.6}$$

Moreover, one can show that

$$V(B) = \frac{n\,b^2}{(n+2)(n+1)}.$$

(15.7)

From (15.6) and (15.7) it follows that the estimator is consistent.

We finally demonstrate the method for r.v.'s with two parameters.

Example 15.11: We want to estimate the two parameters μ and σ of a normally distributed random variable. Let again a sample $(x_1,..,x_n)$ be given. The likelihood function can be written as

$$L(\mu,\,\sigma) = \prod_{i=1}^{n}\left[(2\pi\sigma^2)^{-1/2}\exp\left(-\frac{1}{2}\left(\frac{x_i-\mu}{\sigma}\right)^2\right)\right]$$

Hence,

$$l(\mu,\sigma) = \ln L(\mu,\sigma) = \sum_{i=1}^{n}\left[-\frac{1}{2}\ln(2\pi\sigma^2) - \frac{1}{2}\left(\frac{x_i-\mu}{\sigma}\right)^2\right]$$

$$= -\frac{n}{2}\ln(2\pi\sigma^2) - \frac{1}{2\sigma^2}\sum_{i=1}^{n}(x_i-\mu)^2.$$

We must set both partial derivatives of l equal to zero:

$$\frac{\partial l}{\partial \mu} = -\frac{1}{\sigma^2}\sum_{i=1}^{n}(x_i-\mu)^2 = 0$$

yields $\hat{\mu} = \bar{x}$, and

$$\frac{\partial l}{\partial \sigma} = -\frac{n}{\sigma} + \frac{1}{\sigma^3}\sum_{i=1}^{n}(x_i-\mu)^2 = 0$$

yields

$$\hat{\sigma}^2 = \frac{1}{n}\sum_{i=1}^{n}(x_i-\mu)^2 = \frac{1}{n}\sum_{i=1}^{n}(x_i-\bar{x})^2.$$

We observe that the estimator for σ^2 is biased. We have already seen that in the case of the unbiased estimator one must divide by $(n-1)$.

Example 15.12: We now estimate the two parameters of a gamma distribution, assuming that sample values $(x_1, \ldots, x_n)$ are available. In generalization of Example 15.9, we have the pdf $f(x) = \frac{\alpha}{\Gamma(r)}(\alpha\,x)^{r-1}e^{-\alpha x}$ for $x \geq 0$, where α and r are unknown positive parameters. We obtain

$$L(a, r) = \frac{a^{n\,r}}{\Gamma(r)^n} (x_1 \cdot \, \ldots \, \cdot x_n)^{r-1} e^{-a(x_1 + \ldots + x_n)} \Rightarrow$$

$$l(a, r) = nr \ln a - n \ln(\Gamma(r)) + (r-1)\ln(x_1 \cdot \, \ldots \, \cdot x_n) - a(x_1 + \ldots + x_n).$$

The partial derivatives are

$$\frac{\partial l}{\partial a} = \frac{nr}{a} - n\bar{x} \tag{15.8}$$

and

$$\frac{\partial l}{\partial r} = n \ln a - n\psi(r) + \ln(x_1 \cdot \, \ldots \, \cdot x_n). \tag{15.9}$$

Setting these expressions equal to zero, we get immediately

$$\hat{a} = \frac{r}{\bar{x}}. \tag{15.10}$$

By substituting this for a in (15.9), we obtain

$$n \ln\left(\frac{r}{\bar{x}}\right) - n\psi(r) + \ln(x_1 \cdot \, \ldots \, \cdot x_n) = 0$$

which is equivalent to

$$\ln r - \psi(r) = \ln \bar{x} - \frac{\ln(x_1 \cdot \, \ldots \, \cdot x_n)}{n} \tag{15.11}$$

The estimates are now obtained by solving (15.11) for r, yielding $\hat{r}$ and setting then

$$\hat{a} = \frac{\hat{r}}{\bar{x}}.$$

When the density is complicated or when many parameters exist, a closed-form representation of the estimates is usually not possible. In such a case a numerical method is necessary to maximize the (log) likelihood function.

We conclude this section with a general statement on an *asymptotic property of maximum likelihood estimates*.

Theorem 15.3: If Θ is an *ML* estimator of a parameter θ with value $\hat{\theta}$, based on a sample $(X_1, \ldots, X_n)$ of size n, then for n sufficiently large, Θ has approximately the distribution $n(\theta, 1/t)$, where

$$t = nE\left[\frac{\partial}{\partial \theta} \ln f(X; \theta)\right]^2. \tag{15.12}$$

Here the function f denotes the probability function or the pdf of X, depending on whether X is discrete or continuous.

This property specifies the probabilistic behavior of the *ML* estimator and is therefore considerably stronger then the property of consistency that merely states that the estimate is "close" to θ for sufficiently large n.

Example 15.13: We now analyze the above characteristic for the *ML* estimate of the parameter a of the exponential distribution. Using (15.12), we get

$$\frac{\partial}{\partial a} \ln(a \, e^{-a\,x}) = \frac{\partial}{\partial a} (\ln a - ax) = \frac{1}{a} - x,$$

thus

$$E\left(\frac{1}{a} - x\right)^2 = \int_0^\infty \left(\frac{1}{a} - x\right)^2 a \, e^{-a\,x} dx = \frac{1}{a^2} \Rightarrow t = \frac{1}{a^2},$$

and the approximate distribution of the estimate is

$$N\left(a, \frac{a^2}{n}\right).$$

15.3 Method of moments

This is another universal method to estimate parameters of a distribution. The basic idea is to equate the moments of the distribution (which are assumed to be functions of the parameters) with the corresponding sample moments. Suppose that the probability function or the pdf of a random variable has r unknown parameters $\theta_1, \ldots, \theta_r$ and that the moments about the origin (see Definition 8.4) are given as $\mu'_k = E(X^k) = g_k(\theta_1, \ldots, \theta_r)$ for $k = 1, \ldots, r$. Furthermore, let sample values $(x_1, \ldots, x_n)$ be given with the sample moments $m_k = (1/n)(x_1^k + \cdots + x_n^k)$ for $k = 1, \ldots, r$. Then one tries to solve the system of equations

$$g_1(\theta_1, \ldots, \theta_r) = m_1$$
$$\vdots \qquad\qquad \vdots \qquad\qquad\qquad (15.13)$$
$$g_r(\theta_1, \ldots, \theta_r) = m_r$$

in the variables $\theta_1, \ldots, \theta_r$.

For a small number of parameters, (15.13) can be solved by hand, and otherwise a numerical procedure based on the Newton–Raphson method can be applied. In many cases this procedure is easier than the maximum likelihood method. The estimators are consistent under fairly weak assumptions, but they are often biased. We illustrate the method of moments by means of some examples.

In the case of the Poisson distribution (see Examples 15.5 and 15.7), system (15.13) becomes

$$E(X) = \lambda = m_1 = \bar{x}, \text{ implying } \hat{\lambda} = \bar{x}$$

and the estimate is then the same as previously. Similarly, for the exponential distribution (see Examples 15.4 and 15.8), we obtain

$$E(X) = \frac{1}{\alpha} = m_1 = \bar{x}, \text{ implying } \hat{\alpha} = \frac{1}{\bar{x}}$$

which also agrees with our previous findings.

Example 15.14: Consider the pdf

$$f(x) = \theta x^{\theta-1} \quad \text{for } 0 \le x \le 1, \ \theta > 0.$$

One can easily verify that the expectation is

$$E(X) = \frac{\theta}{\theta+1}.$$

System (15.13) is now

$$\frac{\theta}{\theta+1} - \bar{x} \rightarrow \frac{\theta+1-1}{\theta+1} = 1 - \frac{1}{\theta+1} = \bar{x} \Rightarrow \hat{\theta} = \frac{\bar{x}}{1-\bar{x}},$$

If the sample values are, e.g. $(x_1, \ldots, x_5) = (0.1, 0.2, 0.2, 0.4, 0.6)$, we get

$$\bar{x} = 0.3 \Rightarrow \hat{\theta} = \frac{3}{7}.$$

Example 14.15: The Rayleigh distribution has the pdf

$$f(x) = \frac{x \exp\left(-\frac{x^2}{2s^2}\right)}{s^2} \quad \text{for } x \ge 0, \ s^2 > 0$$

and expectation

$$E(X) = s\sqrt{\frac{2}{\pi}} \tag{15.14}$$

and we obtain

$$s\sqrt{\frac{2}{\pi}} = \bar{X} \Rightarrow \hat{s} = \bar{X}\sqrt{\frac{\pi}{2}}.$$

In particular, the estimator $\hat{s}$ is unbiased, since

$$E(\hat{s}) = E\left(\bar{X}\ \sqrt{\frac{\pi}{2}}\right) = \sqrt{\frac{\pi}{2}}E(X) = s,$$

where the last equation follows from (15.14).

Let as now consider some distributions with two parameters.

For $r = 2$, system (15.13) becomes

$$E(X) = m_1,$$
$$E(X^2) = m_2,$$

(15.15)

and this is of course equivalent to

$$E(X) = m_1 = \bar{X},$$
$$V(X) = S^2,$$

(15.16)

where S^2 denotes the biased sample variance, defined by

$$S^2 = \frac{1}{n}\sum_{i=1}^{n}(x_i - \bar{x})^2.$$

Form (15.16) is advantageous, if a formula for the variance is available from the literature. It is immediately clear that the moment estimators for the normal distributions are the same as the *ML* estimators (see Example 15.11).

Example 15.16: For the gamma distribution, we obtain

$$f(x) = \frac{a}{\Gamma(r)}(ax)^{r-1}e^{-ax}\quad\text{for } x \geq 0; a, r > 0,$$

$$E(X) = \frac{r}{a}, \quad V(X) = \frac{r}{a^2}.$$

Thus, system (15.16) has the form

$$\frac{r}{a} = \bar{X}.$$

$$\frac{r}{a^2} = S^2.$$

Dividing the first equation by the second, we obtain

$$\hat{a} = \frac{\bar{x}}{S^2} \quad \text{and} \quad \hat{r} = \bar{x}\hat{a}.$$

Example 15.17: For the uniform distribution, we get

$$f(x) = \frac{1}{b-a} \quad \text{for } a \le x \le b,$$

$$E(X) = \frac{a+b}{2}, \quad V(X) = \frac{(b-a)^2}{12},$$

and system (15.16) becomes

$$\frac{a+b}{2} = \bar{x},$$

$$\frac{(b-a)^2}{12} = S^2.$$

From this we obtain

$$a + b = 2\bar{x},$$

$$b - a = \sqrt{12S^2},$$

and finally the estimates are

$$\hat{a} = \bar{x} - \sqrt{3S^2},$$
$$\hat{b} = \bar{x} + \sqrt{3S^2}.$$

Example 15.18: The Weibull distribution has the pdf

$$f(x) = a\, b\, x^{b-1} \exp(-a\, x^b) \quad \text{for } x \ge 0;\, a, b > 0,$$

satisfying

$$E(X) = \frac{\Gamma(1/b)}{b\, a^{1/b}}, \quad E(X^2) = \frac{2\Gamma(2/b)}{b\, a^{2/b}}$$

and system (15.15) becomes

$$\frac{\Gamma(1/b)}{b\, a^{1/b}} = \bar{x},$$

$$\frac{2\Gamma(2/b)}{b\, a^{2/b}} = m_2.$$

Isolating the term $a^{1/b}$ results in

$$a^{1/b} = \frac{\Gamma(1/b)}{b\,m_1},$$

$$a^{2/b} = \frac{2\Gamma(2/b)}{b\,m_2},$$ (15.17)

yielding

$$\frac{\Gamma(1/b)^2}{b^2\,m_1^2} = \frac{2\Gamma(2/b)}{b\,m_2} \Rightarrow$$

$$\frac{\Gamma(1/b)^2}{b\,\Gamma(2/b)} = \frac{2\,m_1^2}{m_2}.$$

With the notation $c := 1/b$ we obtain finally

$$\frac{c\,\Gamma(c)^2}{\Gamma(2c)} = \frac{2m_1^2}{m_2}$$ (15.18)

where the left-hand side is a monotone decreasing function in c from which the value of c can be uniquely determined.

Let $\hat{c}$ denote the solution of (15.18), then the estimates of the Weibull distribution are given as (see (15.17))

$$\hat{b} = 1/\hat{c},$$

$$\hat{a} = \left(\frac{\Gamma\left(1/\hat{b}\right)}{\hat{b}\,m_1}\right)^{\hat{b}}.$$ (15.19)

15.4 Linear regression

We now study the estimation of parameters when a r.v. Y can be expressed as a (linear) function of another variable X. The branch of statistics that estimates the relationship between variables is called *regression analysis*. We start with an illustrative example.

Example 15.19: It is a well-known fact that the temperature decreases with the altitude of the locality. The fact is illustrated in Tab. 15.2 and the corresponding scatter plot in Fig. 15.2. It is assumed that the measurements were all carried out at the same time by meteorological stations in different altitudes. (The values are fictive, but basically reflect a simplified realistic situation.) The plotted points suggest that the temperature Y decreases linearly with the altitude X.

Tab. 15.2: Temperature in dependence of the height.

Station	X (height, m)	Y (temperature, °C)
1	250	15
2	300	11
3	450	12
4	650	9
5	700	8
6	900	7
7	1,200	7
8	1,500	4

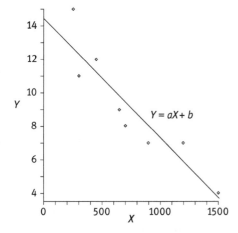

Fig. 15.2: Scatterplot with linear model.

To establish a reasonable model for these data, we assume that Y depends on both X and random fluctuations. Specifically, we suppose that

$$Y = aX + b + \varepsilon, \tag{15.20}$$

where a and b are unknown parameters, called slope and intercept, X is the *known* altitude at which Y is measured, and ε is a random variable.

In general the variables X and Y in (15.20) are called the *independent* and the *dependent* variable, respectively. This relation is known as a *simple linear regression model* in which we assume that the random component satisfies $E(\varepsilon) = 0$ and $V(\varepsilon) = \sigma^2$ for all X. It follows that $E(Y) = aX + b$ and $V(Y) = \sigma^2$, i.e., expectation and variance of ε do not depend on the value of X. Note that the model depends on the three parameters a, b and ε. Before we can define estimates, we must first make clear what we mean by a random sample in the present context. Suppose that n values of X are chosen, say $x_1, \ldots, x_n$ (As already stated above, X is not a random variable here.) For each x_i let y_i be an observation of the r.v. Y; hence $(x_1, y_1), \ldots, (x_n, y_n)$ may be considered as a random sample of Y for given values $x_1, \ldots, x_n$ of X.

Definition 15.6: Given a random sample $(x_1, y_1), \ldots, (x_n, y_n)$ of the r.v. Y in (15.20). The *least square estimates* of the parameters a and b are the values that minimize the sum of squared errors:

$$S(a, b) = \sum_{i=1}^{n} [y_i - (ax_i + b)]^2. \tag{15.21}$$

between the *observed values* y_i and the corresponding *expected values* $ax_i + b$ of Y. Expression (15.21) measures in an intuitive manner, how well the points of a scatter plot are adjusted by the theoretical straight line $y = ax + b$. To minimize the convex function S, we set the partial derivatives equal to zero, yielding

$$\frac{\partial S}{\partial a} = 2\sum_{n}^{i=1} [y_i - (ax_i + b)](-x_i) = 0,$$
$$\frac{\partial S}{\partial b} = 2\sum_{n}^{i=1} [y_i - (ax_i + b)](-1) = 0, \tag{15.22}$$

or equivalently

$$a\sum_{i=1}^{n} x_i^2 + b\sum_{i=1}^{n} x_i = \sum_{i=1}^{n} x_i y_i,$$
$$a\sum_{n}^{i=1} x_i + nb = \sum_{n}^{i=1} y_i. \tag{15.23}$$

These equations are called *normal equations*. System (15.23) can be solved by direct elimination. It can also be transformed into the closed-form representation of the estimates:

$$\hat{a} = \frac{S_{xy}}{S_{xx}},$$
$$\hat{b} = \bar{y} - \hat{a}\,\bar{x}, \tag{15.24}$$

employing the frequently used notations

$$S_{xy} = \sum_{i=1}^{n} y_i(x_i - \bar{x}) = \sum_{i=1}^{n} x_i y_i - n\,\bar{x}\,\bar{y}, \tag{15.25}$$

$$S_{xx} = \sum_{i=1}^{n} (x_i - \bar{x})^2 = \sum_{i=1}^{n} x_i^2 - n\,\bar{x}^2 = \sum_{i=1}^{n} (x_i - \bar{x})x_i, \tag{15.26}$$

and $\bar{x} = \frac{1}{n}\sum_{i=1}^{n} x_i, \bar{y} = \frac{1}{n}\sum_{i=1}^{n} y_i$. System (15.24) has always a unique solution, excluding the trivial case in which all the x_i's are equal.

For the above example we obtain

$$n = 8, \quad \bar{x} = 743.75, \quad \bar{y} = 9.125, \quad S_{xy} = -9693.75, \quad S_{xx} \approx 1.342 \cdot 10^6$$
$$\hat{a} \approx -0.0072, \quad \hat{b} \approx 14.5.$$

The corresponding theoretical straight line $Y = -0.0072X + 14.5$ is indicated in Fig. 15.2. From (15.24) it follows immediately that the theoretical line $Y = aX + b$ always passes through the point $(\bar{x}, \bar{y})$.

Without proof we mention that the estimate of the variance of the error term in (15.20) can be expressed as

$$\frac{1}{n-2} S(\hat{a}, \hat{b}) \tag{15.27}$$

where S is the function in (15.21).

We now study some properties of the estimators. One obtains

$$E(S_{xy}) = E\left[\sum_{i=1}^{n}(x_i - \bar{x})(ax_i + b + \varepsilon_i)\right] = a\sum_{i=1}^{n}(x_i - \bar{x})x_i + b\sum_{i=1}^{n}(x_i - \bar{x}) + E\left(\sum_{i=1}^{n}(x_i - \bar{x})\varepsilon_i\right) = aS_{xx}$$

using (15.2) and observing that all elements of the last sum are zero due to the assumption on the random term in the model (14.20), Thus,

$$E(\hat{a}) = E\left(\frac{S_{xy}}{S_{xx}}\right) = \frac{1}{S_{xx}} a S_{xx} = a,$$

and $\hat{a}$ is unbiased. For the variance, we obtain

$$V(\hat{a}) = V\left(\frac{S_{xy}}{S_{xx}}\right) = \frac{1}{S_{xx}^2} V\left(\sum_{i=1}^{n} y_i(x_i - \bar{x})\right).$$

Since y_i are uncorrelated, having the same variance σ^2, the variance of the sum is the sum of the variances, i.e.,

$$V(\hat{a}) = \frac{1}{S_{xx}^2}\sum_{i=1}^{n} V(y_i(x_i - \bar{x})) = \frac{1}{S_{xx}^2}\sum_{i=1}^{n}(x_i - \bar{x})^2 \sigma^2 = \frac{1}{S_{xx}^2}\sigma^2 S_{xx} = \frac{\sigma^2}{S_{xx}}. \tag{15.28}$$

In a similar way, it can be shown that the estimate for b satisfies

$$E(\hat{b}) = b, \quad V(\hat{b}) = \sigma^2\left(\frac{1}{n} + \frac{\bar{x}^2}{S_{xx}}\right). \tag{15.29}$$

From (15.24), it follows immediately that the estimate $\hat{a}$ is a linear function of the sample values y_i. One can show that this also holds for the estimate $\hat{b}$.

We still investigate some modifications and generalizations of the linear regression model.

Example 15.20: An engineer studies the relation between the resistance of a thread and its diameter. The results of 10 measurements are given in Tab. 15.3.

Tab. 15.3: Diameter and resistance.

Item i	1	2	3	4	5	6	7	8	9	10
X (diameter, mm)	1.2	1.5	1.7	2.0	2.6	3.0	4.0	4.8	5.2	6.0
Y (applied force, kg)	6.0	6.3	9.0	7.9	11.7	12.5	16.5	21.0	21.3	23.1

The corresponding scatter plot (the reader is recommended to create it) suggests that Y increases linearly with X. However, in contrast to the previous example, it is now natural to employ a model of the type

$$Y = aX + \varepsilon \tag{15.30}$$

since for trivial reasons the resistance is zero if the diameter is zero, i.e., the expected resistance is a line passing through the origin of the coordinate system.

The sum of squared errors is now a convex function in one variable

$$S(a) = \sum_{i=1}^{n} [y_i - ax_i]^2. \tag{15.31}$$

Setting the derivative equal to zero yields

$$S'(a) = 2\sum_{i=1}^{n}(y_i - ax_i)(-x_i) = 0 \Leftrightarrow a\sum_{i=1}^{n} x_i^2 = \sum_{i=1}^{n} x_i y_i \tag{15.32}$$

i.e., the estimate of the parameter is given by

$$\hat{a} = \frac{\sum_{i=1}^{n} x_i y_i}{\sum_{i=1}^{n} x_i^2}. \tag{15.33}$$

For the above example, we get $\hat{a} \approx 4.14$.

Evidently model (15.20) can be generalized in two directions. Firstly, one may try to adjust the points in the scatter plot by means of a parabola or a polynomial of higher degree instead of using a linear model, and secondly one may incorporate several independent variables into the model.

We illustrate these possibilities by means of two examples.

Example 15.21: Assume that the random sample in Tab. 15.4 was obtained. The corresponding scatter plot in Fig. 15.3 clearly reveals that the data points cannot be adjusted by means of a straight line, but a quadratic model seems to be adequate.
We consider the model

$$Y = a + bX + cX^2 + \varepsilon, \tag{15.34}$$

Tab. 15.4: Hypothetical observed data.

i	1	2	3	4	5	6	7	8	9	10
x_i	-4	-3	-2	-1.5	-0.5	0	1	2.5	4	5
y_i	15	10	3.5	2	0.5	0.5	1	6	14	17

resulting in the three-dimensional function of squared errors:

$$S(a, b, c) = \sum_{i=1}^{n} [y_i - (a + bx + cx^2)]^2. \tag{15.35}$$

We minimize this convex function by setting the three partial derivatives equal to zero, yielding

$$\frac{S}{\partial a} = 2 \sum_{i=1}^{n} [a + bx_i + cx_i^2 - y_i] = 0,$$

$$\frac{\partial S}{\partial b} = 2 \sum_{i=1}^{n} [a + bx_i + cx_i^2 - y_i] x_i = 0, \tag{15.36}$$

$$\frac{\partial S}{\partial c} = 2 \sum_{i=1}^{n} [a + bx_i + cx_i^2 - y_i] x_i^2 = 0.$$

It is not difficult to transform this system of linear equations into the standard form

$$an + b \sum_{i=1}^{n} x_i + c \sum_{i=1}^{n} x_i^2 = \sum_{i=1}^{n} y_i,$$

$$a \sum_{i=1}^{n} x_i + b \sum_{i=1}^{n} x_i^2 + c \sum_{i=1}^{n} x_i^3 = \sum_{i=1}^{n} x_i y_i, \tag{15.37}$$

$$a \sum_{i=1}^{n} x_i^2 + b \sum_{i=1}^{n} x_i^3 + c \sum_{i=1}^{n} x_i^4 = \sum_{i=1}^{n} x_i^2 y_i.$$

which is a system of three linear equations in the variables a, b and c.
For the data in Tab. 15.4, the solution of (15.37) is

$$\hat{a} \approx 0.6518, \ \hat{b} \approx -0.3299, \ \hat{c} \approx 0.7959.$$

The corresponding parabola $y = \hat{a} + \hat{b}x + \hat{c}x^2$ is indicated in Fig. 15.3.

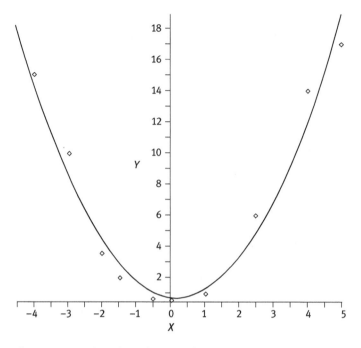

Fig. 15.3: Scatter plot with quadratic model.

Example 15.22: Suppose that the price of a used car of a particular brand depends essentially on the year of production and the mileage. If the depreciation is approximately linear, it is reasonable to use the model

$$Z = a + bX + cY + \varepsilon, \tag{15.38}$$

where X represents the age, Y the traveled distance and Z the price. Suppose that a sample of 10 cars provides the following hypothetical data.

Tab. 15.5: Data for two independent variables.

Car i	1	2	3	4	5	6	7	8	9	10
X (age, years)	1	1	2	3	3	4	5	5	6	6
Y (distance reading, 1,000 km)	10	20	15	30	40	35	40	50	60	70
Z (price, 1,000$)	54	53	48	42	40	36	35	30	24	23

To determine the estimates of the parameter values of (15.38), we minimize the sum of squared errors:

$$S(a, b, c) = \sum_{i=1}^{n} (a + bx_i + cy_i - z_i)^2 \qquad (15.39)$$

yielding the normal equations:

$$na + b \sum_{i=1}^{n} x_i + x \sum_{i=1}^{n} y_i = \sum_{i=1}^{n} z_i,$$

$$a \sum_{i=1}^{n} x_i + \sum_{i=1}^{n} x_i^2 + c \sum_{i=1}^{n} x_i y_i = \sum_{i=1}^{n} x_i z_i, \qquad (15.40)$$

$$a \sum_{i=1}^{n} y_i + b \sum_{i=1}^{n} x_i y_i + c \sum_{i=1}^{n} y_i^2 = \sum_{i=1}^{n} y_i z_i.$$

For the values in Tab. 15.5, we obtain the estimates $\hat{a} = 59.725$, $\hat{b} = -3.83$, $\hat{c} \approx -0.201$.

15.5 Confidence intervals

In all previous developments we have been concerned with obtaining a point estimate for an unknown parameter. For example, the average weight μ of school children of a certain age has been estimated by the average weight $\bar{X}$ of a sample of them. However, this point estimate is a random variable and therefore depends on chance. It is therefore natural to ask how far $\bar{X}$ might be away from μ. By formalizing this question we ask, what is the probability that $|\bar{X} - \mu|$ is smaller than ε or equivalently, for which value of ε it holds

$$P(\mu - \varepsilon \le \bar{X} \le \mu + \varepsilon) = 1 - \alpha. \qquad (15.41)$$

Here $1 - \alpha$ is a specified probability, known as the *confidence coefficient* ($0 < \alpha < 1$). The interval $[\mu - \varepsilon, \mu + \varepsilon]$ is called a $(1 - \alpha) \cdot 100\%$ *confidence interval (for the parameter μ)*.

If X has distribution $N(\mu, \sigma^2)$ where σ^2 is assumed to be known, it is easy to express ε in terms of the confidence coefficient. We know that in this case $\bar{X}$ has distribution $N(\mu, \sigma^2/n)$; thus, we obtain for the probability in (15.41):

$$\Phi\left(\frac{\mu + \varepsilon - \mu}{\sigma/\sqrt{n}}\right) - \Phi\left(\frac{\mu - \varepsilon - \mu}{\sigma/\sqrt{n}}\right) = 2\Phi\left(\frac{\varepsilon}{\sigma}\sqrt{n}\right) - 1 = 1 - \alpha \Rightarrow \Phi\left(\frac{\varepsilon}{\sigma}\sqrt{n}\right) = 1 - \frac{\alpha}{2}.$$

Hence,

$$\varepsilon = \frac{\sigma}{\sqrt{n}} \Phi^{-1}\left(1 - \frac{\alpha}{2}\right), \qquad (15.42)$$

where Φ^{-1} is the inverse distribution function of the standard normal distribution. (Modern statistical software offer options to evaluate diverse inverse distribution functions, also known as *quantile functions*.) Now the confidence interval is

$$\left[\bar{X} - \frac{\sigma}{\sqrt{n}} \Phi^{-1}\left(1 - \frac{\alpha}{2}\right), \quad \bar{X} + \frac{\sigma}{\sqrt{n}} \Phi^{-1}\left(1 - \frac{\alpha}{2}\right) \right]. \tag{15.43}$$

The value ε in (15.42) is often called the *margin of error* (*for the confidence coefficient* $1 - \alpha$).

Example 15.23: Suppose that the height X of the adults of a population is normally distributed with $\sigma = 15$ cm and μ unknown. If a random sample of 200 adults has the average height $\bar{x} = 170$ cm, determine a 95% confidence interval for μ.

Solution: The margin of error is

$$\varepsilon = \frac{15}{\sqrt{200}} \Phi^{-1}\left(1 - \frac{0.05}{2}\right) \approx \frac{15}{\sqrt{200}} \cdot 1.96 \approx 2.08,$$

yielding the confidence interval [167.92, 172.08].

The confidence interval must be interpreted carefully. When the average height encountered in the sample is $\bar{x} = 170$ cm, *then* the parameter μ is contained in that particular interval with probability of 95%. However, one must bear in mind that another sample may result in another sample average $\bar{x}$, yielding another confidence interval $[\bar{x} - \varepsilon, \bar{x} + \varepsilon]$. Thus, the general interpretation of the confidence interval is that μ is contained in (15.43) with $(1 - \alpha) \cdot 100\%$ of probability.

In Section 15.1 we have already seen that the sample proportion $\hat{p}$ is a particular case of a sample mean. Hence, we can estimate the unknown probability p that an element of a population has a certain characteristic by the sample proportion $\hat{p} = \frac{k}{n}$ (see Example 15.3). The probability p can also be interpreted as the proportion of the population, having the respective characteristic. The margin of error is now

$$\varepsilon = \sqrt{\frac{p(1-p)}{n}} \Phi^{-1}\left(1 - \frac{\alpha}{2}\right)$$

where the standard deviation $\sigma/\sqrt{n}$ in (15.42) has been substituted by the corresponding value $\sqrt{(p(1-p))/n}$ (see (15.1)).

However, since p is not known we substitute it for $\hat{p}$, yielding the approximate relation

$$\varepsilon \approx \sqrt{\frac{\hat{p}(1-\hat{p})}{n}} \Phi^{-1}\left(1 - \frac{\alpha}{2}\right). \tag{15.44}$$

Example 15.24: Assume that a sample of 400 articles contains 48 defective items.
(i) Determine a 99% confidence interval for the proportion of defective articles in the population.
(ii) For which sample size the error margin is 0.01.

Solution: The required estimate is given by $\hat{p} = 48/400 = 0.12$ (see (15.1)).

(i) From (15.44) we obtain the error margin

$$\varepsilon \approx \sqrt{\frac{0.12 \cdot 0.88}{400}} \, \Phi^{-1}(0.995) \approx 0.042$$

and the confidence interval is [0.078, 0.162].

(ii) The sample size n in question is given by

$$\sqrt{\frac{0.12 \cdot 0.88}{400}} \, \Phi^{-1}(0.995) \approx 0.01,$$

yielding $n = 7006$.

The last item makes clear that the accuracy of the estimation can only be improved by increasing the sample size.

Before we turn to the question how confidence intervals are calculated, when the variance is not known, we study a generalization and a modification of the concept. In certain situations it might be interesting to set up a confidence interval for a function $h(\theta)$ of a parameter θ. This is not a difficult task, if a confidence interval has already been determined for θ and if h is a monotone function. Let $[L, U]$ be the confidence interval of θ with confidence coefficient $1 - \alpha$. If h is monotone increasing, then the events $L \le \theta \le U$ and $h(L) \le h(\theta) \le h(U)$ are equivalent, i.e., they occur with the same probability. Thus, $[h(L), h(U)]$ is a confidence interval for $h(\theta)$ (with confidence coefficient $1 - \alpha$). Similarly, one obtains the confidence interval $[h(U), h(L)]$ for $h(\theta)$, if h is monotone decreasing.

Example 15.25: The life length X of an electronic device is normally distributed with known variance σ^2. The reliability for a service time of t hours is defined as the probability $R(t; \theta) = P(X > t)$, which for fixed t is a monotone decreasing function of μ. Assume that the average life length of a sample of 50 devices is 580 h, and that the standard deviation is $\sigma = 12$ h. Determine a confidence interval for μ and for the reliability for a service time of 570 h. Choose a confidence coefficient of 95%.

Solution:
(i) From (15.42) we get

$$\varepsilon = \frac{12}{\sqrt{50}} \Phi^{-1}(0.975) = \frac{12}{\sqrt{50}} 1.96 \approx 3.33,$$

yielding the confidence interval [576.67, 583.33] for the average life time.

(ii) The reliability in dependence of μ is

$$R(570, \mu) = P(X > 570) = 1 - \Phi\left(\frac{570 - \mu}{12}\right).$$

The values corresponding to the limits of the above interval are

$$1 - \Phi\left(\frac{570 - 576.67}{12}\right) \approx 0.71 \text{ and } 1 - \Phi\left(\frac{570 - 583.33}{12}\right) \approx 0.87$$

thus the confidence interval for the reliability is [0.71, 0. 87].

In some situations it may be interesting to set up unilateral confidence intervals, i.e., intervals of the form $] - \infty, U]$ or $[L, \infty[$ that contain a certain parameter with probability $1 - \alpha$.

Example 15.26: Suppose that X has distribution $N(\mu, \sigma^2)$ with unknown variance, and let a random sample $(X_1, \ldots, X_n)$ of X be given. We want to determine a unilateral confidence interval $[L, \infty[$ for σ^2, i.e., we are seeking a value L such that $P(\sigma^2 \geq L) = 1 - \alpha$.

Solution: Let

$$S^2 = \frac{1}{n-1} \sum_{i=1}^{n} (X_i - \bar{X})^2$$

be the sample variance. From Theorem 14.2 we know that $((n-1)S^2)/\sigma^2$ has distribution χ^2_{n-1}. Hence, there is a number c such that

$$P\left(\frac{(n-1)\,S^2}{\sigma^2} \leq c\right) = G_{n-1}(c) = 1 - \alpha$$

where G_n denotes the distribution function of χ^2_n. Now we can express c as $c = G^{-1}_{n-1}(1-\alpha)$, yielding

$$P\left(\frac{(n-1)\,S^2}{\sigma^2} \leq G^{-1}_{n-1}(1-\alpha)\right) = P\left(\frac{(n-1)\,S^2}{G^{-1}_{n-1}(1-\alpha)} \leq \sigma^2\right) = 1 - \alpha.$$

Hence, the required lower limit for σ^2 is given by

$$L = \frac{(n-1)\,S^2}{G^{-1}_{n-1}(1-\alpha)}.$$

If a random variable X is not normally distributed, the determination of confidence intervals can be considerably more difficult. We finally show how to obtain a lower limit for the expectation of an exponentially distributed r.v. X. We make use of the moment generating function (Chapter 11) and state the following corollary of Theorem 11.9.

Theorem 15.4: If $(X_1, \ldots, X_n)$ is a random sample of a r.v. X with exponential density $f(x) = \beta e^{-\beta x}$, $x \geq 0$, $\beta > 0$, then the r.v. $W = 2\beta n \bar{X}$ has chi-square distribution χ_{2n}^2, where $\bar{X} = \frac{1}{n}(X_1 + \cdots + X_n)$ is the sample mean.

Proof: From Theorem 11.9 it follows that $n \bar{X} = X_1 + \cdots + X_n$ has gamma distribution with parameters β and n, i.e., the moment generating function is $(\beta/(\beta - t))^n$. From Theorem 11.2 it follows now that W has the mgf

$$\left(\frac{\beta}{\beta - 2\beta\, t}\right)^n = \frac{1}{(1 - 2t)^n},$$

and the latter is the mgf of χ_{2n}^n.

Example 15.27: Assume that the life length X of some electronic equipment is exponentially distributed with pdf as in the last theorem, where $\beta > 0$ is an unknown parameter. Let a random sample with average life length $\bar{x}$ be given that serves to estimate the expectation $1/\beta$ of X. We want to obtain the upper confidence interval $[L, \infty[$, containing the expectation with probability $1 - \alpha$.

Solution: Since $2\beta n \bar{X}$ has distribution χ_{2n}^n, there is a number c, satisfying

$$P(2n \bar{X} \beta \leq c) = G_{2n}(c) = 1 - \alpha,$$

where G_n is the df of χ_{2n}^n (see Example 15.26). From the last relation it follows that $c = G_{2n}^{-1}(1 - \alpha)$ and

$$P\left(\frac{2n \bar{X}}{G_{2n}^{-1}(1 - \alpha)} \leq \frac{1}{\beta}\right) = 1 - \alpha.$$

Hence, the required lower limit is

$$L = \frac{2n \bar{X}}{G_{2n}^{-1}(1 - \alpha)}. \tag{15.45}$$

15.6 Student's *t*-distribution

We finally determine a confidence interval for the expectation of a normally distributed r.v. X, when σ^2 is not known.

For this purpose, we need the random variable

$$T = \frac{(\bar{X} - \mu) \sqrt{n}}{S} \tag{15.46}$$

where S^2 is the sample variance (Definition 14.2(ii)). We make use of the following facts:

(a) $Z = \frac{(\bar{X} - \mu)\sqrt{n}}{\sigma}$ has distribution $N(0,1)$.

(b) $V = \frac{S^2(n-1)}{\sigma^2}$ has a chi-square distribution with $(n-1)$ degrees of freedom (see Theorem 14.2).

(c) The random variables Z and V are independent. (The proof is not easy and will be omitted here.)

It can now be easily seen that the statistic in (15.46) can be written as $T = \left(Z\sqrt{n-1} \right) / \sqrt{V}$, and we obtain its distribution from the following statement.

Theorem 15.5: Suppose that the random variables Z and V are independent and have distributions $N(0,1)$ and χ_k^2, respectively. Then the pdf of

$$T = \frac{Z\sqrt{k}}{\sqrt{V}}$$

is given by

$$h_k(t) = \frac{\Gamma\left[(k+1/2)\right]}{\Gamma(k/2)\,\sqrt{\pi\,k}} \left(1 + \frac{t^2}{k} \right)^{-(k+1)/2} \qquad \text{for } -\infty < t < \infty. \tag{15.47}$$

This distribution is known as *Student's t-distribution with k degrees of freedom*.

The proof can be performed, using the methods of Sections 6.1 and 7.5 (see also Exercise 7.7). The details are left to the reader.

We can now set up a confidence interval for the expectation. In a manner completely analogous to the one of the previous section, we obtain the margin of error:

$$\varepsilon = \frac{S}{\sqrt{n}} H_{n-1}^{-1}\left(1 - \frac{\alpha}{2} \right), \tag{15.48}$$

where H_k is the distribution function corresponding to the density h_k in (15.47). The corresponding confidence interval (for the confidence coefficient $1 - \alpha$) is

$$\left[\bar{X} - \frac{S}{\sqrt{n}} H_{n-1}^{-1}\left(1 - \frac{\alpha}{2} \right), \quad \bar{X} + \frac{S}{\sqrt{n}} H_{n-1}^{-1}\left(1 + \frac{\alpha}{2} \right) \right]. \tag{15.49}$$

In contrast to (15.43), the length

$$L = 2\varepsilon = \frac{2\,S}{\sqrt{n}} H_{n-1}^{-1}\left(1 - \frac{\alpha}{2} \right) \tag{15.50}$$

of the last interval depends of the sample values, since S is a function of the x_i's.

Example 15.28: Determine bilateral confidence intervals for the average weight of school children, given the sample values in Example 15.1. Choose the confidence coefficients $1 - \alpha = 0.9$ and $1 - \alpha = 0.95$.

Solution: From (15.47) we obtain

$$\varepsilon_{90} = \frac{S}{\sqrt{n}} H_{n-1}^{-1}(0.95) \approx \frac{\sqrt{26.04}}{\sqrt{10}} \cdot 1.833 \approx 2.96,$$

$$\varepsilon_{95} = \frac{S}{\sqrt{n}} H_{n-1}^{-1}(0.975) \approx \frac{\sqrt{26.04}}{\sqrt{10}} \cdot 2.262 \approx 3.65,$$

thus the corresponding confidence intervals are [51.44, 57.36] and [50.75, 58.05].

Exercises

15.1 The weight of the adults of a population is normally distributed. In a sample of 20 adults the following weights (in kg) have been observed: 60, 66, 68, 70, 75, 76, 78, 81, 81, 83, 84, 84, 86, 90, 93, 95, 95, 98, 101,105.
    (i) Determine unbiased estimates of expectation and variance.
    (ii) What is the *ML* estimate for the variance?

15.2 Suppose that the length of an object is measured independently by two different devices, and let L_1 and L_2 denote the respective estimates of the true length L. We assume that $E(L_1) = E(L_2) = L$ if both devices are calibrated correctly, but the accuracies of the devices, measured in terms of variance, may be different, i.e., $V(L_1) \neq V(L_2)$.
    Let us use a linear combination $X = \alpha L_1 + (1 - \alpha)L_2$ with $0 < \alpha < 1$ to estimate the length.
    (i) For which values of α the estimator is unbiased?
    (ii) What choice of α minimizes the variance of X?

15.3 Given a sample $(X_1, \ldots, X_n)$ of a random variable X with expectation μ and variance σ^2. Show that $C \sum_{i=1}^{n-1} (X_i - X_{i+1})^2$ is an unbiased estimator of σ^2 for an appropriate choice of C. Find that value of C.

15.4 Prove relation (15.7)

15.5 Determine both estimates of the gamma distribution by the method of Example 15.12, based on the sample values $(x_1,..,x_{10}) = (5, 6, 5, 7, 5.5, 6, 7, 7.5, 6.5, 6)$, see Example 15.9.
    Plot the function on the left-hand side of (15.11). Show that the right-hand side is always positive, which ensures that a solution exists in any case.

15.6 A random variable X has the pdf $f(x) = (\beta + 1)x^\beta$ for $0 \leq x \leq 1$.
    (i) Obtain the ML estimate of β based on a sample of X.
    (ii) Evaluate the estimate if the sample values are (0.3, 0.4, 0.4, 0.5, 0.6, 0.7).

15.7   Suppose that the time to failure T (in hours) of an electronic device has a shifted exponential distribution, i.e., the pdf is $f(t) = a \exp(-a(t - t_0))$ for a known parameter $t_0 > 0$ and $t \geq t_0$. Obtain the ML estimate of a based on a sample $(t_1, \ldots, t_n)$ of recorded failure times.

15.8   Let X be uniformly distributed over an interval $[-a, a]$.
Determine the ML estimate of a based on sample values $(x_1, \ldots, x_n)$.

15.9   The Weibull distribution has the pdf

$$f(x) = a\, b\, x^{b-1}\, \exp(-a\, x^b) \quad for\ x \geq 0;\ a,\ b > 0.$$

Determine for given sample values $(x_1, \ldots, x_n)$ the ML estimate of the parameter a, when b is known.

15.10 The pdf of the Rayleigh distribution can be written as

$$f(x) = \frac{x}{b} \exp\left(-\frac{x^2}{2b}\right) \quad for\ x \geq 0; b > 0.$$

(i)   Determine the ML estimate of b, based on a sample $(x_1, \ldots, x_n)$.
(ii) Show that the estimator is unbiased and consistent.

15.11 Study by means of simulation the expectation and variance of the ML estimator of the Rayleigh distribution, found in the previous exercise (using suitable software). Construct K samples of size n for a r.v., having Rayleigh distribution with parameter $b = 25$ and calculate the corresponding estimates $\hat{\theta}_1, \ldots, \hat{\theta}_K$.
Determine the approximations for expectation and variance of the estimator, given by $\bar{\theta} = 1/K \sum_{i=1}^{K} \hat{\theta}_i$ and $V = \sum_{i=1}^{K} \left(\hat{\theta}_i - \bar{\theta}\right)^2$.
Start the simulation study with $K = 10{,}000$ and $n = 100$ and test, what happens when these values are changed? The simulation confirms the theoretical results of Exercise 15.10?
If possible, realize similar studies for the estimators of other distributions.

15.12 Determine the parameter t (see (15.12)) of the asymptotic distribution of the ML estimate of μ in the normal distribution.

15.13 Determine the moment estimators for the binomial distribution.

15.14 Determine the moment estimates for the Erlang distribution with pdf

$$f(x) = \frac{\lambda^k x^{k-1} e^{-\lambda x}}{\Gamma(k)} \quad for\ x \geq 0;\ \lambda,\ k > 0.$$

15.15 Consider again the Weibull distribution (Exercise 15.9).
(i)   Determine the ML estimate and the moment estimate of the parameter b if $a = 1$.
(ii) Determine the ML estimate of a, if $b = 3$. Simulate the expectation and variance of this estimate as suggested in Exercise 15.11. The estimator is unbiased and consistent?

15.16 Determine the moment estimator of the parameter in the Maxwell–Boltzmann distribution, having pdf

$$f(x) = \frac{\sqrt{2}}{\sqrt{\pi}\, a^3} x^2 \exp\left(-\frac{x^2}{2\,a^2}\right) \quad \text{for } x \geq 0,\ a > 0.$$

The estimator is unbiased and consistent?

15.17 Determine the moment estimators of the parameters in the shifted exponential distribution with pdf

$$f(x) = a\, e^{-a(x-c)} \text{ for } x \geq c;\ a > 0,\ -\infty < c < \infty.$$

15.18 Suppose that the market price of a used car of a certain brand is a linear function of its age. Adjust the simple linear regression model to the following hypothetical data. Create a scatterplot and compare the expected prices with the observed ones.

Car	1	2	3	4	5	6	7	8	9	10
X (age, years)	1	2	3	4	6	8	9	10	12	13
Y (price, 1,000$)	50	45	42	39	38	36	35	33	28	25

15.19 Prove the formulas in (15.29).

15.20 Prove that the estimate $\hat{b}$ in (15.24) is a linear function of the sample values y_i.

15.21 Show that the estimate $\hat{a}$ in (15.33) is unbiased.

15.22 A physicist pretends to check the value of the gravitational constant a, arising in the well-known formula

$$s = \frac{a}{2} t^2,$$

where s indicates the distance of an object freely falling in a vacuum from the starting point at the time t. The following table gives the distances measured at time intervals of 0.1 s. Estimate the value of a by means of a simple regression model.

T (time, s)	0.1	0.2	0.3	0.4	0.5	0.6	0.7	0.8	0.9	1.0
S (distance, m)	0.055	0.19	0.465	0.75	1.3	1.9	2.42	3.51	3.93	5.05

15.23 Compare the observed values in Tab. 15.5 of Example 15.22 with the expected ones provided by model (15.38).

15.24 Formulate the normal equations for the cubic model $Y = a + bX + cX^2 + dX^3 + \varepsilon$.

15.25 Suppose that the estimated vocabulary size Z (measured in adequate units) of schoolchildren depends essentially on the age X (in years) and the intelligence quotient Y measured at school enrolment. Consider the following hypothetical data:

Number of child	1	2	3	4	5	6	7	8	9	10
X	6	6	7	8	9	9	10	11	12	12
Y	100	120	80	90	110	120	130	100	100	90
Z	155	190	125	148	177	190	205	160	167	149

Determine the estimates of the parameters a, b and c, when model (15.39) is applied.

15.26 Assume that the weight X of the pigs of a big farm is normally distributed with $\sigma = 20$ kg and μ unknown. A random sample of 80 animals has the average weight $\bar{x} = 90$ kg. Determine a 90% confidence interval for μ. What happens with the length of the interval if the confidence coefficient increases?

15.27 To predict the results of a presidential election, 800 randomly selected registered voters have been interviewed.

(i) If 350 persons would vote for candidate A, determine a 95% confidence interval for the proportion in the population that would give this candidate their vote.

(ii) For which sample size the error margin is 1%?

15.28 Consider the sample of weights (50, 52, 53, 58, 62, 55, 50, 63, 48, 53) of school children in Example 15.1. Determine the lower limit L for the variance according to Example 15.26, choosing a 99% confidence interval. Determine also an upper limit U.

15.29 Assume that 20 light bulbs with exponentially distributed life lengths have been tested, providing the average life length of $\bar{x} = 1{,}500$ h. Determine the unilateral and bilateral confidence intervals $[L, \infty[,]-\infty, U]$ and $\left[\bar{X} - \varepsilon \le \frac{1}{\beta} \le \bar{X} + \varepsilon\right]$ for the expectation (see Example 15.27).

15.30 Let X have distribution $N(\mu, \sigma^2)$ where the variance is known. Determine the unilateral confidence intervals $[-\infty, U]$ and $[L, \infty]$ for the parameter μ with confidence coefficient α.

15.31 Illustrate Student's t-distribution (15.47) graphically for different values of k and compare the shape with that of the standard normal density.

15.32 Determine the unilateral confidence limits for Example 15.28.

16 Hypothesis tests

This chapter is dedicated to testing hypotheses. This is one of the most useful topics of statistical inference since many practical decision problems may be formulated as a hypothesis test. We will not detail modeling and evaluation of tests but rather focus on their operation, applying diverse probabilistic models. We are now faced with a similar problem as in the previous chapter since we want to win knowledge about the distribution of a r.v. X based on observations in a random sample. But instead of estimating a parameter we want to make a specific decision: we have to decide between accepting and rejecting an affirmative about a parameter.

16.1 Basic concepts and tests

A *statistical hypothesis* is a statement about the distribution of a random variable X involving one or several parameters. In practical applications, X usually represents a certain population, we are interested e.g. in the size of the persons of a natural population or the life length of the "population" of a certain kind of light bulbs. Suppose e.g. that we are interested in the mean size of a population, which is *normally distributed with known variance.* We might be interested in deciding if the mean size μ of adults equals a specific value μ_0 or not, which might have been obtained from previous studies. We can express this formally as

$$H_0 := \mu = \mu_0,$$
$$H_1 : \mu \neq \mu_0. \tag{16.1}$$

The statements H_o and H_1 are called the *null hypothesis* and the *alternative hypothesis,* respectively. There is more than one possibility to state the alternative hypothesis formally. The form chosen in eq. (16.1) is based on the idea that μ can in fact be larger or smaller than μ_0, in this case we speak about a *bilateral alternative hypothesis* (or a *bilateral test*). If there are reasons to believe that it is (almost) impossible that μ is less than or equal to μ_0 one can write

$$H_0 := \mu = \mu_0,$$
$$H_1 : \mu > \mu_0. \tag{16.2}$$

The latter is called a *unilateral alternative hypothesis.* We now want to make a decision about truth or falsity of a hypothesis. Any procedure that leads to such a decision is called a *hypothesis test.* Such a test makes use of a random sample of a random variable (or of the population, respectively). The decision is based on an appropriate statistic. Suppose that a random sample of size n has been selected from the population. As a statistic known to have "good" properties we choose the average height $\bar{X}$ of

https://doi.org/10.1515/9783111332277-016

the sample. If we consider a test of the form (16.1) it is reasonable to accept H_0 if $\bar{X}$ is "close" to μ_0 and to accept H_1 otherwise. Formally, we accept H_0 if and only if $\bar{X} \in [\mu_0 - \varepsilon, \mu_0 + \varepsilon]$, where ε is to be determined. This interval is called *acceptance region*, its end points are known as *critical values* and the complement $]\infty, \mu_0 - \varepsilon \, [\cup] \, \mu_0 + \varepsilon, \infty \, [$ is called *rejection region* or *critical region*. In case of the test (16.2) it is intuitively reasonable to reject H_0 if $\bar{X}$ is considerably larger than μ_0. Formally, the rejection region is $[t, \infty]$ for a critical value $t > \mu_0$ to be determined.

We will now see how to determine the critical values. Any decision is subject to errors. Basically we can distinguish between two types of error:

If H_o is rejected when it is true, we speak of an *error of type* 1, and if H_o is accepted when it is not true, we speak of an *error of type* 2. For example, if adults have the average size μ_0, a random sample may accidentally contain persons which are all considerably larger than μ_0. This situation might yield the error of type 1. Conversely it may happen that the average population size in fact differs from μ_0 and that the sample yields accidentally an average size close to μ_0. In this case the error of type 2 might be committed. We are going to study the probabilities that these errors occur, defining

$$\alpha = P(\text{error } 1) = P(\text{reject } H_0 | H_0 \text{ is true}), \tag{16.3}$$

$$\beta = P(\text{error } 2) = P(\text{accept } H_0 | H_0 \text{ is false}). \tag{16.4}$$

Now we can determine the respective critical intervals for eqs. (16.1) and (16.2) by fixing the probability α to a "small" value, for example $\alpha = 0.05$. The value to which α is fixed is called the *significance level*. In case of the bilateral test we obtain, using the fact that $\bar{X} \sim N(\mu_0, \sigma^2/n)$:

$$\alpha = P(\bar{X} \notin [\mu_0 - \varepsilon, \mu_0 + \varepsilon] | H_0 \text{ is true}) \Rightarrow$$

$$1 - \alpha = P(\bar{X} \in [\mu_0 - \varepsilon, \mu_0 + \varepsilon]) = \Phi\left(\frac{\mu_0 + \varepsilon - \mu_0}{\sigma/\sqrt{n}}\right) - \Phi\left(\frac{\mu_0 + \varepsilon - \mu_0}{\sigma/\sqrt{n}}\right) \tag{16.5}$$

$$= 2\Phi\left(\frac{\varepsilon}{\sigma}\sqrt{n}\right) - 1.$$

By isolating ε we get

$$\varepsilon = \frac{\sigma}{\sqrt{n}}\Phi^{-1}\left(1 - \frac{\alpha}{2}\right), \tag{16.6}$$

yielding the acceptance region

$$\left[\mu_0 - \frac{\sigma}{\sqrt{n}}\Phi^{-1}\left(1 - \frac{\alpha}{2}\right), \mu_0 + \frac{\sigma}{\sqrt{n}}\Phi^{-1}\left(1 - \frac{\alpha}{2}\right)\right]. \tag{16.7}$$

The region (16.7) corresponds to the confidence interval (15.43) found in Section 15.5. An analogous statement holds for all acceptance intervals.

In the case of a unilateral hypothesis (16.2), fixing the probability α gives

$$\alpha = P(\bar{X} > t|H_0 \text{ is true}) = 1 - \Phi \left(\frac{t - \mu_0}{\sigma/\sqrt{n}} \right), \tag{16.8}$$

yielding

$$t = \mu_0 + \frac{\sigma}{\sqrt{n}} \Phi^{-1}(1 - \alpha), \tag{16.9}$$

where $[t, \infty[$ is the rejection region. We are now able to test a hypothesis regarding the parameter μ for any normal distribution with known variance. In order to ensure a good performance of the procedures, the sample size must be large enough. It is frequently recommended that n is at least 30 in practical applications. In this case moderate deviations from normality have only a small impact on the validity of the tests.

It should be intuitively clear that the test H_0: $\mu = \mu_0$, H_1: $\mu < \mu_0$ is completely analogous to eq. (16.2) because of the symmetry of the normal distribution. The rejection region for H_0 is then

$$\left[-\infty, \mu_0 = \frac{\sigma}{\sqrt{n}} \Phi^{-1}(1 - \alpha) \right]. \tag{16.10}$$

In the following we assume that the reader is aware of this fact. Moreover, further test formulations are possible, for example, H_0: $\mu \leq \mu_0$, H_1: $\mu > \mu_0$ which will not be detailed here.

We illustrate the hypothesis tests by means of examples.

Example 16.1: During a geological study of the continental drift, the question arises, from which civilization descended the inhabitants of a certain island. It is assumed that the islanders could have descended from any of several civilizations A, B, C, D, . . . and it is known that the size X of adults from A has the expectation $\mu = 170$ cm and the standard deviation $\sigma = 15$ cm. Further the average size of a sample of 150 persons of the island resulted in 172.2 cm. We perform a hypothesis test to decide about the ancestry of the islanders for a significance level of 5%.

(i) Suppose no information about the size of the civilizations B, C, D, . . . is known. This means in particular that the average sizes of these populations can be both larger or smaller than 170 cm. Therefore we perform a bilateral test with the hypotheses (see eq. (16.1)):

H_0: The islanders derive from the civilization A $\Leftrightarrow \mu = 170$
H_1: The islanders do *not* derive from the civilization A $\Leftrightarrow \mu \neq 170$.

From eq. (16.6) we get

$$\varepsilon = \frac{15}{\sqrt{150}} \Phi^{-1}\left(1 - \frac{0.05}{2}\right) \approx \frac{15}{\sqrt{150}} \cdot 1.96 \approx 2.40,$$

thus the acceptance interval (16.7) is [167.6, 172.4]. Since the sample average falls inside of this interval, we accept the hypothesis H_0.

(ii) We now assume that A is the smallest civilization with respect to the body size, i.e., all other civilizations B, C, D, . . . have a larger average size. This assumption results in the unilateral test (16.2) with the hypotheses

H_0: The islanders derive from the civilization A $\Leftrightarrow \mu = 170$,
H_1: The islanders do *not* derive from the civilization A $\Leftrightarrow \mu > 170$.

From eq. (16.9) we get $t = 170 + \left(\frac{15}{\sqrt{150}}\right)\Phi^{-1}(1 - 0.05) \approx 172.01$. Since the sample mean 172.2 now falls into the critical region [172.01, ∞], the null hypothesis must be rejected.

In the previous chapter we have already seen that the proportion is a specific case of a sample mean (Sections 15.1, 15.5). We now consider the tests

$$H_0 : p = p_0$$
$$H_1 : p \neq p_0 \tag{16.11}$$

and

$$H_0 : p = p_0$$
$$H_1 : p > p_0, \tag{16.12}$$

where p is the proportion of the population and p_0 a specific value of p. In analogy to the reasoning (16.5)–(16.7) the bilateral test (16.11) has the acceptance region $[p_0 - \varepsilon, p_0 - \varepsilon]$ with

$$\varepsilon = \sqrt{\frac{p_0(1 - p_0)}{n}} \Phi^{-1}\left(1 - \frac{\alpha}{2}\right). \tag{16.13}$$

The standard deviation σ in eq. (16.5) now assumes the specific form $\sqrt{p_0(1 - p_0)}$. For the unilateral test (16.12) the rejection region is $[t, \infty[$ with

$$t = p_0 + \sqrt{\frac{p_0(1 - p_0)}{n}} \Phi^{-1}(1 - \alpha). \tag{16.14}$$

Example 16.2: A manufacturer of glass articles claims that 5% of the production of a new type of candle holder is defective. In a random sample of 1200 candle holders 73 defective have been encountered. We want to test whether the statement of the manufacturer is credible, considering two situations.

(i) Suppose that no further information about the real proportion of defective items in the "population of candle holders" is given; thus we perform a bilateral test with the hypotheses

H_0: $p = 0.05$ (the statement of the producer is correct),
H_1: $p \neq p_0$.

For $\alpha = 0.05$, eq. (16.13) yields $\varepsilon = \sqrt{\dfrac{0.05 \cdot 0.95}{1,200}} \Phi^{-1}(0.975) = 0.0123$.

The acceptance region is therefore [0.0377, 0.0621]. Since the sample proportion $\dfrac{73}{1,200} \approx 0.0608$ is contained in this interval, the null hypothesis is accepted.

(ii) We now suppose that the manufacturer could have exaggerated with respect to the error rate in order to make a good advertisement for his products. Thus we assume that the real portion of defective candle holders is larger than the stated value of 5%, yielding the unilateral test H_0: $p = 0.05$, H_1: $p > 0.05$. From eq. (16.13) we obtain $t = 0.05 + \sqrt{\dfrac{0.05 \cdot 0.95}{1,200}} \Phi^{-1}(0.95) \approx 0.0603$. Now the sample proportion 0.0608 is contained in the rejection region [0.0603, ∞[and H_0 must be rejected.

16.2 OC-function, power function and *p*-value

As we have seen earlier it is easy to determine the critical values resulting in a prescribed probability of the error of type 1, when dealing with the tests (16.1), (16.2) or some of its variants, discussed before. However, controlling the error of type 2 is more complicated since it is not related to a single parameter value (see eq. (16.4)). When H_0 is false, the parameter can usually assume an infinite number of values (an exception is the case of two simple hypotheses, see Exercise 16.3). The following concepts are useful when trying to keep the errors under control.

Definition 16.1: The *operating characteristic function* (OC-function) of a hypothesis test is given by

$$\beta(\mu) = P(\text{accept } H_0 | \mu). \tag{16.15}$$

The *power function* of the test is defined as

$$P(\mu) = 1 - \beta(\mu) = P(\text{reject } H_0|\mu). \tag{16.16}$$

These functions represent the probability to accept or reject the null hypothesis as a function of the unknown parameter μ. In principle, one of these two concepts would be sufficient because any of them can be easily expressed by the other. It depends on the area of application, which function is more popular.

For the bilateral test we get from eq. (16.7)

$$\beta(\mu) = P(\text{accept } H_0|\mu) = P(\bar{X} \in [\mu_0 - \varepsilon, \mu_0 + \varepsilon]),$$

where ε is given by eq. (16.6). Thus,

$$\beta(\mu) = \Phi\left(\frac{\mu_0 + \varepsilon - \mu}{\sigma/\sqrt{n}}\right) - \Phi\left(\frac{\mu_0 - \varepsilon - \mu}{\sigma/\sqrt{n}}\right)$$

$$= \Phi\left((\mu_0 - \mu)\frac{\sqrt{n}}{\sigma} + \Phi^{-1}\left(1 - \frac{\alpha}{2}\right)\right) - \Phi\left((\mu_0 - \mu)\frac{\sqrt{n}}{\sigma} - \Phi^{-1}\left(1 - \frac{\alpha}{2}\right)\right). \tag{16.17}$$

In particular, functions (16.15) and (16.16) are useful to compare the functioning of two tests.

Example 16.3: Let us reconsider the decision problem in Example 16.1 for which we have $\mu_0 = 170$, $\sigma = 15$, $n = 150$, $\alpha = 0.05$, $\varepsilon \approx 2.40$ The solid line in Fig. 16.1 illustrates the power function $P(\mu) = 1 - \beta(\mu)$ for this case. It can be observed that the function is symmetric with respect to $\mu = 170$. The value of $P(\mu)$, i.e., the probability of rejecting H_0 correctly, increases with the distance of μ from the value 170. Corresponding to our intuition the falsity of the null hypothesis can be detected the easier, the farther is μ away from the value 170.

The dotted line in Fig. 16.1 represents the power function in the case when the sample size is raised to $n = 300$. This line always runs above the solid one, i.e., for any value of the parameter μ, the test with $n = 300$ leads with higher probability to the right decision than that with $n = 150$. Generalizing this observation one can say that a hypothesis test with a larger sample size is superior to one with a smaller sample size. This again corresponds to our intuition according to which larger samples permit "safer" statements.

For the unilateral test (16.2) we obtain the OC-function $\beta(\mu) = P(\text{accept } H_0|\mu) = P(\bar{X} \le t|\mu)$, where t is given by eq. (16.9). Thus

$$\beta(\mu) = \Phi\left(\frac{t - \mu}{\sigma/\sqrt{n}}\right) = \Phi\left((\mu_0 - \mu)\frac{\sqrt{n}}{\sigma} + \Phi^{-1}(1 - \alpha)\right). \tag{16.18}$$

Let us continue studying the last example, assuming now that the civilization A is the smallest with respect to the body size. The appropriate hypothesis test is then the unilateral test (16.2), having the OC-function (16.18). For the values

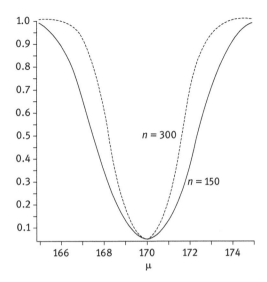

Fig. 16.1: Power functions of Example 16.3 (bilateral test).

$$\mu_0 = 170, \ \sigma = 15, \ n = 150, \ \alpha = 0.05 \qquad\qquad (16.19)$$

of this example the power function $P(\mu) = 1 - \beta(\mu)$ is given by the solid line in Fig. 16.2. Similar to the bilateral case, the value of $P(\mu)$, i.e., the probability of rejecting H_0 correctly, increases with the distance of μ from the value 170.

For example, we obtain $P(173) \approx 0.79$, i.e., for the parameter value 173 the null hypotheses is recognized as false with a probability of about 79%. This "quality" would definitely not be considered as sufficient for practical purposes. But as in the bilateral case, we obtain a superior hypothesis test by augmenting the sample size n. The upper dotted line in Fig. 16.2 represents the power function for $n = 300$. It is very instructive to examine how changes of the values (16.19) influence the shape of the power function. Setting, e.g., $\sigma = 25$ (where the other values of eq. (16.19) remain unchanged) yields the lower dotted curve in Fig. 16.2. Hence, for a larger variance in the population size the test results become more uncertain.

It is to be hoped that the reader is motivated to examine the impact of other changes in the values (16.19).

Before we discuss the concept of p-value we want to compare the performances of the bilateral and the unilateral hypothesis test. The power functions (16.17) and (16.18) corresponding to these tests are plotted in Fig. 16.3 (choosing the unknowns as in eq. (16.19)). The curve of the unilateral test runs throughout above that of the bilateral, i.e., for parameter values greater than $\mu_0 = 170$ the probability to make the right decision is always larger for the unilateral test. The same observation is made for other values of μ_0, σ, n and $\alpha = 0.05$.

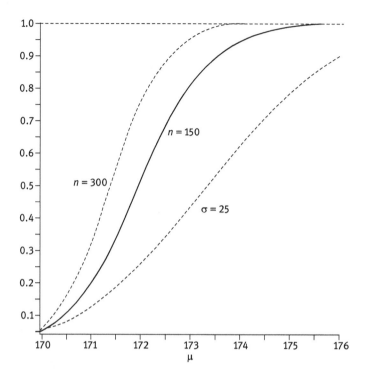

Fig. 16.2: Power functions of Example 16.3 (unilateral test).

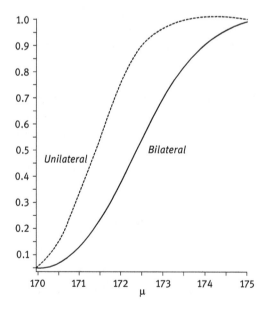

Fig. 16.3: Bilateral versus unilateral power function.

In Section 16.1 we have introduced the significance level as an a priori given value in which the probability of the type 1 error is fixed. Indeed, one can proceed in that manner and accept or reject the null hypothesis depending on the observed sample. However, one can approach the test problem in another way and ask: For which values of the significance level the null hypothesis would be accepted? (see Exercise 16.2).

Definition 16.2: Suppose that a statistic has been calculated for a random sample which is used to perform a hypothesis test. The *p*-value is the probability to obtain a value of the statistic which is "at least as extreme" as the actually observed one, when the null hypothesis is true.

It can be observed that the *p*-value also represents the largest value of the significance level for which H_0 would be accepted.
We illustrate the concept by means of two examples.

Example 16.4: A bus operator is considering the possibility of introducing a new bus route for an interurban connection. According to a preliminary study the duration of a trip is normally distributed with $\mu = 300$ min and $\sigma = 30$ min. The first 10 test drives along this route resulted in an average duration of 314 min. Determine the *p*-value for the "sample of test drives," assuming that the normality assumption and the value of σ are correct.

We perform a bilateral test H_0: $\mu = 300$, H_1: $\neq 300$. A value of the test statistic $\bar{X}$ is "at least as extreme" as the observed value, if $|\bar{X} - 300| \geq 14$, i.e., in the bilateral case, deviations in both directions are possible. Since $\bar{X} \sim N(300, 90)$, the *p*-value is

$$P\left(|\bar{X} - 300| \geq 14\right) = 2P(\bar{X} \leq 286) = 2\Phi\left(\frac{286 - 300}{\sqrt{90}}\right) \approx 0.14.$$

Now the hypothesis H_0 is accepted if and only if

$$\varepsilon > 0.14 \Leftrightarrow P\left(|\bar{X} - 300| > 0.14\right) > P\left(|\bar{X} - 300| > \varepsilon\right) = \alpha,$$

where α is the significance level (see eq. (16.5)). Therefore H_0 is accepted if and only if α is smaller than the *p*-value of 14%.

Example 16.5: For the sample in Example 16.2, we have distinguished between the bilateral and the unilateral test. In the second case the *p*-value is

$$P\left(\hat{p} \geq \frac{73}{1,200}\right) = 1 - \Phi\left(\frac{73/1,200 - 0.05}{\sqrt{\frac{0.05 \cdot 0.95}{1,200}}}\right) \approx 0.0425.$$

For the bilateral test we obtain the *p*-value

$$P\left(\hat{p} \ge \frac{73}{1{,}200}\right) + P\left(\hat{p} \le \frac{47}{1{,}200}\right) = 2P\left(\hat{p} \ge \frac{73}{1{,}200}\right) \approx 0.085.$$

In the bilateral test H_0 is accepted for any significance level of at most 8.5%, and in the unilateral test any significance level of at most 4.25% results in accepting H_0.

16.3 Tests for the mean of a normal distribution with unknown variance

Let us study again the tests (16.1) and (16.2) with the only difference that the variance σ^2 of X is now not known. We first consider the bilateral test

$$H_0 : \mu = \mu_0,$$
$$H_1 : \mu \ne \mu_0. \tag{16.20}$$

We use the same test statistic $\bar{X}$ and make use of the fact that the transformation

$$T = \frac{(\bar{X} - \mu_0)\sqrt{n}}{S} \tag{16.21}$$

has Student's distribution with $(n-1)$ degrees of freedom, if the null hypothesis is true (see Section 15.6). In analogy to Section 16.1 the acceptance region is now $[\mu_0 - \varepsilon, \mu_0 + \varepsilon]$, where

$$\varepsilon = \frac{S}{\sqrt{n}} H_{n-1}^{-1}\left(1 - \frac{\alpha}{2}\right) \tag{16.22}$$

and H_k denotes the Student distribution function with k degrees of freedom (compare with eqs. (15.48) and (16.6)). For the unilateral test

$$H_0 : \mu = \mu_0,$$
$$H_1 : \mu > \mu_0 \tag{16.23}$$

the rejection region is $[t, \infty[$ with

$$t = \mu_0 + \frac{\hat{\sigma}}{\sqrt{n}} H_{n-1}^{-1}(1 - \alpha) \tag{16.24}$$

(compare with eq. (16.9)).

Example 16.6: According to the information of the leaflet of a pain reliever the average duration of action is 6 h. Assume that 15 test patients reported the following duration times of the medicament (in hours): (6.5, 6, 5.5, 7, 4, 6, 6.5, 5.5, 7, 6, 6.5, 5.5, 7, 4.5,

6.5). The statement can be confirmed by the sample result for a significance level of 5%?

(i) We perform the bilateral test (16.20). For the given sample we obtain $\bar{x} = 5.6$, $S^2 = \frac{1}{14}\sum_{i=1}^{15}(x_i - \bar{x})^2 \approx 0.7929$ and $H_{14}^{-1}(0.975) \approx 2.1448$. Hence, $\varepsilon = \sqrt{\frac{0.7929}{15}} \cdot 2.1448 \approx 0.493$ and $\bar{x} = 5.6$ belongs to the acceptance region $[6 - 0.493,\ 6 + 0.493] = [5.507,\ 6.493]$.

(ii) Assuming that the mean duration of the analgesic effect cannot exceed 6 h, the unilateral test H_0: $\mu = 6$, H_1: $\mu < 6$ is appropriate. The rejection region is then $[-\infty, t]$ with $t = \mu_0 - \frac{\hat{\sigma}}{\sqrt{n}}H_{14}^{-1}(1 - 0.05) = 6 - \sqrt{\frac{0.7929}{15}} \cdot 1.7613 \approx 5.595$ and H_0 is accepted again.

16.4 Tests for the variance of a normal distribution

Suppose that a random variable X has distribution $N(\mu, \sigma^2)$ where both parameters μ and σ^2 are unknown. We want to test the hypothesis that σ^2 equals a specific value σ_0^2. We first consider the bilateral test

$$H_0 : \sigma^2 = \sigma_0^2,$$
$$H_1 : \sigma^2 \neq \sigma_0^2.$$

According to Theorem 14.2(ii), the used statistic

$$V = \frac{(n-1)\ S^2}{\sigma_0^2} \tag{16.25}$$

has distribution χ_{n-1}^2, if H_0 is true. The acceptance region for a significance level a is defined as

$$\left[G_{1-n}^{-1}\left(\frac{a}{2}\right),\ G_{n-1}^{-1}\left(1 - \frac{a}{2}\right) \right], \tag{16.26}$$

where G_k again denotes the distribution function of χ_k^2, i.e., V exceeds the upper limit of eq. (16.26) with probability of $a/2$ and falls below the lower limit with the same probability.

For the unilateral test

$$H_0 : \sigma^2 = \sigma_0^2,$$
$$H_1 : \sigma^2 > \sigma_0^2.$$

the rejection region is, $\left[G_{n-1}^{-1}(1 - a), \infty \right[$ and for the test

$$H_0 : \sigma^2 = \sigma_0^2,$$

$$H_1 : \sigma^2 < \sigma_0^2.$$

the rejection region is $\left[0, \ G_{n-1}^{-1}(\alpha)\right]$.

Example 16.7: We want to test whether the variance of the duration of the action time of the pills in Example 16.6 could be equal to 0.9. We perform a bilateral test with $\alpha = 0.05$. The value of the test statistic

$$V = \frac{14 \cdot 0.7929}{0.9} = 12.337$$

is contained in the acceptance interval (16.26), given by [5.629, 26.119], hence the null hypothesis is accepted.

16.5 Comparing the means of two normal distributions

Assume that there are two random variables X, Y with distributions $N(\mu_1, \sigma_1^2)$ and $N(\mu_2, \sigma_2^2)$, respectively, where the expectations are unknown and the variances are known. We first consider the bilateral test:

$$H_0 : \mu = \mu_0,$$

$$H_1 : \mu \neq \mu_0.$$

We use the test statistic $\bar{X} - \bar{Y}$, where $\bar{X}$ is the mean of a sample $(X_1, \ldots, X_n)$ of X and $\bar{Y}$ the mean of a sample $(Y_1, \ldots, Y_m)$ of Y, having the distribution

$$N\left(\mu_1 - \mu_2, \ \frac{\sigma_1^2}{n} + \frac{\sigma_2^2}{m}\right).$$

Under the null hypothesis the reduced statistic

$$Z_0 = \frac{\bar{X} - \bar{Y}}{\sqrt{\frac{\sigma_1^2}{n} + \frac{\sigma_2^2}{m}}} \tag{16.27}$$

has distribution $N(0, 1)$. The acceptance region for Z_0 is therefore

$$[-\varepsilon, \varepsilon], \text{ with } \varepsilon = \Phi^{-1}\left(1 - \frac{\alpha}{2}\right). \tag{16.28}$$

For this choice of ε it holds that Z_0 exceeds ε with probability $\alpha/2$ and falls below $-\varepsilon$ with probability $\alpha/2$.

A similar reasoning yields the rejection region

$$\left[\Phi^{-1}(1-\alpha),\ \infty\right[\tag{16.29}$$

for the unilateral test H_0: $\mu_1 = \mu_2$, H_1: $\mu_1 > \mu_2$ and the rejection region

$$\left]-\infty,\ \Phi^{-1}(\alpha)\right] \tag{16.30}$$

for the test H_0: $\mu_1 = \mu_2$, H_1: $\mu_1 < \mu_2$.

Example 16.8: According to the advertisement of an automotive company the new version of a popular car type has a lower gasoline consumption than the previous model. Test drives with 60 cars of the newer and 50 cars of the older model resulted in average consumptions of $\bar{x} = 7.8$ and $\bar{y} = 8.1$ L/100 km. Due to experiences with the respective injection systems it is known that the standard deviations are $\sigma_1 = 1.2$ and $\sigma_2 = 1.5$, respectively. We perform the test H_0: $\mu_1 = \mu_2$, H_1: $\mu_1 < \mu_2$ for a significance level of 5%. The value

$$Z_0 = \frac{7.8 - 8.1}{\sqrt{\frac{1.2^2}{60} + \frac{1.5^2}{50}}} \approx -1.1421$$

of the statistic (16.27) is not contained in the rejection region $]-\infty, -1.6449]$; thus the null hypothesis is not rejected.

We have repeatedly noted that a proportion is an important specific case of the mean. In this case the test is, for example, H_0: $p_1 = p_2$, H_1: $p_1 > p_2$, where the p_i are the real proportions in the two populations. The other two test formulations are analogous.

The test statistic is now the difference between the sample proportions $\hat{p}_1 - \hat{p}_2$, having distribution

$$N\left(p_1 - p_2,\ \frac{p_1(1-p_1)}{n} + \frac{p_2(1-p_2)}{m}\right)$$

and under the null hypothesis H_0: $p = p_1 = p_2$ the variance is $p(1-p)\left(\frac{1}{n} + \frac{1}{m}\right)$.

Usually p is estimated by the "pooled sample," i.e., one substitutes p for

$$\hat{p} = \frac{k+l}{n+m} = \frac{n\,\hat{p}_1 + m\,\hat{p}_2}{n+m}, \tag{16.31}$$

where k and l are the number of elements in the two samples having the considered characteristic for which the proportion is calculated. The reduced test statistic is therefore

$$Z_0 = \frac{\hat{p}_1 - \hat{p}_2}{\sqrt{\hat{p}(1-\hat{p})\left(\frac{1}{n} + \frac{1}{m}\right)}} \tag{16.32}$$

having approximately the distribution $N(0,1)$. The acceptance and rejection regions for the three tests are the same as in eqs. (16.28)–(16.30).

Example 16.9: An institute of tropical medicine has developed an improvement of a vaccine against an infectious disease. The improved and the conventional version of the drug has been tested on control groups of 50 and 40 voluntary patients in which the effect has proved to be satisfactory for $k = 43$ and $l = 30$ patients, respectively. Assuming that the new vaccine cannot be worse than the previously used one, we perform the unilateral test $H_0: p_1 = p_2$, $H_1: p_1 > p_2$ for $\alpha = 5\%$. We get $\hat{p} = \frac{43+30}{50+40} \approx 0.8111$ and the value of the test statistic

$$Z_0 = \frac{0.86 - 0.75}{\sqrt{0.8111 \cdot 0.1889 \left(\frac{1}{50} + \frac{1}{40}\right)}} \approx 1.3247$$

is not contained in the critical region $[1.6449, \infty]$, thus H_0 is accepted. Hence, one assumes that the difference in the effects observed in the two patients groups is due to random variations.

We finally study tests to compare the mean of two normal distributions in the case of unknown variances σ_1^2, σ_2^2. We have to distinguish the two cases $\sigma_1^2 = \sigma_2^2$ and $\sigma_1^2 \neq \sigma_2^2$.

Case 1: $\sigma_1^2 = \sigma_2^2 = \sigma^2$: We first consider the bilateral test

$$H_0 : \mu = \mu_0,$$

$$H_1 : \mu \neq \mu_0.$$

We use a test statistic analogous to that of the beginning of the section, which is obtained by substituting the variances σ_1^2 and σ_2^2 in eq. (16.27) – which are now equal – by one and the same *combined estimator* S_c. Similar to eq. (16.31) one defines

$$S_c^2 = \frac{(n-1)S_1^2 + (m-2)S_2^2}{n+m-2}, \tag{16.33}$$

yielding the statistic

$$T_0 = \frac{\bar{X} - \bar{Y}}{S_c \sqrt{\frac{1}{n} + \frac{1}{m}}} \tag{16.34}$$

which has Student's distribution (15.47) with $n + m - 2$ degrees of freedom under the null hypothesis. The acceptance region for T_0 is therefore

$$[-\varepsilon, \varepsilon], \text{ with } \varepsilon = H_{n+m-2}^{-1}\left(1 - \tfrac{\alpha}{2}\right), \tag{16.35}$$

where H_k is the distribution function corresponding to eq. (15.47). In a straightforward manner we obtain the rejection region

$$\left[H_{n+m-2}^{-1}(1-a),\ \infty\right[\tag{16.36}$$

for the unilateral test $H_0\colon \mu_1 = \mu_2$, $H_1\colon \mu_1 > \mu_2$ and the rejection region

$$\left]-\infty,\ H_{n+m-2}^{-1}(a)\right] \tag{16.37}$$

for the test $H_0\colon \mu_1 = \mu_2$, $H_1\colon \mu_1 < \mu_2$.

The last test is usually called *combined t*-test since the sample variances are combined to estimate the common variance. Another name is *independent t-test* since it is assumed that the random variables X and Y are independent.

We illustrate the procedure by the following modification of Example 16.8.

Example 16.10: Test drives with 60 cars of the newer and 50 cars of the older model resulted in average consumptions of $\bar{x} = 7.8$ and $\bar{y} = 8.1$ L/100 km. The sample variances of the consumptions were $s_1^2 = 1.8$ and $s_2^2 = 2$. It is assumed that the real variances are equal.

(i) We perform the test $H_0\colon \mu_1 = \mu_2$, $H_1\colon \mu_1 < \mu_2$ for a significance level of 5%. The combined estimator for the variance is

$$S_c^2 = \frac{59 \cdot 1.8 + 49 \cdot 2}{108} \approx 1.8907$$

and the value of the statistic

$$T_0 = \frac{7.8 - 8.1}{\sqrt{1.8907}\sqrt{\frac{1}{60} + \frac{1}{50}}} \approx -1.1394$$

is not contained in the rejection region $]-\infty, -1.6591]$. Hence the null hypothesis is accepted.

(ii) In the case of the bilateral test the acceptance region for the null hypothesis is $[-1.9822, 1.9822]$, and H_0 is again accepted.

Case 2: $\sigma_1^2 \neq \sigma_2^2$: The test procedures are as in the previous case with the only difference that the test statistic is now

$$T_0^* = \frac{\bar{X} - \bar{Y}}{\sqrt{\frac{S_1^2}{n} + \frac{S_2^2}{m}}}. \tag{16.38}$$

When the null hypothesis is true, T_0^* follows approximately a Student's distribution, where the number of degrees of freedom is obtained by rounding the value

$$v = \frac{\left(\frac{s_1^2}{n} + \frac{s_2^2}{m}\right)^2}{\frac{\left(s_1^2/n\right)^2}{n+1} + \frac{\left(s_2^2/m\right)^2}{m+1}} - 2 \tag{16.39}$$

to the next integer. The test problem of case 2 is called the *Behrens-Fisher problem.*

Example 16.11: A nutritionist wants to compare the effectiveness of two diet plans that have been developed for weight reduction. Both methods have been applied for 2 months and for each of the 25 patients the weight reduction X was determined. Ten patients adopted the plan 1 and had a mean weight reduction of $\bar{x} = 10.2\,\text{kg}$ with the sample variance of $s_1^2 = 14$. Fifteen other patients adopted plan 2 and achieved a mean weight reduction of $\bar{y} = 12.1\,\text{kg}$ with the sample variance of $s_2^2 = 20$.

We perform the bivariate test for $\alpha = 5\%$:

$$H_0 : \mu = \mu_0,$$

$$H_1 : \mu \neq \mu_0.$$

The test statistic assumes the value $T_0^* = \dfrac{10.2 - 12.1}{\sqrt{\frac{14}{10} + \frac{20}{15}}} \approx -1.1492$ and the number of degrees of freedom is

$$v = \frac{\left(\frac{14}{10} + \frac{20}{15}\right)^2}{\frac{(14/10)^2}{11} + \frac{(20/15)^2}{16}} - 2 \approx 23.8 \approx 24.$$

Thus the acceptance region for H_0 is $[-\varepsilon, \varepsilon]$, with $\varepsilon = H_{24}^{-1}(0.975) \approx 2.0639$, containing the observed value of the statistic. The null hypotheses is accepted, i.e., there is not sufficient evidence to believe that the diet plans are significantly different.

16.6 Comparing the variances of two normal distributions

For the following test it is necessary to introduce the F-distribution which plays an important role in statistical applications.

Theorem 16.1: Let X and Y be random variables with distributions χ_n^2 and χ_m^2. Then the density of the random variable $F = \frac{X/n}{Y/m} = \frac{mX}{nY}$ is given by

$$f(x) = \frac{\Gamma\left(\frac{n+m}{2}\right)}{\Gamma\left(\frac{n}{2}\right) \cdot \Gamma\left(\frac{m}{2}\right)} \left(\frac{n}{m}\right)^{n/2} x^{n/2-1} \left(1 + \frac{n}{m}x\right)^{-(n+m)/2} \text{ for } x \geq 0. \qquad (16.40)$$

This is called (*Snedecor's*) *F-distribution* with n and m degrees of freedom, also known as *F-ratio distribution*.

Suppose that the random variables X and Y have distributions $N(\mu_1, \sigma_1^2)$ and $N(\mu_2, \sigma_2^2)$, respectively, where all parameters $\mu_1, \sigma_1^2, \mu_2, \sigma_2^2$ are unknown. We first consider the bilateral test

$$H_0 : \sigma_1^2 = \sigma_2^2,$$
$$H_1 : \sigma_1^2 \neq \sigma_2^2.$$

We make use of the test statistic

$$F_0 = \frac{S_1^2}{S_2^2}, \qquad (16.41)$$

where S_1^2 and S_2^2 are the sample variances of the random samples $(X_1, \ldots, X_n)$ of X and $(Y_1, \ldots, Y_m)$ of Y. From Theorems 14.2(ii) and 16.1, it follows now that this statistic has Snedecor's F-distribution with $n-1$ and $m-1$ degrees of freedom if the null hypothesis is correct.

Similar to previous reasoning we obtain the acceptance region as $\left[K^{-1}\left(\frac{\alpha}{2}\right),\right.$ $\left. K^{-1}\left(1 - \frac{\alpha}{2}\right)\right]$, where K denotes the distribution function corresponding to eq. (16.40) (see Exercise 16.15).

Without loss of generality we may assume that X is the variable that may have the larger variance. Then we obtain the only unilateral test

$$H_0 : \sigma_1^2 = \sigma_2^2,$$
$$H_1 : \sigma_1^2 > \sigma_2^2.$$

where the rejection region for H_0 is $[K^{-1}(1-\alpha), \infty[$.

Example 16.12: We test whether the variances in the two samples of Example 16.11 could be different, choosing a significance level of 5%. The test statistic has the value $F_0 = \frac{14}{20} = 0.7$.

(i) For the bilateral test the acceptance region is $[0.2002, 4.9949]$, hence H_0 is accepted.

(ii) After changing the notations of the variances we get for the unilateral test $F_0 = \frac{20}{14} = 1.4286$, which is not contained in the rejection interval $[3.7870, \infty[$. Again, H_0 is accepted.

16.7 Chi-square goodness-of-fit test

In the above tests the distribution of the random variable X is assumed to be known and the hypotheses involve their parameters. We now turn to another important situation: we do not know the distribution of X and we want to test whether X follows a certain distribution, e.g. the normal or the Poisson distribution. Thus we have the hypotheses H_0: X follows that distribution and H_1: X does not follow that distribution.

   We describe a test for the goodness of the fit which is based on the chi-square distribution. It is assumed that a sample of size n of the r.v. X is given. This sample is given in "grouped form," i.e., the range is partitioned into k classes (intervals) $C_1, \ldots,$ C_k. We denote by O_i the observed number of sample values falling within C_i and by E_i the expected number of values falling within C_i, assuming that the null hypothesis is true $(i = 1, \ldots, k)$. The test statistic is

$$\chi_0^2 = \sum_{i=1}^{k} \frac{(O_i - E_i)^2}{E_i}, \tag{16.42}$$

which has approximately a chi-square distribution with $k - p - 1$ degrees of freedom, where p is the number of parameters in the distribution of X. Similar to the value (15.21) in linear regression, the statistic (16.42) attempts to express the amount of deviations between observed and expected frequencies in form of a single number. Small values support the null hypothesis and large values suggest that H_0 might be wrong. Accordingly, the rejection region is $\left[G_{n-p-1}^{-1}(1 - \alpha), \infty \right[$, where G_r denotes the distribution function of the chi-square distribution with r degrees of freedom (see Section 10.4).

   In practical applications, the statistic χ_0^2 is modified, when very small expected frequencies occur. In this case the corresponding components in eq. (16.42) are subject to strong random fluctuations. There is no common agreement among all statisticians, but a frequent rule is to combine adjacent classes with small expected frequencies, until the total frequency exceeds a minimum value, say 3. The values O_i and E_i of the new class are obtained by adding the observed and expected frequencies of the original classes. The total number of classes in eq. (16.42) – which need not have the same length – reduces accordingly (see Example 16.14).

Example 16.13: (Absence of parameters): We want to test first whether a certain die is fair. Suppose that it has been rolled 6000 times. The hypothetical observed frequencies of the possible results 1, . . ., 6 are given in Tab. 16.1.

According to the null hypothesis, in any throw of the die the result i appears with probability $p_i = 1/6$ for $i = 1, \ldots, 6$. The expectation of each number is therefore 1,000 for each result. The value of the statistic is

Tab. 16.1: Hypothetical results in tossing a die.

Result	1	2	3	4	5	6
Observed frequency	954	1,050	890	1127	942	1,037
Expected frequency	1,000	1,000	1,000	1,000	1,000	1,000

$$\chi_0^2 = \frac{(954 - 1{,}000)^2}{1{,}000} + \cdots + \frac{(1{,}037 - 1{,}000)^2}{1{,}000} = 37.578.$$

For the probabilistic model, i.e., a uniform discrete distribution, no parameters have to be estimated. The number of degrees of freedom is therefore $k - p - 1 = 6 - 0 - 1 = 5$ and for a significance level of $\alpha = 5\%$ the critical value is $G_5^{-1}(0.95) \approx 11.07$ Since the observed value of the statistic exceeds this value, the null hypothesis must be rejected.

Example 16.14 (Discrete distribution): A small subsidiary of a car insurance company studied the frequency of serious accidents in which their clients were involved during a period of 5 years. The results for the $n = 122$ clients are given in the second column of Tab. 16.2.

Tab. 16.2: Frequencies of road accidents.

Number of accidents, X	Observed frequency, O_i	Expected frequency, E_i
0	53	48.32
1	40	44.75
2	18	20.73
3	7	6.399
4	3	1.482
≥5	1	0.319

It is assumed that X follows a Poisson distribution, i.e., $p_i = P(X = i) = \frac{e^{-\lambda}\lambda^i}{i!}$ for $i = 0, 1,$ .... The unknown parameter, representing the expected number of accidents must be estimated from the data, yielding $\hat{\lambda} = (0 \cdot 53 + 1 \cdot 40 + 2 \cdot 18 + 3 \cdot 7 + 4 \cdot 4)/122 \approx 0.9262$. The expected frequencies $E_i = np_i$ are listed in the last column of Tab. 16.2. In particular, the last value is $n - E_0 - \ldots - E_4 = 0.319$. Since this value is smaller then 3, we combine the two last classes, resulting in

$$\chi_0^2 = \frac{(53 - 48.32)^2}{48.32} + \frac{(40 - 44.75)^2}{44.75} + \frac{(18 - 20.73)}{20.73} + \frac{(7 - 6.399)^2}{6.399} + \frac{(4 - 1.801)^2}{1.801} \approx 4.06.$$

The number of degrees of freedom is now $5 - 1 - 1 = 3$; thus for a significance level of 5% the critical value is $G_3^{-1}(0.95) \approx 7.815$, which is much larger than the above value of the test statistic. Thus H_0 is accepted, i.e., the Poisson model seems to be well suited for the data.

Example 16.15 (Continuous distribution): In control examinations realized by the Ministry of Health, the weights of $n = 100$ test participants have been determined. The average weight and the sample variance are 75 kg and 100 kg^2, respectively. We test whether the weights follow the distribution $N(75, 100)$. Therefore we subdivide the sample into $k = 8$ classes $[a_{i-1}, a_i]$ for $i = 1, \ldots, 8$ and compare the corresponding observed and expected frequencies. According to a common practice the limits of the classes are determined such that the theoretical frequencies $E_i = np_i$ for all classes are equal, i.e., the a_i must satisfy

$$p_i = \int_{a_{i-1}}^{a_i} f(x)\ dx = \frac{1}{8}, \quad i = 1, \ldots, 7 \quad a_0 = -\infty, \ a_8 = \infty, \tag{16.43}$$

where f is the pdf of the distribution $N(75,100)$. The solution of (16.43) is $(a_1, \ldots, a_7) =$ (63.50, 68.26, 71.81, 75, 78.19, 81.75, 86.50). The observed number of persons with a weight inside of the respective classes is given in the second column in Tab. 16.3.

Tab. 16.3: Observed and expected frequencies.

Classes	Observed frequency, O_i	Expected frequency, E_i
$x < 63.50$	11	12.5
$63.50 \leq x < 68.26$	15	12.5
$68.26 \leq x < 71.81$	10	12.5
$71.81 \leq x < 75.00$	13	12.5
$75.00 \leq x < 78.19$	12	12.5
$78.19 \leq x < 81.75$	15	12.5
$81.75 \leq x < 86.50$	11	12.5
$86.50 \leq x$	13	12.5

The value of the test statistic is

$$\chi_0^2 = \frac{(11 - 12.5)^2}{12.5} + \frac{(15 - 12.5)^2}{12.5} + \cdots + \frac{(13 - 12.5)^2}{12.5} = \frac{24}{12.5} = 1.92.$$

Since there are $k - p - 1 = 8 - 2 - 1 = 5$ degrees of freedom, the critical value for a significance level of 5% is $G_5^{-1}(0.95) \approx 11.07$. The hypothesis that the weight of the test persons has distribution $M(75, 100)$ cannot be rejected.

16.8 Contingency table chi-square test

In many practical applications arises the question whether two characteristics of a population are statistically independent or not. For example, one can ask whether there is any relation between smoking and lung cancer, or between frequent drinking of alcohol and liver cirrhosis or whether the choice of the subject in university studies is gender specific.

Given two variables X and Y which are usually qualitative/categorical, i.e., their values are not numbers but categories between which no order relation needs to exist. For example, if X is the party affiliation of a voter, then the categories are e.g. Social Democratic Party, Christian Union, etc. between which does no natural order relation exists. The data of a sample are usually presented in the form of Tab. 16.4, where the x_i and y_i represent the categories of X and Y. The $O_{i,j}$ denote the observed number of sample elements falling into the ith category of X and the jth category of Y; the s_i and t_j denote the total of the lines and columns and $N = s_1 + \cdots + s_n = t_1 + \cdots + t_n$ is the sample size. The totals null hypothesis states that there is no dependence between the variables X and Y.

Tab. 16.4: Contingency table.

	Y			
X	y_1	$\cdots$	y_n	
x_1	$O_{1,1}$	$\cdots$	$O_{1,n}$	s_1
$\vdots$	$\vdots$		$\vdots$	$\vdots$
x_m	$O_{m,1}$	$\cdots$	$O_{m,n}$	s_m
	t_1	$\cdots$	t_n	N

As will be illustrated in the following example, the expected frequencies under the null hypothesis are given by

$$E_{i,j} = \frac{s_i \, t_j}{N} \text{ for } i = 1, \ldots, m, j = 1, \ldots, n. \tag{16.44}$$

The test statistic

$$\chi_0^2 = \sum_{i=1}^{n} \sum_{j=1}^{m} \frac{(O_{i,j} - E_{i,j})^2}{E_{i,j}}, \tag{16.45}$$

that again measures the relative deviations between observed and expected frequencies, is approximately distributed as $\chi^2_{(n-1)(m-1)}$. The number of degrees of freedom corresponds to the number of freely selectable frequencies in Tab. 16.4, when the s_i and t_j are given. Similar to the previous tests, the rejection region for the independency hypothesis is $[G^{-1}_{(n-1)(m-1)}(1-\alpha), \infty[$.

Example 16.16: A sample of 500 persons of different age groups has been interviewed with respect to the drinking habits. The results are given in Tab. 16.5.

Tab. 16.5: Hypothetical data for alcohol consumption and age group.

X, age group	Y, alcohol consumption				
	Never	Rarely	Moderately	Frequently	
20–40	14	42	81	63	200
40–60	32	49	51	38	170
>60	15	35	62	18	130
	61	126	194	119	500
	12.2%	25.2%	38.8%	23.8%	

The additional last line of this contingency table contains the percentage of the whole sample, falling into the corresponding category of Y. For example, 12.2% of the sample never drink alcohol and 25.2% drink rarely. Assuming now that there is no relation between age group and drinking habits, then the same percentages of drinking habits must occur in all age groups, i.e., 12.2% of each age group never drink, 25.2% of each age group drink rarely, etc. In particular, 12.2% of the class of oldest persons never drink, i.e., the expected number of never drinking persons in this age group is $0.122 \cdot 130 = \frac{61 \cdot 130}{500} = \frac{s_3 \, t_1}{N}$ which illustrates formula (16.44). The complete matrix of expected values is

$$E = \begin{pmatrix} 24.4 & 50.4 & 77.6 & 47.6 \\ 20.74 & 42.84 & 65.96 & 40.46 \\ 15.86 & 32.76 & 50.44 & 30.94 \end{pmatrix},$$

yielding

$$\chi_0^2 = \sum_{i=1}^{3} \sum_{j=1}^{4} \frac{(O_{i,j} - E_{i,j})^2}{E_{i,j}} = \frac{(14 - 24.4)^2}{24.4} + \cdots + \frac{(18 - 30.94)^2}{30.94} \approx 29.77.$$

Let the significance level be 1%. Then the critical value is $G_6^{-1}(0.99) \approx 16.81$, and the null hypothesis must be rejected. It can be concluded that the drinking habits in the hypothetical example depend on the age group. However, the considered test does not allow to specify the form of dependence. For example, it cannot prove that the alcohol consumption increases or decreases with the age. The test result merely states that there exists some kind of dependence between age and alcohol consumption.

Exercises

16.1 Show step by step that for a test H_0: $\mu = \mu_0$, H_1: $\mu > \mu_0$ the critical value is given by eq. (16.9).

16.2 Perform the calculations in Example 16.1 for a significance level of $\alpha = 1\%$. How far influences the reduction of α the test decision? What happens with the acceptance region or the critical region, respectively, if α is reduced.

16.3 Consider the following modification of Example 16.1: The islanders could have descended from *exactly two* possible civilizations A or B. The size X of the population A has distribution $N(170, 400)$ and the size Y of B has distribution $N(180, 900)$. The average size $\bar{x}$ of a random sample of $n = 50$ islanders is given. We obtain the following hypotheses H_0: $\mu = 170$, H_1: $\mu = 180$.

In such a case one speaks about *simple hypotheses* since the parameter is completely specified in each case. Consider the following decision rule for the hypothesis test: Accept H_0 if $\bar{x} \leq t$ (where t is a critical value to be determined) and accept H_1 if $\bar{x} > t$.

(i) Express the error probabilities α and β in terms of t (see eqs. (16.3), (16.4) and (16.8)).

(ii) Illustrate the functions $\alpha(t)$, $\beta(t)$ and $\alpha(t) + \beta(t)$ graphically over the interval $[170, 180]$.

(iii) Determine t and β for $\alpha = 0.05$.

(iv) Determine t and α for $\beta = 0.05$.

(v) Determine t such that $\alpha = \beta$. What is the common value of α and β?

(vi) Determine t such that $\alpha + \beta$ is minimized. What are the values of α and β in this case?

16.4 A popular automotive company has launched a new version of a sports car on the market, which supposedly has less gasoline consumption than the predecessor model. The producer states that the new model has an average consumption of 8 L/100 km. The standard deviation of the consumption is assumed to be $\sigma = 1.2$ L and the average consumption in 50 test drives has been 8.5 L/100 km.

(i) Decide whether the information provided by the producer is credible. Use the hypotheses test H_0: $\mu = 8$, H_1: $\mu > 8$ with a significance level of 1%.

(ii) Would the test decision be the same if the standard deviation of the consumption were $\sigma = 1.6$ L?

16.5 A well-known television station claims that 70% of the population had viewed a particular program of this transmitter. In interviews of 300 randomly selected families, 192 confirmed that they have seen this program. Decide whether the assertion of the TV station is credible. Perform for a significance level of 2%

(i) the bilateral test H_0: $p = 0.7$, H_1: $p \neq 0.7$,

(ii) the unilateral test H_0: $p = 0.7$, H_1: $p < 0.7$.

16.6 How large must the sample size n in Example 16.3 be, such that for $\mu \geq 171$ the error of type 2 is at most 0.07?

16.7 Determine the p-value for the two tests in Example 16.1.

16.8 In testing whether a given coin is fair, it has been tossed 1000 times, where 532 heads and 468 tails occurred. Determine the p-value for a bilateral test
(i) using the binomial distribution,
(ii) using the normal distribution.

16.9 (i) Determine the p-value for the two tests in Example 16.6.
(ii) For which sample size the acceptance region in Example 16.6 (i) would have the length 0.2?
(iii) Determine the critical values for Example 16.6, applying the tests in Section 16.1 and assuming that $\sigma^2 \approx 0.7929$. Compare with the critical values found previously.

16.10 Perform the two unilateral hypothesis tests
(i) $H_0 : \sigma_1^2 = 1$,
$H_1 : \sigma^2 > 1$
and
(ii) $H_0 : \sigma_1^2 = 2$,
$H_1 : \sigma^2 < 2$
for the situation in Example 16.7.

16.11 (i) How low must be the average gasoline consumption $\bar{x}$ in Example 16.8 so that H_0 would have to be rejected (assuming a significance level of 5% and 1%, respectively).
(ii) What is the probability that the correct test decision is taken, if the real reduction of consumption is 0.5 L/100 km (for $\alpha = 5\%$)?
(iii) Assuming that both samples are of the size n. For which value of n, the probability of the previous item is 0.95?

16.12 Consider the test of Example 16.9.
(i) Determine the p-value for the observed sample.
(ii) Perform a bilateral test for $\alpha = 5\%$. Which hypothesis is accepted?

16.13 How large must be the difference in the average weight reductions in Example 16.11, such that H_0 would be rejected?

16.14 Prove the Theorem 16.1.

16.15 Show that $K^{-1}(1 - \alpha) = \dfrac{1}{K^{-1}(\alpha)}$, where K denotes the distribution function corresponding to eq. (16.40).

16.16 Assume that the variance $S_1^2 = 14$ in Example 16.12 remains unchanged. For which values of S_2^2 the null hypothesis would be rejected
(i) in the bilateral test,
(ii) in the unilateral test?

16.17 (i) Determine 60,000 values of the statistic in Example 16.13 by means of a random number generator, simulating the tossing of a fair die. Create a histogram by means of suitable software and compare it with the density of the distribution χ_5^2.

(ii) What is the p-value of the observed sample in Example 16.13?

16.18 In a study of the health ministry the body sizes of 60 adults have been determined. The values in non-decreasing order (in cm) are

140, 142, 142, 143, 144, 145, 145, 147, 149, 150,

150, 151, 152, 153, 153, 154, 155, 156, 156, 158,

159, 160, 162, 163, 163, 164, 165, 166, 167, 167,

168, 168, 169, 169, 169, 170, 172, 174, 174, 175,

175, 176, 176, 177, 178, 179, 179, 180, 181, 182,

182, 183, 184, 185, 186, 186, 187, 188, 189, 190.

Test in analogy to Example 16.14, if the body size could be normally distributed. Choose 5 and 6 classes, respectively.

16.19 Calculate the Example 16.16 again, for the case when the drinking categories "never" and "rarely" are combined into a single category "never or rarely." Does this alteration change the observed value of the statistic and the critical value considerably?

16.20 Former long-time members of a large bodybuilding club have been interviewed with respect to possible health problems, particularly spine problems. The fictive outcomes of the consultation of 200 members are summarized in the following table:

X, training hours per week	Y, degree of spine problems		
	No problems	Moderate problems	Severe problems
<4	16	10	4	30
4–8	30	25	15	70
8–12	25	40	15	80
>12	10	7	3	20
	81	82	37	200

Is there a relation between training intensity and spine problems? Perform a contingency table chi-square test for a significance level of 2%.

17 Exploratory data analysis

This last chapter is dedicated to the analysis of sample data. In the first two sections we restrict ourselves to data of a single random variable. These topics are also known as descriptive statistics. In Section 17.3, we study in detail the relation between two quantitative random variables, making use of the regression analysis introduced in Chapter 15.

17.1 Introduction

Some of the most important concepts of statistics are population, sample parameter, statistic and variable, which are detailed as follows:

Population: It is the set of individuals, animals, objects or observations to be studied. Examples are the natural populations of Germans, English or Brazilians, the pigs of a farmer, the light bulbs available in a store or the temperatures observed in a certain region over a period of time. A population can be finite or infinite.

Sample: It is a representative subset of the population.
This means that the sample must reflect the characteristics of the larger group. For example, a sample of a natural population should contain similar proportions of males and females and similar proportions of young and old people as the population.

Parameter: It is a characteristic of the population.
Examples are the average life expectancy, the most preferred kind of food and the average gross income of a natural population.

Statistic: It is a characteristic of the sample.
To any parameter one can construct a corresponding statistic, see Fig. 17.1.

Assume that a sample of the English population has been determined by randomly selecting 500 adults. The parameter we are interested in is the average size of the English adults. The corresponding statistic is the average size of persons in the sample.

Population	Sample
population of England	500 randomly selected adults
Parameter	**Statistic**
average size of adults of England	$\Leftarrow$ $\begin{cases} \text{average size} \\ \text{of sample} \\ \text{members} \end{cases}$

Fig. 17.1: Introduced concepts and examples.

https://doi.org/10.1515/9783111332277-017

The reasoning in estimating the parameter is now very simple. The statistic is used as an estimator for the parameter. In particular, the average size of the persons in the random sample is used to estimate the average size of the whole population. A beginner in statistics might ask why not study the whole population instead of a sample. However, it should be noted that populations are either far too large to be fully studied, or the researcher does not have access to all members of the population. But there are further reasons that might preclude studying the entire population. Suppose that the lifespan of light bulbs offered for sale is to be determined. If the entire light-bulb population were tested, there would be nothing left to sell. Another important issue is how to determine a sample. If a complete list of the population is available, simple random sampling is a good option. This method guarantees that the sample is representative for the population. For other cases sampling methods are available that will not be detailed here.

A **variable** is any characteristic or quantity that can be measured or counted. We distinguish between **quantitative** and **qualitative** variables. Quantitative variables are divided into **discrete** and **continuous** variables (see Fig. 17.2). For exact definitions we refer to Chapter 5. As a rule, however, one can say that the values of discrete variables are natural numbers (including zero), and continuous variables are characterized by the fact that they can assume any value within an entire interval. For example, the number of children of a family is a discrete variable and the weight of a person is a continuous variable. Qualitative variables are divided into **ordinal** and **nominal** variables. The former have outcomes (values) that are ranked in a specific order, while there is no order between the outcomes of a nominal variable. For example, the values of the variable "smoking habit" may be "not at all," "a little," "moderately" and "a lot." Though these outcomes are not objectively defined, there is an order between them: "not at all" is less than "a little," "a little" is less than "moderately," etc. Thus the "smoking habit" is an ordinal variable. Finally, the variable "political party" is a nominal variable since there is no order between outcomes like "Labor Party," "Socialist Party," etc.

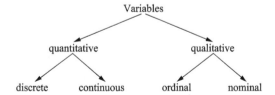

Fig. 17.2: Types of variables.

There are several kinds of tables and graphics in descriptive statistics that can be found in most introductory textbooks. At the moment we limit ourselves to the most important ones, namely the frequency distribution and the histogram. The simple concepts are illustrated by means of an example. Table 17.1 contains the fictitious results of determining the body weight of a class of 90 students. In a frequency distribution the data are grouped into classes. For example, the class $45 \vdash 50$ is the set of weights x such that $45 \leq x < 50$. The absolute frequency F_i denotes the number of students falling in a certain class. The accumulated frequency is the number of students of all preceding classes up to the present one. Finally, the relative frequency is defined by $f_i = F_i/n$, where $n = F_1 + F_2 + \cdots$ is the sample size.

Tab. 17.1: Frequency distribution of weights of 90 students.

Classes (in kg)	Absolute frequencies, F_i	Accumulated frequencies, F_{ac}	Relative frequencies, f_i
45 ⊢ 50	6	6	0.067
50 ⊢ 55	8	14	0.089
55 ⊢ 60	10	24	0.111
60 ⊢ 65	15	39	0.117
65 ⊢ 70	20	59	0.222
70 ⊢ 75	18	77	0.200
75 ⊢ 80	8	85	0.089
80 ⊢ 85	5	90	0.056

An important issue in constructing a frequency distribution is determining the number K of classes. Some textbooks recommend to choose

$$K = 5 \qquad \text{for } n \leq 25, \tag{17.1}$$

$$K \approx \sqrt{n} \quad \text{for } n > 25.$$

Another recommendation is **Sturge's formula**

$$K = \lceil 1 + 3.322 \cdot \log(n) \rceil, \tag{17.2}$$

where log and $\lceil X \rceil$ denote the decadic logarithm and the smallest integer greater than or equal to x, respectively. For Tab. 17.1, formula (17.1) would result in 9 or 10 classes since $\sqrt{90} \approx 9.48$, while Sturge's formula results in $K = \lceil 1 + 3.322 \cdot 1.954 \rceil = \lceil 7.49 \rceil = 8$, which is the number of classes actually chosen in this table.

The length or amplitude of the classes is approximately given by $(M - m)/K$, where M and m denote the **maximum** and **minimum** value of the data set, respectively. Of course, the end points of the classes are frequently rounded. The histogram of the considered frequency distribution is illustrated in Fig. 17.3.

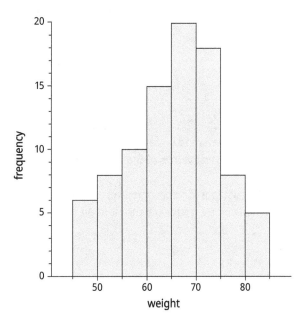

Fig. 17.3: Histogram for Tab. 17.1.

17.2 Summary measures

In order to summarize a data set in some way, we present "typical" data points of a numerical data set. The first three well-known concepts are also called **measures of central tendency**. We consider first the case of ungrouped data, where $x_1, \ldots, x_n$ are the elements of a set. The average number or **mean** is defined by $\bar{x} = (1/n) \sum_{i=1}^{n} x_i$. The most frequent number is called the **mode**. For example, if $n = 10$ and $(x_1, \ldots, x_{10}) = (2, 3, 5, 4, 6, 1, 6, 6, 2, 1)$ we have $\bar{x} = \frac{1}{10} \cdot 36 = 3.6$ and the mode is $Mo = 6$. Of course it is possible that several modes exist. For example, in the case $(x_1, \ldots, x_{10}) = (1, 2, 2, 2, 4, 5, 6, 6, 6, 7)$ we have the two modes 2 and 6. If the data points are ordered, the **median** is defined as the middle number. The exact definition is as follows:

$$\tilde{x} = \begin{cases} x_{(n+1)/2} & \text{if } n \text{ is odd,} \\ \dfrac{x_{n/2} + x_{n/2+1}}{2} & \text{if } n \text{ is even.} \end{cases} \tag{17.3}$$

The idea behind this concept is to define a value such that (approximately) half of the data is smaller and (approximately) half of the data is larger than this value. Obviously, $\tilde{x}$ has this characteristic, as the following example illustrates.

Example 17.1:

Consider the ordered sequences

(i)   4, 5, 7, 10, 12, 17, 20, 25, 30,

and

(ii) 8, 13, 20, 38, 36, 47, 58, 75, 90, 105.

In sequence (i) the number of elements $n = 9$ is odd, thus the median is $x_{(9+1)/2} = x_5 = 12$, in the second sequence $n = 10$ is even and the median is $\bar{x} = \frac{x_{10/2} + x_{10/2+1}}{2} = \frac{36+47}{2} = 41.5$.

The example shows that it is not the numerical values of the x_i that determine the median, but only their positions.

In the following we illustrate how to calculate the above measures when observations of continuous data are given in the form of a frequency distribution.

Example 17.2:

We consider the fictive distribution in Tab. 17.2.

Tab. 17.2: Frequency distribution of heights of 300 students.

Classes (in kg)	Absolute frequencies, F_i
140 ⊢ 150	50
150 ⊢ 160	80
160 ⊢ 170	70
170 ⊢ 180	60
180 ⊢ 190	40

In fact, the exact heights of the students are not known. It is only known that $F_1 = 50$ students have a height between 140 and 150 cm, $F_2 = 80$ students have a height between 150 and 160 cm, etc. Thus the mean height can only be estimated. Assuming that the heights inside of an interval are approximately uniformly distributed, then the sum of weights of the persons in each interval is about $F_i\, x_i$, where x_i is the midpoint of the ith interval. This results in the following approximation for the mean weight:

$$\bar{x} = \frac{1}{n}\sum_i F_i x_i = \frac{1}{300}(50 \cdot 145 + 80 \cdot 155 + 70 \cdot 165 + 60 \cdot 175 + 40 \cdot 185) = 163.67 \text{ kg.}$$

The mode can also only be approximately determined. The following **formula of Czuber** is frequently used for this purpose:

$$\text{Mo} = l + \frac{d_1}{d_1 + d_2}\, h, \tag{17.4}$$

where l is the lower limit of the modal class, i.e., the class with the highest frequency, h is the amplitude of a class, F_1 is the frequency of the **modal class**, F_0 and F_2 are the frequencies of the classes preceding and succeeding this class, respectively. Finally, $d_1 = F_1 - F_0$ and $d_2 = F_1 - F_2$.

The modal class of Tab. 17.2 is the class 150 ⊢ 160, having amplitude $h = 10$. Its lower limit is $l = 150$. The considered frequencies are $F_1 = 80$, $F_0 = 50$ and $F_2 = 70$, implying $d_1 = 30$ and $d_2 = 10$. Thus, eq. (17.4) yields the mode $\text{Mo} = 150 + (30/(30+10)) \cdot 10 = 157.5$.

The formula (17.4) is based on the idea of estimating the real mode of the theoretical density, see Exercise 17.5.

To calculate the median in the continuous case it is most intuitive to illustrate it geometrically in the histogram, see Fig. 17.4. The frequency F_i of a class is proportional to the area of the rectangle over this class.

For example, the area of the rectangle over the class 140 ⊢ 150 is $10 \cdot 50$, and the total area of the histogram corresponds to $10 \cdot \sum_i F_i = 10 \cdot n = 3{,}000$. The median $\tilde{x} = Q_2 = 162.857$ is the point on the x-axis so that the dashed vertical line $y = \tilde{x}$ bisects the total area of 3,000 of the histogram. We say that $\tilde{x}$ has the area of 1,500 on the left (and the area of 1,500 on the right).

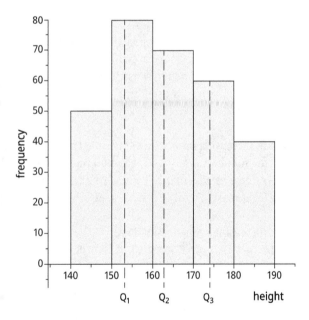

Fig. 17.4: Histogram with quartiles for Tab. 17.2.

The value of $\tilde{x}$ is easily determined as follows: First one observes that the class 160 ⊢ 170 contains the median since the endpoints of this class leave the areas of $500 + 800 = 1{,}300$ and $1{,}300 + 700 = 2{,}000$ on the left, respectively. Since $\tilde{x}$ must have the area of 1500 on the left, we get

$$500 + 800 + (\bar{x} - 160) \cdot 70 = 1{,}500, \tag{17.5}$$

implying $\bar{x} = 162.857$.

Now it is obvious how to generalize the concept of median for continuous data. The **quartiles** Q_1, Q_2 and Q_3 divide the area of the histogram in four equal areas, i.e., Q_1, Q_2 and Q_3 have the area of 25%, 50% and 75% of the histogram at the left, respectively. In particular, Q_2 equals the median. With a reasoning analogous to eq. (17.5) we get

$$500 + (Q1 - 150) \cdot 80 = 750$$

and

$$(180 - Q3) \cdot 60 + 400 = 750$$

since Q_3 has the area of 750 on the right. The last two equations imply $Q_1 = 153.125$ and $Q_3 = 174.167$, see Fig. 17.4. The **deciles** $D_1, \ldots, D_9$ divide the area of the histogram in 10 equal areas, i.e., D_i has $10 \cdot i$ % of the area of the histogram on the left, $i = 1, \ldots, 9$. Finally, the **percentiles** $p_1, \ldots, p_{99}$ divide the area of the histogram in 100 equal areas, where p_i has i % of the area on the left ($i = 1, \ldots, 99$). In principle, the area of the histogram can be divided in any number q of equal parts. This results in the **q-quantiles** $Q_1, \ldots, Q_{q-1}$, where Q_i has $100 \cdot (i/q)$ % of the area of the histogram on the left ($i = 1, \ldots, q - 1$).

We now define the quantiles for ungrouped data. We emphasize that various definitions are possible. The following concept might seem a bit artificial in contrast to the continuous case. In practical applications, it makes little sense to divide, for example, a sample of seven data into four parts. However, as usual in mathematics, a definition is formulated as generally as possible.

Definition 17.1: Let $x_1, \ldots, x_n$ be data in increasing order. The q-quantile Q_i is defined as follows for $i = 1, \ldots, q - 1$. Consider the numbers $S_i = i \cdot \frac{n+1}{q}$. If S_i is an integer we set $Q_i = x_{S_i}$. Otherwise we define the greatest integer less than S_i by $m = \lfloor S_i \rfloor$ and the corresponding fractional part by $t = m - S_i$. Then we set $Q_i = x_m + t(x_{m+1} - x_m)$.

Example 17.3: Consider the following increasing sequences of data:
(a) Determine the quartiles for $n = 12$: $(x_1, \ldots, x_{12}) = (3, 5, 8, 10, 15, 20, 35, 50, 60, 65, 70, 80)$
We get $S_i = \frac{13}{4} i$, thus

$$S_1 = \frac{13}{4} \to m = 3, \; t = \frac{1}{4} \to Q_1 = x_3 + \frac{1}{4}(x_4 - x_3) = 8.5,$$

$$S_2 = \frac{13}{2} \to m = 6, \; t = \frac{1}{2} \to Q_2 = x_6 + \frac{1}{2}(x_7 - x_6) = 27.5,$$

$$S_3 = \frac{39}{4} \to m = 6, \; t = \frac{3}{4} \to Q_3 = x_9 + \frac{3}{4}(x_{10} - x_9) = 63.75.$$

(b) Determine the quartiles for

    i. $n = 7$: $(x_1, \ldots, x_7) = (8, 12, 15, 20, 35, 50, 65)$

        We get $S_i = 2i$,

$$S_1 = 2 \rightarrow Q_1 = x_2 = 12,$$
$$S_2 = 4 \rightarrow Q_2 = x_4 = 20,$$
$$S_3 = 6 \rightarrow Q_3 = x_6 = 50.$$

(c) Determine the deciles for $n = 15$:

    $(x_1, \ldots, x_{15}) = (9, 11, 16, 25, 30, 40, 55, 60, 76, 80, 84, 89, 95, 100, 110)$

    We get $S_i = \frac{16}{10} i$,

$$S_1 = \frac{8}{5} \rightarrow m = 1, \ t = \frac{3}{5} \rightarrow D_1 = x_1 + \frac{3}{5}(x_2 - x_1) = 10.2,$$

$$S_2 = \frac{16}{5} \rightarrow m = 3, \ t = \frac{1}{5} \rightarrow D_2 = x_3 + \frac{1}{5}(x_4 - x_3) = 17.8,$$

$$S_3 = \frac{24}{5} \rightarrow m = 4, \ t = \frac{4}{5} \rightarrow D_3 = x_4 + \frac{4}{5}(x_5 - x_4) = 29.$$

The remaining deciles are $(D_4, \ldots, D_9) = (46, 60, 78.4, 85, 93.8, 104)$.

So far we have examined measures of position, which represent individual values relative to other values in the data set. In the remainder of this section, we examine measures characterizing the shape of the distribution, namely measures of dispersion, symmetry and kurtosis.

Given the ungrouped values $x_1, \ldots, x_n$ in increasing order. Then **range** and **interquartile range** are defined by

$$R = x_n - x_1 \tag{17.6}$$

and

$$IQR = Q_3 - Q_1. \tag{17.7}$$

The following three measures aim to describe the deviations of the single values from the mean $\bar{x}$. The **variance** and the **mean deviation** are given by

$$V = \sigma^2 = \frac{1}{n} \sum_i (x_i - \bar{x})^2 \tag{17.8}$$

and

$$D_M = \frac{1}{n} \sum_i |x_i - \bar{x}|, \tag{17.9}$$

and the **standard deviation** is defined by

$$\sigma = \sqrt{V}. \tag{17.10}$$

Example 17.4:

Let $(x_1, \ldots, x_{10}) = (1{,}000,\ 1{,}300,\ 1{,}450,\ 1{,}600,\ 1{,}730,\ 1{,}840,\ 2{,}050,\ 2{,}500,\ 3{,}150,\ 4{,}660)$ denote the net incomes of 10 persons in \$.

We obtain $R = 3{,}660$, $IQR = 2{,}662.5 - 1{,}412.5 = 1{,}250\$$ (see the previous example), $\bar{x} = 2{,}128\$$, $V = 1{,}053{,}776\$^2$, $D_M = 785.2\$$ and $\sigma = 1{,}026.54\$$.

In order to decide whether the value of a measure is large or small, one has to compare it with a reference value (e.g., with the average value). For this reason, the following relative measures, called **coefficients of dispersion**, are also considered:

$$\text{Coefficient of range} \quad \frac{x_n - x_1}{x_n + x_1} \tag{17.11}$$

$$\text{Coefficient of interquartile range} \quad \frac{Q_3 - Q_1}{Q_1 + Q_3} \tag{17.12}$$

$$\text{Coefficient of variance (or of variation)} \quad \frac{V}{\bar{x}} \tag{17.13}$$

$$\text{Coefficient of standard deviation} \quad \frac{\sigma}{\bar{x}} \tag{17.14}$$

$$\text{Coefficient of mean deviation} \quad \frac{D_M}{\bar{x}} \tag{17.15}$$

Example 17.4 (Continuation):

For the above data we obtain the following coefficients of dispersion:

$$\frac{3660}{5660} \approx 0.65, \quad \frac{1250}{4075} \approx 0.31, \quad 495, \quad 0.48 \text{ and } 0.37.$$

We now turn to the case of grouped data, illustrating the calculations using Tab. 17.2.

Analogous to eqs. (17.6)–(17.10) we define **range** and **interquartile range** by $R = M - m$ and $IQR = Q_3 - Q_1$, where M is the upper limit of the maximum interval and m the lower limit of the minimum interval and Q_3 and Q_1 are the quartiles previously defined for the continuous case. For the example given by Tab. 17.2 we get $R = 190 - 140 = 50$ and $IQR = 174.167 - 153.125 = 21.042$. As we did in Example 17.2, we make the hypothetical assumption that all values in the classes are equal to the midpoint of the interval. Therefore the **variance** and the **mean deviation** are now given by

$$V = \sigma^2 = \frac{1}{n} \sum_i (x_i - \bar{x})^2 \cdot F_i \tag{17.16}$$

and

$$D_M = \frac{1}{n}\sum_i |x_i - \bar{x}| \cdot F_i,$$ (17.17)

where x_i and F_i are the midpoints and frequencies of the classes, respectively. The **standard deviation** is again defined by

$$\sigma = \sqrt{V}.$$

For the distribution in Tab. 17.2 we have $n = 300$, $(x_1, \ldots, x_5) = (145,155,165,175,185)$, $(F_1, \ldots, F_5) = (50,80,70,60,40)$ and $\bar{x} = 163.67$, thus $V = 164.89$, $D_M = 10.84$ and $\sigma = 12.84$.

The definition of the relative measures is analogous to eqs. (17.11)–(17.15) and will not be further discussed here.

Before we turn to shape characteristics (see Section 8.4), we introduce the underlying concept of sample moments. In Definition 8.4, we defined the moments for arbitrary random variables. In particular, for a discrete variable X the **moments about the origin** are defined as

$$\mu_k' = E(X^k) = \sum_{i=1}^{\infty} x_i^k p_i$$ (17.18)

and the central moments as

$$\mu_k = E\left[\left(X - E(X)^k\right)\right] = \sum_{i=1}^{\infty} (x_i - E(X))^k \, p_i$$ (17.19)

where x_i and p_i denote the values of X and the corresponding probabilities.

For a sample $(x_1, \ldots, x_n)$ we define the **sample moments** as

$$m_k' = \frac{1}{n}\sum_{i=1}^{n} x_i^k$$ (17.20)

and

$$m_k = \frac{1}{n}\sum_{i=1}^{n} (x_i - \bar{x})^k$$ (17.21)

Formulas (17.20) and (17.21) are obtained from eqs. (17.18) and (17.19), substituting the probabilities p_i by $1/n$, i.e., one interprets the sample elements as the possible results of a discrete distribution that all appear with the same probability.

Analogously to eqs. (8.5) and (8.6) we obtain the **(sample) coefficients of skewness and kurtosis**

$$\alpha_3 = \frac{m_3}{\sigma^3}$$ (17.22)

$$\alpha_4 = \frac{m_4}{\sigma^4},\tag{17.23}$$

where σ is the standard deviation. We illustrate the shape measures (17.22) and (17.23) by means of some fictive examples.

Example 17.5:
Consider a group of 10 persons with weights in kilogram $(x_1, \ldots, x_{10}) = (68, 69, 79, 82, 84, 84, 98, 105, 109, 110)$. We obtain $\bar{x}=88.8$, $\sigma = 14.89$, $m_3 = 456.264$, $m_4 = 79{,}782.79$, $\alpha_3 = 0.1382$, and $\alpha_4=1.622$. The distribution is therefore slightly positively skewed and platykurtic (see Section 8.4)

Example 17.6:
Given the following frequency distribution of the number of children of 350 families in a certain district

Number of children, x_i	Number of families with i children, F_i
0	56
1	73
2	107
3	88
4	26

In the case of grouped data, formula (17.21) for the sample moments becomes

$$m_k = \frac{1}{n} \sum_i (x_i - \bar{x})^k F_i,\tag{17.24}$$

where n is the sample size. For the considered distribution we get $\bar{x}=1.871$, $\sigma = 1.175$, $m_3 = -0.1082$, $m_4 = 4.0157$, $\alpha_3 = -0.0667$, $\alpha_4=2.1068$. The distribution is slightly negatively skewed and platykurtic.

Example 17.7:
The weight distribution of 480 athletes of different disciplines (sprinter and weight-lifters, among others) who took part in an international competition is as follows.

Classes (in kg)	Absolute frequencies, F_i
40 ⊢ 50	52
50 ⊢ 60	122
60 ⊢ 70	75
70 ⊢ 80	62
80 ⊢ 90	51
90 ⊢ 100	40
100 ⊢ 110	35
110 ⊢ 120	23
120 ⊢ 130	12
130 ⊢ 140	8

As in the variance calculation for frequency distributions of continuous variables, we theoretically assume that all weights are equal to the midpoint of their class. Therefore, we use eq. (17.24) for the calculation of the sample moments, where the x_i are the midpoints of the classes. We obtain $\bar{x}=74.19$, $\sigma = 22.87$, $m_3 = 8{,}693.81$, $m_4 = 724{,}538$, $a_3=0.7271$, $a_4=2.6501$. According to the given criteria the distribution is positively skewed and platykurtic, see Fig. 17.5. For later use we note that $Mo = 55.98$, $Q_1 = 55.57$, $Q_2 = 68.8$, $Q_3 = 89.5$, $D_1 = 49.23$, $D_9 = 108.57$.

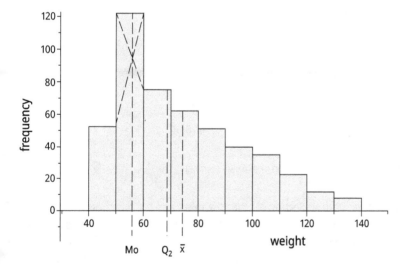

Fig. 17.5: Histogram of the weight distribution of 480 athletes.

Example 17.8:
The net income distribution of the 400 members of a bowling club is shown in the following table.

Classes (in $)	Absolute frequencies, F_i
1,000 ⊢ 1,500	10
1,500 ⊢ 2,000	20
2,000 ⊢ 2,500	22
2,500 ⊢ 3,000	59
3,000 ⊢ 3,500	113
3,500 ⊢ 4,000	110
4,000 ⊢ 4,500	35
4,500 ⊢ 5,000	19
5,000 ⊢ 5,500	12

We obtain $\bar{x}=3{,}352.50$, $\sigma = 822.80$, $\alpha_3 = -0.2320$, $\alpha_4 = 3.3878$. The distribution is slightly negatively skewed and leptokurtic, see Fig. 17.6. Furthermore, we get $Mo = 3473.68$, $Q_1 = 2906.78$, $Q_2 = 3393.81$, $Q_3 = 3845.45$, $D_1 = 2227.27$, $D_9 = 4371.43$.

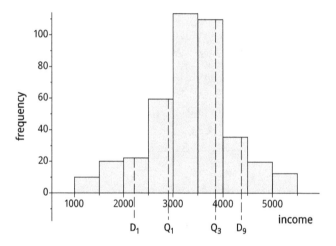

Fig. 17.6: Histogram of the income distribution of 400 club members.

We have already pointed out in Section 8.4 that the measures for skewness and kurtosis of a distribution cannot always represent a geometric property perfectly. The same also applies to the analog concepts (17.22) and (17.23) for a sample. Due to these problems, attempts have often been made to fix the weak points of the known concepts or to develop alternative measures. For example, it turned out that the measure (17.22) is imprecise for a small sample size n. To compensate for this inaccuracy, the **adjusted coefficient of skewness**, defined by $\dfrac{\sqrt{n\,(n-1)}}{n-2}\,\alpha_3$ is often used. For larger values of n this factor is approximately 1. The so-called **excess kurtosis** is defined by $\alpha_3 - 3$. This value is positive if the distribution is leptokurtic, and negative if the distribution is platykurtic. For a (theoretical or empirical) distribution usually applies

Mo $< \tilde{x} < \bar{x}$ when the distribution is positively skewed,
$\bar{x} < \tilde{x} <$ Mo when the distribution is negatively skewed and $\qquad$ (17.25)
$Mo = \tilde{x} = \bar{x}$ in the case of symmetry.

We verify these relations by means of the last two examples, see Fig. 17.5.
Based on eq. (17.25) one gets the following two coefficients of Pearson

$$P_1 = \frac{\bar{x} - Mo}{\sigma}, \qquad (17.26)$$

$$P_2 = 3 \frac{\bar{x} - x}{\sigma}, \qquad (17.27)$$

which are generally positive, negative or zero when the distribution is positively skewed, negatively skewed or symmetric, respectively. The **Galton skewness** is given by

$$G_{\text{skew}} = \frac{(Q_3 - Q_2) - (Q_2 - Q_1)}{Q_3 - Q_1} = \frac{Q_1 + Q_3 - 2\,Q_2}{Q_3 - Q_1}. \qquad (17.28)$$

A quantile-based measure like this is much more intuitive than one based on moments. The last expression compares the right side of the interquartile range $(Q_3 - Q_2)$ with its left side $(Q_2 - Q_1)$. There are innumerable variants of this formula conceivable, for example, one could substitute Q_3 and Q_1 by the deciles D_9 and D_1. There are also quantile-based measures of the kurtosis which can also be easily interpreted by means of intuitive geometrical considerations, for example, the coefficient of kurtosis:

$$K = \frac{Q_3 - Q_1}{2(D_9 - D_1)}, \qquad (17.29)$$

which compares the interquartile range $(Q_3 - Q_1)$ with the interdecile range $(D_9 - D_1)$, see Fig. 17.6. For the standardized normal distribution we get $K = 0.2632$; consequently, a distribution is said to be leptokurtic if $K > 0.2632$ and platykurtic if $K < 0.2632$.

One must be aware of the fact that all measures of skewness and kurtosis are just "rules of thumb". Values of different measures can deviate significantly from one another. Two measures of skewness may even have different signs, i.e., one of the measures characterizes a distribution as positively skewed and the other as negatively skewed. Table 17.3 summarizes all measures introduced in this chapter for the last two examples. The interpretation is left to the reader.

Tab. 17.3: Shape characteristics for the distributions in Examples 17.7 and 17.8.

Example	Skewness				Kurtosis	
	a_3	P_1	P_2	G_{skew}	a_4	K
17.7	0.7271	0.7962	0.7070	0.2202	2.6501	0.2859
17.8	−0.2320	−0.1473	−0.1506	−0.0377	3.3878	0.2189

17.3 Bidimensional analyses

So far in this chapter we have dealt with data from one random variable only. In this section we study the relationship between two variables. The following measure compares two quantitative variables X and Y. We assume that we have a sample of paired data $(x_1, y_1), \ldots, (x_n, y_n)$. For example, Tab. 17.4 shows the heights and corresponding weights of 10 randomly chosen persons.

Tab. 17.4: Data of a sample of $n = 10$ adults.

Person i	Height, x_i (in cm)	Weight, y_i (in kg)	Theoretical weights $z_i = \hat{a}x_i + \hat{b}$
1	158	53	53.19
2	162	57	57.76
3	169	69	65.76
4	172	75	69.19
5	176	63	73.76
6	181	80	79.47
7	184	85	82.90
8	192	88	92.04
9	195	93	95.47
10	195	102	95.47

$\hat{a} = 1.1427$, $\hat{b} = -127.4$, $\rho = 0.9489$.

The data pairs (x_i, y_i) are illustrated as points in Fig. 17.7, where the straight line has been determined by linear regression, see Section 15.4.

We define a correlation coefficient analogously to Section 8.6. From eqs. (8.17) and (8.18) it follows that the population correlation coefficient is defined by

$$\rho = \frac{E(X, Y) - E(X)E(Y)}{\sqrt{V(X)\ V(Y)}}.$$

Substituting $E(XY)$, $E(X)$ and $V(X)$ by the corresponding sample moments $\frac{1}{n}\sum_{i=1}^{n} x_i y_i$, $m_1' = \frac{1}{n}\sum_{i=1}^{n} x_i = \bar{x}$ and $m_2 = \frac{1}{n}\sum_{i=1}^{n}(x_i - \bar{x})^2 = \frac{1}{n}\sum_{i=1}^{n} x_i^2 - \bar{x}^2$, see eqs. (17.20) and (17.21) we obtain the **sample (linear) correlation coefficient**

$$\rho = \frac{\frac{1}{n}\sum_{i=1}^{n} x_i y_i - \bar{x}\bar{y}}{\sqrt{\left(\frac{1}{n}\sum_{i=1}^{n} x_i^2 - \bar{x}^2\right)\left(\frac{1}{n}\sum_{i=1}^{n} y_i^2 - \bar{y}^2\right)}} = \frac{\sum_{i=1}^{n} x_i y_i - n\bar{x}\bar{y}}{\sqrt{\left(\sum_{i=1}^{n} x_i^2 - n\bar{x}^2\right)\left(\sum_{i=1}^{n} y_i^2 - n\bar{y}^2\right)}} \quad (17.30)$$

For the data in Tab. 17.4, we obtain $\bar{x} = 178.4$, $\bar{y} = 76.5$, $\sum_{i=1}^{n} x_i^2 = 319{,}860$, $\sum_{i=1}^{n} y_i^2 = 60{,}835$ and $\sum_{i=1}^{n} x_i y_i = 138{,}298$, resulting in $\rho = \frac{138{,}298 - 10 \cdot 76.5 \cdot 178.4}{\sqrt{(319{,}860 - 10 \cdot 178.4^2)(60{,}835 - 10 \cdot 76.5^2)}} = 0.9489$.

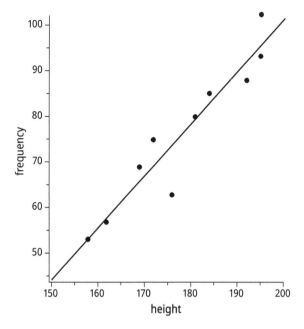

Fig. 17.7: Scatterplot for the data in Tab. 17.4.

The parameters of the straight line $\hat{a}x + \hat{b}$ in Fig. 17.7 and the value of ρ are presented at the bottom of Tab. 17.4. The last column in Tab. 17.4 contains the theoretical values. In order to understand the interpretation of the correlation coefficient we consider several examples. In Tab. 17.5 we relate the time that insurance brokers operate in a given city with the number of clients acquired. The relation is illustrated in Fig. 17.8.

Tab. 17.5: Time of activity and number of clients.

Broker, i	Time of activity, x_i (in years)	Number of clients, y_i
1	1	5
2	3	60
3	4	25
4	6	90
5	7	30
6	9	150
7	12	100
8	13	180
9	15	120
10	20	210

$\hat{a} = 9.9844$, $\hat{b} = 7.1406$, $\rho = 0.8658$.

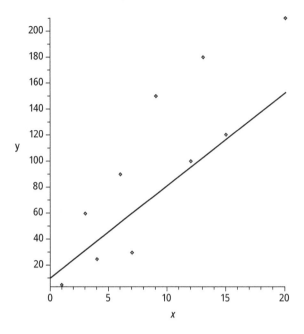

Fig. 17.8: Activity time versus client number.

The correlation coefficient is now $\rho = 0.8658$.

In general, the sample correlation coefficient satisfies

$$-1 \leq \rho \leq 1 \tag{17.31}$$

(see Theorem 8.8). In the case $\rho = 1$ a perfect positive linear relationship between the variables X and Y exists, i.e., all points (x_i, y_i) lie on an increasing straight line. If ρ is "close to 1," then the points in the scatterplot lie close to an increasing straight line, and the larger ρ is, the closer the points are to the line (see the examples above). For negative values of ρ the interpretation is analogous, except that the data points are now scattered around a descending straight line. For "small" absolute values of ρ no straight line can be identified around which the data points are scattered.

In Tab. 17.6 the average values of cars of a certain brand depending on age are listed. The graphical illustration is given in Fig. 17.9.

Tab. 17.6: Wear and tear of a car brand.

Car, i	Age, x_i (in years)	Mean value, y_i (in 1000 \$)
1	1	100
2	2	90
3	5	70
4	6	50
5	8	70
6	8	60
7	11	45
8	13	35
9	15	20
10	16	40

$\hat{a} = -4.4124$, $\hat{b} = 95.505$, $p = -0.9174$.

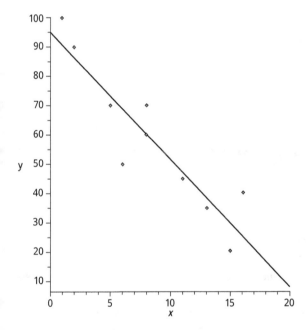

Fig. 17.9: Mean car value by age.

The reader might ask when a correlation is strong and when it is weak. But it is hardly possible to answer this question objectively. Only a very rough classification is possible. Table 17.7 presents recommendations usually found in textbooks.

Tab. 17.7: Classification of the correlation coefficient.

Range of values	Strength of correlation
$0.9 < |\rho| \le 1$	Very strong
$0.7 < |\rho| \le 0.9$	Strong
$0.5 < |\rho| \le 0.7$	Moderate
$0.3 < |\rho| \le 0.5$	Weak
$|\rho| \le 0.3$	Inexistent

Example 17.9: Assume that we have nine data pairs (x_i, y_i) with $(x_1, x_2, \ldots, x_9) = (-4, -3, \ldots, 3, 4)$ and $y_i = x_i^2$ for $i = 1, \ldots, 9$. It can be easily verified that $\rho = 0$, i.e., the coefficient of correlation indicates that there is no (linear) correlation between the variables X and Y.

The example shows that nonlinear relationships may not be discovered, using the measure ρ.

Another way to decide whether a correlation is strong is to study the distribution of the correlation coefficient. It can be shown that the test statistics $t = \rho\sqrt{(n-2)/(1-\rho^2)}$ has approximately a t-Student distribution with $n - 2$ degrees of freedom, when the data x_i and y_i are normally distributed.

Example 17.10: The data in Tab. 17.4 result in $t = 0.9489\sqrt{(10-2)/(1-0.9489^2)} = 8.505$. The probability to observe such a high correlation, assuming that the heights and weights are normally distributed, is $P(T \ge 8.505) = \int_{8.505}^{\infty} t_8 \, dt = 0.000014$, where t_8 is the Student's density with 8 degrees of freedom. Thus the null hypothesis H_0 that such a high value of ρ is coincidental, must be rejected with very high evidence.

Let us now assume for comparison that the correlation coefficient were $\rho = 0.5$. This would result in $t = 0.5\sqrt{(10-2)/(1-0.5^2)} = 1.632$ and $P(T \ge 1.632) = \int_{1.632}^{\infty} t_8 \, dt = 0.07$, and H_0 could not be rejected for a significance level of 5%.

In all of the previous examples in this section, we started with a sample of paired data (x_i, y_i) and calculated a measure that showed the strength of the correlation between the values x_i and the corresponding values y_i.

We now turn to a similar problem. We have a one-dimensional sample of data $x_1 \ldots x_n$ of a random variable X and a model to describe these data. We now have to compare each empirical value x_i with a corresponding theoretical value z_i. We consider three types of squared differences:
Total sum of squares:

$$SS_{tot} = \sum_{i=1}^{n} (x_i - \bar{x})^2. \tag{17.32}$$

Residual sum of squares (unexplained sum of squares):

$$SS_{res} = \sum_{i=1}^{n} (x_i - z_i)^2. \tag{17.33}$$

Regression sum of squares (explained sum of squares):

$$SS_{reg} = \sum_{i=1}^{n} (z_i - \bar{x})^2. \tag{17.34}$$

Now we define the **coefficient of determination** by

$$R^2 = 1 - \frac{SS_{res}}{SS_{tot}}. \tag{17.35}$$

satisfying

$$0 \le R^2 \le 1. \tag{17.36}$$

This is a measure for the goodness of fit that may be interpreted as the proportion of variance explained by the model. According to this measure, a model describes the data x_i the better, the larger R^2 is. In the ideal case, the values x_i and z_i are identical, and then it holds $SS_{res} = 0$ and thus $R^2 = 1$. In many applied areas a value $R^2 > 0.9$ is interpreted as a very good model fit.

Example 17.11: Suppose that the data $(x_1, \ldots, x_5) = (4, 5, 8, 16, 18)$ are given and we want to model them by a curve ax^b. To fit the model as well as possible, we determine the parameters a and b such that the value R^2 is maximized. Since SS_{tot} does not contain parameter values, this is equivalent to minimizing

$$SS_{res}(a, b) = \sum_{i=1}^{5} \left(x_i - a \cdot i^b \right)^2.$$

With the help of an appropriate software we obtain the optimal parameter values $\hat{a} = 2.3849$, $\hat{b} = 1.2761$, resulting in $SS_{res} = 10.47$, $SS_{tot} = 164.8$ and $R^2 = 0.9365$. The model is therefore well suited for the given data x_i. Table 17.8 compares empirical and theoretical values.

Tab. 17.8: Fitting results of Example 17.11.

x_i	$z_i = \hat{a}\, i^b$
4	2.3849
5	5.7761
8	9.6906
16	13.9891
18	18.5977

In principle, any continuous or discrete function can be fitted to given data $x_1 \ldots x_n$ in this way. In any case, the sum of squared deviations between empirical and theoretical data $SS_{res} = \sum_{i=1}^{n} (x_i - z_i)^2$ has to be minimized. We present an application with real data, see the site www.worldometers.info.

Example 17.12: The growth of the Brazilian population since 1955 is shown in Tab. 17.9. Since it is well known that a population grows in general exponentially, we fit the model ae^{bx} to the values. The fitting process yields $\hat{a} = 70{,}943$, $\hat{b} = 0.08364$, thus $z_i = \hat{a}e^{\hat{b} i}$. For example, the population in 1980 was $x_6 = 120{,}694{,}000$ persons, the model prediction is $z_6 = 117{,}183{,}000$ persons. Moreover, we get $SS_{res} = 1.233 \cdot 10^9$, $SS_{reg} = 3.072 \cdot 10^{10}$, $SS_{tot} = 3.372 \cdot 10^{10}$ and $R^2 = 0.9634$, i.e., the model fits the observed values well. However, a closer look at Fig. 17.10 reveals an interesting trend. The population growth accelerates from 1955 ($i = 1$) to about 1990 ($i = 8$) and declines thereafter. This may indicate a changed attitude toward family planning. To account for this, more complex models are required.

Tab. 17.9: Population growth in Brazil.

i	Year	Obs. value x_i (in thousands)	Predicted value z_i (in thousands)
1	1955	62,534	77,132
2	1960	72,179	83,861
3	1965	83,374	91,177
4	1970	95,113	99,132
5	1975	107,216	107,780
6	1980	120,694	117,183
7	1985	135,274	127,406
8	1990	149,003	138,521
9	1995	162,020	150,606
10	2000	174,790	163,745
11	2005	186,127	178,030
12	2010	195,714	193,561
13	2015	204,472	210,448
14	2020	212,559	228,808

When modeling by linear regression, interesting relations can be observed between the squared differences (17.32)–(17.34).

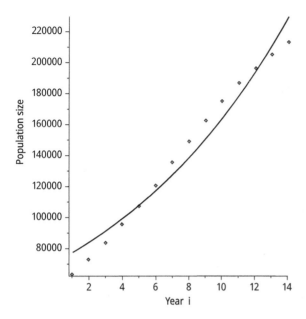

Fig. 17.10: Population growth in Brazil (1955–2020).

Proposition 17.1: Given paired data $(x_1, y_1), \ldots, (x_n, y_n)$, representing outcomes of two related random variables X and Y. Let $\hat{a}X + \hat{b}$ be the regression line, where

$$\hat{a} = \frac{\sum\limits_{i=1}^{n} x_i y_i - n\bar{x}\bar{y}}{\sum\limits_{i=1}^{n} x_i^2 - n\bar{x}^2} \quad \text{and} \quad \hat{b} = \bar{y} - \hat{a}\bar{x},$$

see Section 15.4. Thus, $z_i = \hat{a}\,x_i + \hat{b}$ are the theoretical values that the model assigns to the empirical values y_i. Then the following relations hold

(i) $SS_{tot} = SS_{res} + SS_{reg}$, equivalent to

$$\sum\limits_{i=1}^{n} (y_i - \bar{y})^2 = \sum\limits_{i=1}^{n} (y_i - z_i)^2 + \sum\limits_{i=1}^{n} (z_i - \bar{y})^2.$$

(ii) $\rho^2 = R^2$

i.e., the coefficient of determination (17.35) is the square of the coefficient of linear correlation (17.30).

The relations are very important for practical applications. The proofs are not trivial, and the interested reader is referred to books on Linear Regression Analysis for more details. We limit ourselves to verifying the relationships using the above examples:

For the observed weights in Tab. 17.4 we obtain $SS_{tot} = \sum_{i=1}^{n} (y_i - \bar{y})^2 = 2{,}312.5$. The corresponding theoretical weights found by linear regression are $z_i = \hat{a}x_i + \hat{b} = 1.1427$

$x_i - 127.4$, yielding $SS_{res} = \sum_{i=1}^{n}(y_i - z_i)^2 = 230.41$, and $SS_{reg} = \sum_{i=1}^{n}(z_i - \bar{y})^2 = 2{,}082.09$, verifying item (i) of Proposition 17.1. By using eq. (17.35) we obtain $R^2 = 1 - \frac{230.41}{2,312.5} = 0.9004$. Comparing with the value ρ in Tab. 17.4 yields $R^2 = \rho^2 = 0.9489^2$, verifying item (ii) of Proposition 17.1.

For the nonlinear model in Example 17.12 we have found $SS_{tot} = 3.372 \cdot 10^{10}$, $SS_{res} = 1.233 \cdot 10^9$, $SS_{reg} = 3.072 \cdot 10^{10}$ and $SS_{res} + SS_{reg} = 3.195 \cdot 10^{10}$, thus property (i) of the proposition holds only approximately.

Exercises

17.1 Given the ordered data $(x_1, \ldots, x_{30}) = (3, 7, 9, 14, 25, 29, 35, 35, 35, 45, 65, 68, 72, 80, 90, 98, 103, 110, 115, 120, 126, 128, 137, 137, 140, 155, 160, 160, 170, 170)$.
Determine mean, variance, mode, the quartiles, the deciles and the 7-quantiles. Write a program to determine these characteristics.

17.2 Given strictly increasing values $x_1, \ldots, x_n$, where n is divisible by 20. Show that the quartiles Q_i fall between $x_{ni/4}$ and $x_{ni/4+1}$ for $i = 1, 2, 3$ and that the deciles D_i fall between $x_{ni/10}$ and $x_{ni/10+1}$ for $i = 1, \ldots, 9$.

17.3 Determine mean, variance, mode, the quartiles and the deciles for the frequency distribution in Tab. 17.1.

17.4 (i) The age distribution of 30 members of a swimming club is as follows:

Classes (in years)	Absolute frequencies, F_i
20 ⊢ 30	6
30 ⊢ 40	7
40 ⊢ 50	8
50 ⊢ 60	5
60 ⊢ 70	4

Calculate the mean age by means of the approximation formula, see Example 17.2.
(ii) Assume now that the exact ages are given by
$(x_1, \ldots, x_{30}) = (20, 21, 23, 25, 27, 29, 31, 32, 34, 36, 37, 37, 39, 40, 41, 42, 44, 45, 47, 47, 49, 50, 52, 55, 58, 59, 62, 64, 67, 68)$.
Calculate the mean with the formula for ungrouped data and compare with the result of part (i).
(iii) What is the theoretically largest possible error in calculating the mean by the approximation?

17.5 Let us imagine that a car insurance company has registered the damage claims of $n = 100$ policyholders. In our hypothetical case we assume that the amount of damage X (with unit of $\$1{,}000$) is described by a gamma distribution with parameters $a = 0.5$ and $r = 2.75$. For the amount of damage we consider the seven classes $0 \vdash 2$, 2

⊢ 4, . . ., 12 ⊢ 14. The probability that a value of X falls in the class C_i in the table below is $p_i = P(2i - 2 \le X \le 2i) = \int_{2(i-1)}^{2i} f(x)dx$, where the gamma density is $f(x) = \frac{a}{\Gamma(r)} (ax)^{r-1} e^{-ax}$. The theoretically expected frequencies of the classes C_i are np_i. Calculating these expressions and rounding to the closest integer results in the values F_i of the table. (One can also interpret X as a multinomial distributed variable, where the classes are the possible events, and apply Theorem 9.9 to get the expectations.). We obtain the following frequency distribution of the amount of damage.

Classes (in 1,000$)	Absolute frequencies, F_i
$C_1 = 0 \vdash 2$	11
$C_2 = 2 \vdash 4$	27
$C_3 = 4 \vdash 6$	25
$C_4 = 6 \vdash 8$	17
$C_5 = 8 \vdash 10$	10
$C_6 = 10 \vdash 12$	5
$C_7 = 12 \vdash 14$	3

Verify the frequencies in the table. Illustrate the data by a histogram and the theoretical curve $nf(x)$. Determine the mode by the Czuber formula and compare with the maximum point of $nf(x)$.

17.6 Given the following frequency distribution of the number of children of 400 families in a certain district. Calculate the mean, the mode and the median. In the case of the median, convert the data into ungrouped data.

Number of children, i	Number of families with i children, F_i
0	56
1	70
2	104
3	88
4	56
5	26

17.7 Consider the histogram of Example 17.2 (Fig. 17.4). Let $A(x)$ denote the area on the left of x. Realize that $F(x) = (1/A_{tot}) A(x)$ is a distribution function, called the empirical distribution function, where $A_{tot} = 3{,}000$ is the total area of the histogram. Give the formula for F and illustrate the function geometrically. Describe the relationship between F, the density and the quantile function in general.

17.8 Discuss the measures (17.8) and (17.9) and formulate a possible generalization.

17.9 Consider the two data sets $(x_1, x_2, x_3) = (50, 100, 150)$ and $(y_1, y_2, y_3) = (4{,}950, 5{,}000, 5{,}050)$ which might be interpreted as the hypotethical weights of three persons and three elephants in kilogram, respectively. Intuitively one can say that the persons have very different weights while the weights of the elephants are almost identical.

Apply all of the measures (17.6)–(17.15) to the two data sets and verify that only the coefficients (17.11)–(17.15) are useful for comparison.

17.10 Show that the variance in eq. (17.16) can be written as $V = \sum_i x_i^2 \cdot f_i - \bar{x}^2$. Is there an analogous transformation for eq. (17.17)?

17.11 Compare the variances and standard deviations of the data in parts (i) and (ii) of Exercise 17.4.

17.12 Determine skewness a_3 and kurtosis a_4 for
(i)   the data in Exercise 17.1,
(ii) the data in Exercise 17.6,
(iii) and the frequency distribution in Exercise 17.5.

17.13 Determine a_3, a_4 and the measures (17.26)–(17.29) for the following distribution of heights of 500 persons. Write a program to determine the quantiles of a continuous distribution given in form of a frequency distribution.

Classes (in cm)	Absolute frequencies, F_i
140 ⊢ 150	80
150 ⊢ 160	120
160 ⊢ 170	150
170 ⊢ 180	90
180 ⊢ 190	60

17.14 Determine the coefficient P_1 (see 17.26) for the gamma distribution. Show that this coefficient also depends only on the parameter r (see Section 8.4). Compare P_1 and a_3 for the values of r indicated in Fig. 8.3.

17.15 Determine the quantile function of the Weibull distribution (inverse of the distribution function). Use the results to determine Q_1, Q_2, Q_3, D_1 and D_9. Determine the measures (17.28) and (17.29) and show that they are independent of the parameter a (compare with Exercise 8.23).

17.16 Show that the coefficients a_3 and a_4 for grouped data have the following invariance properties. Both coefficients do not change by
(i)   multiplying the frequency vector $(F_1, \ldots, F_k)$ by a constant $c > 0$,
(ii) adding a constant c to each component of the vector $(x_1, \ldots, x_k)$ of interval midpoints (which corresponds to a shift in the intervals).
(iii) multiplying the vector $(x_1, \ldots, x_k)$ by any constant $c > 0$.

17.17 Assume that paired data (x_i, y_i) are given with $y_i = cx_i$ for $i = 1, \ldots, n$. Show that $\rho = 1$ when c is positive and $\rho = -1$ when c is negative.

17.18 Consider the following modification of the wear and tear data in Tab. 17.6:

Car, i	Age, x_i (in years)	Mean value, y_i (in $1,000)
1	1	125
2	2	60
3	4	70
4	6	100
5	6	55
6	8	30
7	12	25
8	12	60
9	14	20
10	15	40

Determine ρ and the linear regression line. Create a scatter plot. Would you call this a "strong" correlation (see Tab. 17.7)? Verify Proposition 17.1 for the presented example.

17.19 Determine the correlation coefficient $\rho(n)$ for the paired data $(x_i, y_i) = (i, i^3)$ for $i = 1, \ldots, n$. How does ρ change for increasing values of n? Determine $\rho(10)$ and $\rho(\infty)$.

17.20 Assume that normally distributed paired values (x_i, y_i) are given for $i = 1, \ldots, 20$. What is the highest value of ρ for which the null hypothesis H_0: $\rho = 0$ (there is no correlation) may be accepted. Use the test statistics of Example 17.10 and a significance level of 5%.

17.21 Perform a linear regression for the population growths data of Example 17.12. Determine the predicted populations as in Tab. 17.9 and create a scatter plot. Determine ρ and the quantities of Proposition 17.1. Interpret the results.

17.22 In practice, data are often uncertain. Assume that the weight of the second person in Tab. 17.4 is $57 + c$. Express the regression results at the bottom of this table as a function of c.

Appendix

Solutions to exercises

Chapter 1

1.1.2   (a) $\{4,5,6,9\}$ (b) $\{4,5,9\}$ (c) $\{1, \ldots, 5,7, \ldots, 10\}$ (d) $\{7,8,10\}$

1.1.3   There are four ways.

1.1.5   Not necessarily, since a set $A_i \cap B$ may be empty.

1.1.8 2^{mn}

1.1.9   (a) No, since $A \cap B$ may be finite, for example if $A = \{1,3,5,7, \ldots\}$ and $B = (1,2,4,6,8, \ldots\}$

      (b) Yes, let $A = \{a_1, a_2, \ldots\}$, $B = \{b_1, b_2, \ldots\}$. Define the function in Definition 1.3 by $f(n) = b_{n/2}$ for n even and $f(n) = a_{(n+1)//2}$ for n odd.

      (c) No. For example, if $C = [1,2]$ and $D = \{2,3\}$, then $C \cap D$ is finite.

      (d) Yes, since $C \cup D$ contains the uncountably infinite set C.

1.1.10   For $n = 2$ construct the function f in a geometric manner similar to Example 1.2. Use induction.

1.2.1   There are 8! ways.

1.2.2   10! words

1.2.3 $26^3 \cdot 10^5$.

1.2.5   (a) $\begin{pmatrix} 18 \\ 6 \end{pmatrix}$ (b) $\begin{pmatrix} 8 \\ 3 \end{pmatrix} \begin{pmatrix} 10 \\ 4 \end{pmatrix}$

1.2.6 $\begin{pmatrix} 200 \\ 3 \end{pmatrix} \begin{pmatrix} 300 \\ 5 \end{pmatrix} \begin{pmatrix} 400 \\ 4 \end{pmatrix}$

1.2.8   Suppose that an urn contains n_i balls of the color i for $i = 1, \ldots r$. Then there exist $\begin{pmatrix} n_1 \\ i_1 \end{pmatrix} \cdots \begin{pmatrix} n_r \\ i_r \end{pmatrix}$ k-combinations of balls, using i_j balls of color j. Hence it holds

$$\begin{pmatrix} n_1 + \cdots + n_r \\ k \end{pmatrix} = \sum_{(i_1, \ldots i_r)} \begin{pmatrix} n_1 \\ i_1 \end{pmatrix} \cdots \begin{pmatrix} n_r \\ i_r \end{pmatrix},$$

      where the sum runs over all r-tuples $(i_1, \ldots, i_r)$ with $i_1 + \cdots + i_r = k$ and $0 \le i_j \le n_j$.

1.2.11   The recurrence formula follows immediately from Proposition 1.8.

1.2.12   The assertions follow from Proposition 1.6 for $a_1 = \cdots = a_k = 1$ and $a_1 = \cdots = a_{k-1} = -1$, $a_k = k - 1$.

1.2.13 $\begin{pmatrix} 18 \\ 3, 4, 5, 6 \end{pmatrix} = 514,594,080$

1.2.14 $\begin{pmatrix} n \\ k \end{pmatrix} D_{n-k}$

1.2.15 $4! - D_4 = 24 - 9 = 15$

https://doi.org/10.1515/9783111332277-018

1.2.16 (a) The assertion follows from Proposition 1.4 for $n = m = k$,

(b) $\begin{pmatrix} -1/2 \\ n \end{pmatrix} = \frac{(-\frac{1}{2})(-\frac{3}{2})\cdots(-\frac{2n-1}{2})}{n!} = \frac{1\cdot3\cdot5\cdots(2n-1)}{(-2)^n n!}$

$= \frac{(2n)!}{(-2)^n n!\cdot2\cdot4\cdots(2n)} = \frac{(2n)!}{(-2)^n n!\cdot2^n n!} = \frac{1}{(-4)^n}\begin{pmatrix} 2n \\ n \end{pmatrix}$.

1.3.1 From formula (1.7), it follows that $\Gamma(x) = 0$ is impossible.

1.3.3 The relation follows by induction, using eq. (1.5).

1.3.4 $\Gamma\left(-\frac{n}{2}\right) = \frac{(-2)^{(n+1)/2}\sqrt{\pi}}{n\cdot(n-2)\cdot(n-4)\cdots1}$ for $n \in \mathbb{N}$, n odd.

1.3.5 The assertion follows from (1.19), setting $z = 1$ and observing that $\psi(1) = -\gamma$.

1.3.7 (a) set $t = x^b$ in (1.26)

(b) Set $t = \frac{x}{1+x}$ in (1.26)

(c) Set $t = bx^c$ and $z = \frac{a}{c}$ in (1.3).

1.3.8 It holds, e.g., $B(x,y) = \dfrac{1}{x\cdot\begin{pmatrix} x+y-1 \\ x \end{pmatrix}}$

1.4.1 (a) $-3/4$ (b) $x(\ln(x)-1)$ and $\frac{1}{2}\ln(y) - \frac{1}{4}$

1.4.2 (a) 31.2240 (b) 6.6404

1.4.3 (a) $\int_0^1\int_{-x}^x 3y^2 e^x\,dy\,dx + \int_1^2\int_{x-2}^{2-x} 3y^2 e^x\,dy\,dx = 12(1-3e+e^2)$,

(b) $\int_0^1\int_0^1 3(v-u^2)^2 e^{v+u^2} 4u\,dv\,du = 12(1-3e+e^2)$,

$G = \{(u,v)\mid 0 \le u,\ v \le 1\},\ J = 4u.$

1.4.4 (a) $\int_{-\sqrt{2}}^{\sqrt{2}}\int_{-\sqrt{1-x^2/2}}^{\sqrt{1-x^2/2}} (x^2+5xy^2)^2\,dy\,dx$

$= \int_{-\sqrt{2}}^{\sqrt{2}}\left[x^2\sqrt{4-2x^2} + \frac{5}{12}x(4-2x^2)^{3/2}\right]dx = \frac{\pi}{\sqrt{2}}.$

(b) The relation $x^2 + 2y^2 \le 2$ is equivalent to $(r\cos(\Phi))^2 + 2(r\sin(\Phi))^2 \le 2 \Leftrightarrow r^2[\cos^2(\Phi)$

$+ 2\sin^2(\Phi)] \le 2 \Leftrightarrow r \le \sqrt{\dfrac{2}{1+\sin^2(\Phi)}}.$

Thus, $G = \left\{(\Phi, r)\mid 0 \le \Phi \le 2\pi,\ 0 \le r \le \sqrt{\dfrac{2}{1+\sin^2(\Phi)}}\right\}.$

$\int_0^{2\pi}\int_0^{\sqrt{2/(1+\sin^2(\Phi))}} r\left[(r\cos(\Phi))^2 + 5r\cos(\Phi)(r\sin(\Phi))^2\right]dr\,d\phi = \frac{\pi}{\sqrt{2}}.$

1.4.5 176.4727.

1.4.6 $\frac{1}{12}(d^4-c^4)(b^3-a^3)$.

1.4.7 $\frac{2}{3}(c^{3/2}-d^{3/2})[a(\ln(a)-1)-b(\ln(b)-1)]$.

1.4.8 The volume of the ellipsoid is given by $\frac{4}{3}\pi abc$.

1.4.9 (a) $\int_{-a}^a\int_{-b\sqrt{1-x^2/a^2}}^{b\sqrt{1-x^2/a^2}} (x^2+y^2)\,dy\,dx = \frac{ab\pi}{4}(b^2-a^2)$.

(b) By proceeding as in Example 1.8 we obtain

$$\int_0^{2\pi} \int_0^{f(\phi)} r^3 \ dr \ d\phi = \frac{ab\pi}{4} \left(b^2 - a^2 \right),$$

where $f(\phi) = \sqrt{\left(\cos^2(\phi)/a^2 \right) + \left(\sin^2(\phi)/b^2 \right)}^{-1}$ is the maximum distance of an ellipse point from the origin, for a given angle ϕ. Adequate software can calculate this integral; however, the introduction of polar coordinates does not "simplify" the integration.

Chapter 2

2.3 The number of chosen balls is between 8 and 24 balls, thus $S = \{8, \ldots, 24\}$.

2.5 (i) $A \cup B \cup C$; (ii) $[(A \cap B) \cup (A \cap C) \cup (B \cap C)]\backslash(A \cap B \cap C)$; (iii) $(A \cup B)\backslash C$; (iv) $S\backslash(A \cap B \cap C)$

2.6 (i) $S = \{1, \ldots, 6\} \times \{1, \ldots, 6\}$, 36 outcomes

(ii) S consists of the 2-subsets of $\{1, \ldots, 5\}$, 10 outcomes

(iii) $S = \{1, \ldots, 5\} \times \{1, \ldots, 5\}$, 25 outcomes

2.7 (i) $(4,6),(5,5),(6,4)$.

(ii) $(1,1),(1,2),(2,1),(1,3),(2,2),(3,1),(1,4),(2,3),(3,2),(4,1)$.

(iii) The event is empty.

2.9 The sample space is $S = R_+^4$ and the considered event $E = [70, \infty[\times [100, \infty[\times \{[170, \infty[\times [0, \infty[\cup [0, \infty[\times [160, \infty[\}$.

2.10 (i) $S = \{0,1, \ldots, 6\} \times \{0,1, \ldots, 6\}$

(ii) $B_1 = \{(x,y) \mid 0 \le x < y \le 6\}$, $B_2 = \{(x,y) \mid 0 \le x,y \le 2\}$, $B_3 = \{(x,y) \mid 1 \le x,y \le 6\}$,

(iii) In at least one branch none of the cash machines is in use.

2.11 The results cannot be correct since they would imply

$$P(A \cup B) = P(A) + P(B) - P(A \cap B) = 0.8 + 0.7 - 0.4 = 1.1 > 1.$$

2.12 $P(A \cup B \cup C) = 1/5 + 1/5 + 1/5 - 0 - 0 - 1/10 + 0 = \frac{1}{2}$. Hence, the desired probability is $P(\bar{A} \cap \bar{B} \cap \bar{C}) = 1 - P(A \cup B \cup C) = 1/2$.

2.13 (a) $1 - z$ (b) $x - z$.

2.16 $P(\bar{A} \cap B) = P(B\backslash A) = P(B) - P(A \cap B)$,

$P(\bar{A} \cap \bar{B}) = 1 - P(A \cup B) = 1 - P(A) - P(B) + P(A \cap B)$.

2.17 The desired probability is $P(A) - P(A \cap B) - P(A \cap C) + P(A \cap B \cap C) = 0.5 - 0.25 - 0.2 + 0.15 = 0.2$.

Chapter 3

3.1 $\dfrac{\binom{5}{3}\binom{4}{1}+\binom{5}{4}\binom{4}{0}}{\binom{9}{4}} = \dfrac{5}{14}$

3.3 (i) We obtain the desired probability

$$\left[\binom{7}{2}\binom{6}{2}\binom{5}{4}+\binom{7}{2}\binom{6}{4}\binom{5}{2}+\binom{7}{4}\binom{6}{2}\binom{5}{2}+\binom{7}{3}\binom{6}{3}\binom{5}{2}\right.$$
$$\left.+\binom{7}{3}\binom{6}{3}\binom{5}{2}+\binom{7}{3}\binom{6}{2}\binom{5}{3}+\binom{7}{2}\binom{6}{3}\binom{5}{3}\right]/\binom{18}{8} = \dfrac{26{,}425}{43{,}758}.$$

(ii) Consider the event R = {the selection contains less than two red balls} and define the events B and G analogically for blue and green balls. The complement of the event considered in the exercise is $R \cup B \cup G$. We obtain

$$P(R) = \left[\binom{7}{0}\binom{11}{8}+\binom{7}{1}\binom{11}{7}\right]/\binom{18}{8} = \tfrac{26{,}425}{43{,}758} \text{ and in the same way } P(B) = \tfrac{5{,}247}{43{,}758},$$

$$P(G) = \tfrac{9{,}867}{43{,}758}. \text{ Furthermore, } P(R \cap B) = 0, \ P(R \cap G) = \binom{7}{1}\binom{6}{6}\binom{5}{1}/\binom{18}{8} = \tfrac{35}{43{,}758},$$

$$P(B \cap G) = \left[\binom{7}{7}\binom{6}{0}\binom{5}{1}+\binom{7}{1}\binom{6}{1}\binom{5}{1}+\binom{7}{6}\binom{6}{1}\binom{5}{1}\right]/\binom{18}{8} = \tfrac{221}{43{,}758}$$

and $P(R \cap B \cap G) = 0$. By means of the inclusion-exclusion principle we obtain now $P(R \cup B \cup G) = \tfrac{26{,}425 + 5{,}247 + 9{,}867 - 0 - 35 - 221 + 0}{43{,}758} = \tfrac{17{,}333}{43{,}758}.$ Hence the desired probability is $1 - \tfrac{17{,}333}{43{,}758} = \tfrac{26{,}425}{43{,}758}$ as in part (i).

3.4 (i) $\dfrac{\binom{700}{140}\binom{200}{160}}{\binom{900}{300}} \approx 5.91 \cdot 10^{-55}$. (ii) $1 - \dfrac{1}{\binom{900}{300}}\sum_{i=0}^{69}\binom{200}{i}\binom{700}{300-i} \approx 0.3135.$

3.5 $D_n = \Gamma(n+1, -1)e^{-1}$

3.6 (a) The probability is $\dfrac{\binom{n}{k}D_{n-k}}{n!} = \dfrac{1}{k!}\sum_{i=2}^{n-k}\dfrac{(-1)^i}{i!}.$

Numerical values: 0.3679, 0.3679, 0.1840, 0.0611, 0.0156, 0.0028, 0.0007, 0, 0 (b) 0.0190

3.7 $\dfrac{1}{ek!}$

3.10 (i) $1 - P(\bar{J} \cap M) = 1 - 10/60 = 5/6$

(ii) $1 - P(J \cap \bar{M}) = 1 - 17/60 = 43/60$

3.11 Define $p(x) := \dbinom{1{,}100}{200-x}\dbinom{300}{x} \Big/ \dbinom{1{,}400}{200}$

(i) $p(38) \approx 0.0506$

(ii) $1 - \sum_{x=0}^{49}p(x) \approx 0.1092$

3.12 $\frac{99}{100} \cdot \frac{98}{100} \cdot \ldots \cdot \frac{61}{100} \approx 0.1122 \cdot 10^{-3}$.

3.13 (i) 7^5

(ii) $7 \cdot 6 \cdot 5 \cdot 4 \cdot 3$

3.14 (i) By means of a tree diagram, we obtain for $n > 2$:

$$\frac{1}{n^2} + 2\frac{n-2}{n^2} + \frac{1}{n^2} = 2\frac{n-1}{n^2}$$

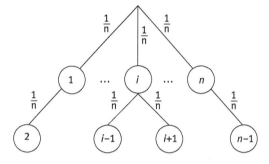

(ii) Similar to part (i) we get the probability $2/n$.

3.15 $2^{30} - \binom{30}{0} - \binom{30}{1} - \cdots - \binom{30}{5} = 1,073,567,387$.

3.16 (i) $\dfrac{1}{\binom{n}{k}} \displaystyle\sum_{(k_1 \cdots k_r)} \binom{n_1}{k_1} \cdots \binom{n_r}{k_r}$

where the sum runs over all r-tuples $(k_1, \ldots, k_r)$, satisfying $k_1 + \cdots + k_r = n$ and $m \leq k_i \leq n_i$ for $i = 1, \ldots, r$.

(ii) $\dfrac{1}{\binom{45}{40}} \displaystyle\sum_{k_1=6}^{8} \sum_{k_2=6}^{10} \sum_{k_3=6}^{12} \binom{8}{k_1}\binom{10}{k_2}\binom{12}{k_3}\binom{15}{40-k_1-k_2-k_3} = \frac{1,705}{1,763} \approx 0.9671$.

Note that the value $40 - k_1 - k_2 - k_3$ in the last binomial coefficient ranges from 10 to 22. If it is larger than 15, this coefficient is zero.

Chapter 4

4.1 Let X denote the number of good tubes drawn. We ask for $P(X = 2|X \geq 1) =$

$$\frac{P(X=2)}{P(X\geq1)} = \frac{\binom{10}{2}/\binom{15}{2}}{1 - \binom{5}{2}/\binom{15}{2}} = \frac{5}{13}.$$

4.2 Consider the events $A = \{$the first ball is black$\}$ and $B = \{$the second ball is black$\}$. The required probability is $P(B|A)P(A) + P(B|\bar{A})P(\bar{A}) = \frac{8}{14} \cdot \frac{2}{3} + \frac{7}{14} \cdot \frac{1}{3} = \frac{23}{42}$.

4.3 $P(B) = \frac{P(A \cup B) - P(A)}{1 - P(A)}$.

4.4 (i) $\frac{12}{30} \cdot \frac{11}{29} \cdot \frac{10}{28} = \frac{11}{203}$,

  (ii) $\frac{12}{30} \cdot \frac{11}{29} \cdot \frac{10}{28} + 3 \cdot \frac{18}{30} \cdot \frac{12}{29} \cdot \frac{11}{28} = \frac{352}{1,015}$

4.5 $= \frac{4}{5}$

4.6 It holds $P(A \cap B) = P(A) + P(B) - P(A \cup B) = p - 0.3$.
  Mutually exclusive for $p - 0.3 = 0 \Rightarrow p = 0.3$,
  Independent for $p - 0.3 = 0.5p \Rightarrow p = 0.6$.

4.7 Let X denote the number of tails obtained by a person:

$$P(X=0)^2 + P(X=1)^2 + P(X=2)^2 + P(X=3)^2 = \frac{5}{16}.$$

4.8 $\frac{10}{36 - 6} = \frac{1}{3}$

4.9 $\frac{3}{16}$

4.10 It holds $P(A \mid B) > P(A) \Leftrightarrow P(A \cap B) > P(A)P(B) \Leftrightarrow P(B \mid A) > P(B)$.

4.11 (i) $\frac{15}{25} \cdot \frac{14}{24} \cdot \frac{13}{23} = \frac{91}{460}$,

  (ii) $\dfrac{\binom{10}{2}\binom{15}{2}}{\binom{25}{4}} = \dfrac{189}{506}$,

  (iii) $\frac{15}{28}$.

4.13 (ii) $3 \cdot \frac{5}{12} \cdot \frac{4}{11} \cdot \frac{7}{10} \cdot \frac{3}{9} = \frac{7}{66}$

4.14 $\frac{25}{45} \cdot \frac{20}{44} \cdot \frac{24}{43} = \frac{20}{1,419}$

4.15 (i) $\frac{1}{15}$,

  (ii) $\frac{36}{6 \cdot 15} = \frac{2}{5}$,

  (iii) $\frac{P(X < Y, Z)}{P(X < Y)} = \frac{25 + 16 + 9 + 4 + 1}{6 \cdot (5 + 4 + 3 + 2 + 1)} = \frac{11}{18}$.

4.16 (ii) $P(A) = \frac{4.5 - 0.03 \cdot a}{100}$, $P(B_3 \mid A) = \frac{2a}{450 - 3a}$.

4.17 $\dfrac{\frac{5}{19} \cdot \frac{6}{18}}{\frac{5}{19} \cdot \frac{6}{18} + \frac{6}{19} \cdot \frac{5}{18} + \frac{8}{19} \cdot \frac{6}{18}} = \frac{5}{18}$

4.18 (i) $0.6 \cdot 0.2 + 0.3 \cdot 0.3 + 0.1 \cdot 0.15 = 0.225$,

  (ii) $\frac{0.6 \cdot 0.2}{0.225} = \frac{8}{15}$.

4.19 $\dfrac{0.05 \cdot 0.2}{0.05 \cdot 0.2 + 0.95 \cdot 0.01} = \frac{20}{39}$.

4.20 $\dfrac{\frac{1}{2} \cdot \frac{1}{3}}{\frac{1}{2} \cdot \frac{1}{3} + \frac{1}{2} \cdot 1} = \frac{1}{4}$.

4.21 (i) $\frac{35}{92}$

(ii) $\dfrac{\frac{2}{5}\cdot\frac{9}{24}\cdot\frac{8}{23}+\frac{2}{5}\cdot\frac{15}{24}\cdot\frac{9}{23}+\frac{3}{5}\cdot\frac{10}{24}\cdot\frac{9}{23}}{1-\frac{3}{5}\cdot\frac{14}{24}} = \frac{342}{897}$

4.22 $P(A\cap J) = \frac{1}{5} \neq P(A)\ P(J) = \frac{3}{10}\cdot\frac{2}{5} \Rightarrow$ dependence.
Independence occurs for $f_{1,1} = 84$, $f_{1,2} = 36$, $f_{2,1} = 56$, $f_{2,2} = 24$.

4.23 Let $S = \{1, \ldots, 8\}$ be the sample space, then the events $A = \{1,2,3,4\}$, $B = \{3,4,5,6\}$, $C = \{2,3,5,7\}$ and $D = \{2,4,5,8\}$ have the required characteristics.

4.24 (i) $0.999^{500} \approx 0.6064$

(ii) $(1-p)^{500} = 0.99 \Rightarrow p = 1 - 0.99^{(1/500)} \approx 2.01 \cdot 10^{-5}$.

4.25 The required probability is

$$abcde + (1-a)bcde + a(1-b)cde + ab(1-c)de + abc(1-d)e + abcd(1-e)$$
$$+ (1-a)(1-b)cde + (1-a)b(1-c)de + (1-a)bc(1-d)e + (1-a)bcd(1-e)$$
$$+ a(1-b)(1-c)de + a(1-b)c(1-d)e + a(1-b)cd(1-e) + ab(1-c)(1-d)e$$
$$+ ab(1-c)d(1-e) + abc(1-d)(1-e) \approx 0.9958.$$

4.26 (i) $P(E) = p_1p_2 + p_1p_3 + p_4p_5p_6 - p_1p_2p_3 - p_1p_2p_4p_5p_6 - p_1p_3p_4p_5p_6 + p_1\ldots p_6 = 0.56924$,

(ii) $A_2 \cap E = (A_1 \cap A_2) \cup (A_1 \cap A_2 \cap A_3) \cup (A_2 \cap A_4 \cap A_5 \cap A_6)$
$$= (A_1 \cap A_2) \cup (A_2 \cap A_4 \cap A_5 \cap A_6)$$
$$\Rightarrow P(A_2 \cap E) = p_1p_2 + p_2p_4p_5p_6 - p_1p_2p_4p_5p_6$$
$$\Rightarrow P(A_2|E) = \frac{p_1p_2 + p_2p_4p_5p_6 - p_1p_2p_4p_5p_6}{P(E)} \approx 0.7919.$$

4.27 $E = (A_1 \cap A_2 \cap A_3) \cup (A_6 \cap A_7 \cap A_8) \cup (A_1 \cap A_2 \cap A_5 \cap A_8) \cup$
$(A_1 \cap A_4 \cap A_7 \cap A_8) \cup (A_3 \cap A_5 \cap A_6 \cap A_7) \cup (A_2 \cap A_3 \cap A_4 \cap A_6)$
$(A_1 \cap A_3 \cap A_4 \cap A_5 \cap A_7) \cup (A_2 \cap A_4 \cap A_5 \cap A_6 \cap A_8)$

4.29 (i)

Secretary present					All languages represented?	Probability
1	2	3	4	5		
×	×	×	×	×	Yes	p^5
×	×	×	×		Yes	$p^4(1-p)$
×	×	×		×	Yes	$p^4(1-p)$
×	×		×	×	Yes	$p^4(1-p)$
×		×	×	×	Yes	$p^4(1-p)$
	×	×	×	×	Yes	$p^4(1-p)$
×	×	×			Yes	$p^3(1-p)^2$
×	×		×		Yes	$p^3(1-p)^2$
×	×			×	Yes	$p^3(1-p)^2$
×		×	×		No	
×		×		×	No	
×			×	×	No	
	×	×	×		No	
	×	×		×	Yes	$p^3(1-p)^2$
	×		×	×	Yes	$p^3(1-p)^2$
		×	×	×	No	

The required probability is $p^5 + 5p^4(1-p) + 5p^3(1-p)^2 = p^5 - 5p^4 + 5p^3$.

(ii) We obtain

$$E = (A_1 \cap A_2 \cap A_3) \cup (A_1 \cap A_2 \cap A_4) \cup (A_1 \cap A_2 \cap A_5) \cup (A_2 \cap A_3 \cap A_5)$$
$$\cup (A_2 \cap A_4 \cap A_5) \cup (A_1 \cap A_3 \cap A_4 \cap A_5)$$

where the events are defined by $E = \{$all languages are represented by the secretaries present at work$\}$, $A_i = \{$secretary i is present at work$\}$ and the intersections correspond to the minimum working sets.

The inclusion-exclusion principle yields

$$P(E) = \left(5p^3 + p^4\right) - \left(8p^4 + 7p^5\right) + \left(2p^4 + 18p^5\right) - \left(15p^5\right) + \left(6p^5\right) - \left(p^5\right)$$
$$= p^5 - 5p^4 + 5p^3.$$

The expressions in parenthesis refer to the elements of the sum in Theorem 2.6.

4.30 The components of the sample space partition are $C_1 = B_1 \cap B_2$, $C_2 = B_1 \cap \bar{B}_2$, $C_3 = \bar{B}_1 \cap B_2$, $C_4 = \bar{B}_1 \cap \bar{B}_2$. In Example 4.5, we applied Theorem 4.2 with C_i instead of B_i and $m = 4$.

4.31 From the tree diagram we obtain nine paths from the starting point 0 to a realization of the event A with corresponding probabilities:

Path	Probability
$0 \to r \to r \to r$	$P_{rr} = (7 \cdot 6 \cdot 5)/(15 \cdot 14 \cdot 13)$
$0 \to r \to b \to r$	$P_{rb} = (7 \cdot 5 \cdot 6)/(15 \cdot 14 \cdot 13)$
$0 \to r \to g \to r$	$P_{rg} = (7 \cdot 3 \cdot 6)/(15 \cdot 14 \cdot 13)$
$0 \to b \to r \to r$	$P_{br} = (5 \cdot 7 \cdot 6)/(15 \cdot 14 \cdot 13)$
$0 \to b \to b \to r$	$P_{bb} = (5 \cdot 4 \cdot 7)/(15 \cdot 14 \cdot 13)$
$0 \to b \to g \to r$	$P_{bg} = (5 \cdot 3 \cdot 7)/(15 \cdot 14 \cdot 13)$
$0 \to g \to r \to r$	$P_{gr} = (3 \cdot 7 \cdot 6)/(15 \cdot 14 \cdot 13)$
$0 \to g \to b \to r$	$P_{gb} = (3 \cdot 5 \cdot 7)/(15 \cdot 14 \cdot 13)$
$0 \to g \to g \to r$	$P_{gg} = (3 \cdot 2 \cdot 7)/(15 \cdot 14 \cdot 13)$

(i) The total probability of the event A is the sum of the above probabilities, yielding $P(A) = \frac{7}{15}$,

(ii) $P(B|A) = \frac{P(A \cap B)}{P(A)} = \frac{P_{rb} + P_{bb} + P_{gb}}{P(A)} = \frac{1/6}{7/15} = \frac{5}{14}$,

(iii) $P(C|A) = \frac{P(A \cap C)}{P(A)} = \frac{P_{rr} + P_{bb} + P_{gg}}{P(A)} = \frac{28/195}{7/15} = \frac{4}{13}$.

Chapter 5

5.1 $P(X=0)=0.064$, $P(X=1)=0.288$, $P(X=2)=0.432$, $P(X=3)=0.216$.

5.2 (i) $P(X=k)=\dfrac{\binom{7}{k}\binom{23}{5-k}}{\binom{30}{5}}$,

(ii) $P(X=k)=\binom{5}{k}\left(\tfrac{7}{30}\right)^k\left(\tfrac{23}{30}\right)^{5-k}$

5.3 (i) 2/3, (ii) 255/256, (iii) 1/31

5.4 $P(X=k)=\tfrac{1}{2^{10}}\binom{10}{k}$ for $k=0,1,\ldots,10$.

5.5 (i) 9/10, (ii) ½, (iii) 3/5.

5.6 $\sum_{k=0}^{10}\binom{10}{k}0.8^k0.2^{10-k}\approx0.6778$.

5.7 $\tfrac{1}{6^n}\sum_{j=k}^{n}\binom{n}{j}5^{n-j}$

5.8 (i) $X:=$ number of steps to the right, $Y:=$ position of the marker, $Y=2X-n$, thus

$$P(Y=k)=P\left(X=\frac{n+k}{2}\right)=\binom{n}{\frac{n+k}{2}}p^{(n+k)/2}(1-p)^{(n-k)/2}\quad\text{for }k$$

$$=-n,-n+2,\ldots,n$$

(ii) $P(|Y|\ge30)=P(|2X-100|\ge30)=P(X\ge65\text{ or }X\le35)=2\sum_{k=0}^{35}\binom{100}{k}\tfrac{1}{2^{100}}\approx0.0035$

5.9 $\sum_{k=4}^{5}\binom{5}{k}\left(\tfrac14\right)^k\left(\tfrac34\right)^{5-k}=\tfrac{1}{64}$

5.10 $p_n(k+1)=p_n(k)\tfrac{p(n-k)}{(1-p)(k+1)}$

5.11 (i) From the recurrence relation of Exercise 5.10, it follows that the $p_n(k)$ are increasing for $\tfrac{p(n-k)}{(1-p)(k+1)}>1\Leftrightarrow k<pn+p-1$ and decreasing for $k>pn+p-1$.
(ii) Set, e.g., $n=6$ and $p=1/2$ in case (a) and $n=6$ and $p=4/7$ in case (b)

5.14 Let $p^{(i)}$ denote the probability of the event A in the ith realization of E. Then

$$P(X=k)=\sum_{|I|=k}\prod_{i\in I}p^{(i)}\prod_{j\in\{1,\ldots,n\}\setminus I}(1-p^{(j)})$$

where the sum runs over all k-subsets of $\{1,\ldots,n\}$ and the products over all $i\in I$ and all $j\in\{1,\ldots,n\}\setminus I$, respectively (see Example 4.9).

5.15 (i) $F(x) = \begin{cases} 0 & \text{for } x < 0 \\ \dfrac{x}{4} - \dfrac{x^2}{8} & \text{for } 0 \le x < 1, \\ \dfrac{3}{8} - \dfrac{x}{2} + \dfrac{x^2}{4} & \text{for } 1 \le x < 2, \\ \dfrac{-5}{8} + \dfrac{x}{2} & \text{for } 2 \le x < 3, \\ \dfrac{-41}{8} + \dfrac{7 \cdot x}{2} - \dfrac{x^2}{2} & \text{for } 3 \le x < 3.5, \\ 1 & \text{for } x \ge 3.5 \end{cases}$

(ii) 7/16 and 5/8

5.16 (i) $C = 3/4$

(ii) $F(x) = \begin{cases} 0 & \text{for } x < 0 \\ \dfrac{x^{3/2}}{2} & \text{for } 0 \le x < 1, \\ \dfrac{3x^2}{4} - \dfrac{x^3}{4} & \text{for } 1 \le x < 2, \\ 1 & \text{for } x \ge 2 \end{cases}$

5.17 37/56.

5.18 Proof by induction on n.

5.19 (i) 1/5, (ii) $\binom{3}{1} \left(\frac{1}{4}\right) \left(\frac{3}{4}\right)^2 = \frac{27}{64}$

5.20 $F(x) = \frac{1}{2} + \frac{1}{\pi} \arctan(x)$ for $x \in \mathrm{IR}$,

$$f(x) = F'(x) = \frac{1}{\pi(1+x^2)} \text{ for } x \in \mathrm{IR} \text{ (Cauchy distribution).}$$

5.22 (i) $F(x) = 1 - e^{-3x}$ for $x \ge 0$ and 0 elsewhere.

(ii) $F(5) - F(2) = e^{-6} - e^{-15}$, 0, $F(10) = 1 - e^{-30}$.

5.23 $f(2) = 1/5$, $f(5) = 2/5$, $f(10) = 2/5$.

5.24 $f(x) = \begin{cases} (3/4)x^3 & \text{for } -1 \le x < 1, \\ 1/2 & \text{for } 1 \le x < 2, \\ 0 & \text{elsewhere.} \end{cases}$

5.25 $b = 6$,

$$F(x) \begin{cases} 0 & \text{for } x < 2, \\ \dfrac{-3}{x} + \dfrac{3}{2} & \text{for } 2 \le x < 6, \\ 1 & \text{for } x \ge 6, \end{cases}$$

5.26 (i) We proof the statement by induction, using integration by parts, obtaining

$$G(k) = (n-k)\binom{n}{k}\left\{\left[\frac{y^{n-k}}{n-k}(1-y)^2\right]_{y=0}^{y=1-p} + \int_0^{1-p}\frac{y^{n-k}}{n-k}k(1-y)^{k-1}dy\right\}$$

$$= (n-k)\binom{n}{k}\left\{\frac{(1-p)^{n-k}}{n-k}p^k + \int_0^{1-p}\frac{y^{n-k}}{n-k}k(1-y)^{k-1}dy\right\}$$

$$= \binom{n}{k}(1-p)^{n-k}p^k + \binom{n}{k}k\int_0^{1-p}y^{n-k}(1-y)^{k-1}dy$$

$$= \binom{n}{k}(1-p)^{n-k}p^k + G(k-1).$$

Therefore, $G(k)$ and $F(k)$ satisfy the same recursive relation, i.e. $G(k) - G(k-1) = F(k) - F(k-1) = \binom{n}{k}(1-p)^{n-k}p^k$. Moreover, $G(0) = F(0)$.

(ii) From part (i) we obtain $F(k) = G(k) = (n-k)\binom{n}{k}\int_0^{1-p}y^{n-k-1}(1-y)^k dy = (n-k)\binom{n}{k}B(n-k,k)I_{1-p}(n-k,k)$, where $I_{1-p}(\ldots)$ is the incomplete beta function defined by eq. (1.24).

Chapter 6

6.1 $g(y) = 2y\alpha e^{-\alpha y^2}$ for $y \geq 0$.

6.2 The range space of Y is $\left[e^{-3},e^{-2}\right]$, $g(y) = -\frac{4}{65y}(\ln y)^3$,
$G(y) = -\frac{1}{65}(\ln y)^4 + \frac{81}{65}$.

6.3 $g(y) = \frac{1}{a}e^{-y/a}$ for $y \geq 0$.

6.4 $g(y) = \frac{1}{\pi\sqrt{1-y^2}}$ for $-1 \leq y \leq 1$.

6.5 (i) $a = 4\sqrt{\frac{b^3}{\pi}}$, (ii) $g(w) = a\sqrt{\frac{2w}{m^3}}e^{(-2bw)/m}$ for $w \geq 0$.

6.6 $g(y) = \frac{e^{-y}}{3}$ for $-\ln 5 \leq y \leq -\ln 2$.

6.7 $g(y) = \frac{1}{2\sqrt{y}}\left[f(\sqrt{y}) + f(-\sqrt{y})\right]$ for $y \geq 0$.

6.8 Similar to Example 6.6.

6.9 The range space of Y is $[-1/4, 2]$.

$$G(y) = \begin{cases} F\left(\dfrac{1}{2} + \sqrt{\dfrac{1}{4} + y}\right) - F\left(\dfrac{1}{2} - \sqrt{\dfrac{1}{4} + y}\right) & \text{for } -1/4 \le y \le 3/4, \\[3mm] F\left(\dfrac{1}{2} + \sqrt{\dfrac{1}{4} + y}\right) & \text{for } 3/4 \le y \le 2. \end{cases}$$

6.11 Let Y denote the cost in question. This r.v. has the possible values 0, 50, 100, . . ., 250. It holds $P(Y = 50k) = P(X = 2k) + P(X = 2k - 1)$ for $k = 0, 1, \ldots, 5$, where X is binomially distributed with parameters $n = 10$ and $p = 0.7$.

6.12 $P(X = k) = P(Y = 10{,}000 - 1{,}000k) = 0.7 \cdot 0.3^k$ for $k = 1, 2, \ldots$.

Chapter 7

7.2 (i) 0.51, (ii) 0.86, (iii) 0.28, (iv) 0.61/0.8.

7.3 (i) 0.457, (ii) 0.805, (iii) 0.258, (iv) 0.536/0.8.

7.4 (i) Joint distribution:

X	0	1	2
Y			
0	3/66	9/66	3/66
1	12/66	18/66	6/66
2	6/66	8/66	1/66

7.5 (i) 1/60, (ii) 8/9,

(iii) $g(x) = \frac{x^2}{10} + \frac{1}{30}$ for $0 \le x \le 3$. $h(y) = \frac{y}{20} + \frac{9}{20}$ for $0 \le y \le 2$,

(iv) $g(x|y) = \frac{1}{3}\frac{3x^2 + y}{y + 9}$, $h(y|x) = \frac{3x^2 + y}{6x^2 + 2}$,

(v) $F(x, y) = \frac{1}{60}x^3 y + \frac{1}{120}xy^2$ for $0 \le x \le 3$, $0 \le v \le 2$.

7.7 Similar to Example 7.12, one obtains $h(u) = \int_{-\infty}^{\infty} f_1(uv)f_2(v)|v|dv$.

7.8 (ii) $P(B_1) = 1/2$, $P(B_2) = \frac{43 + 36\pi}{144\pi}$,

(iii) $g(x) = \frac{(3x + 6)\sqrt{1 - x^2}}{3\pi}$ for $-1 \le x \le 1$, $h(y) = \frac{(2y + 6)\sqrt{1 - y^2}}{3\pi}$ for $-1 \le y \le 1$,

(v) $F(x, y) = \int_{s=-1}^{x} \int_{t=\sqrt{1-s^2}}^{\min\left(y, \sqrt{1-s^2}\right)} f(s, t)dtds$ for the case $x^2 + y^2 \le 1$ and $x, y \ge 0$.

7.9 The required density is $h(u) = \int_{\max(1000, 1000/u)}^{\infty} \frac{10^6}{u^2 v^3} dv$ for $u > 0$. Hence,

$$h(u) = \frac{1}{2} \text{ for } 0 < u \le 1, \quad h(u) = \frac{1}{2u^2} \text{ for } u \ge 1.$$

7.10 The pdf of D^2: $g(y) = \frac{1}{2\sqrt{y}}e^{-\sqrt{y}}$ for $y > 0$,

pdf of I: $h(i) = \int_{2/i}^{3/i} \frac{1}{2\sqrt{y}}e^{-\sqrt{y}}y\,dy$ for $i > 0$,

7.11 Joint pdf of (W, V): $\frac{1}{12}\left(2 - \sqrt{\frac{w}{v}}\right)$ for $0 \le w \le 4v$, $0 \le v \le 3$ (see Theorem 7.1), pdf of power W:

$$h(w) = \int_{w/4}^{3} \frac{1}{12}\left(2 - \sqrt{\frac{w}{v}}\right)dv = \frac{1}{2} + \frac{w}{24} - \frac{\sqrt{3w}}{6} \text{ for } 0 \le w \le 12.$$

7.12 (ii) 0.38, (iii) 0.16.

7.13 (ii) $F(x, y) = \int_{1-y}^{x}\int_{1-s}^{y} 8s(s + t - 1)dtds$

$$= -1/3 + 2x^2 - (8/3)x^3 + x^4 + (4/3)y - 2y^2 + (4/3)y^3 - (1/3)y^4$$
$$- 4yx^2 + 2y^2x^2 + (8/3)yx^3 \text{ for } (x, y) \in R.$$

(iii) $G(x) = F(x, 1) = x^4$, $H(y) = F(1, y) = (4y^3 - y^4)/3$

(v) (a) $1 - G(3/4) = 175/256$, (b) $H(1/3) = 11/243$, (c) $F(3/4,1/2) = 11/768$

(d) $G(1/3) - F(1/3, 4/5) = 1/81 - 16/16{,}875 = 577/50{,}625$.

7.14 For example: $F(x, y) = F(a, y)$ for $x > a$, $0 < y < b$,

$F(x, y) = F(x, b)$ for $0 < x < a$, $y > b$.

7.15 (i) $\dfrac{P(2X \le Y \le 1)}{P(Y \ge 2X)} = \dfrac{\frac{5}{6} - \frac{25}{12}e^{-1}}{\frac{5}{6} - \frac{11}{6}e^{-2}} \approx 0.1143$

(ii) $\dfrac{P(X + Y \le 2, X \le 1/2)}{P(X \le 1/2)} = \dfrac{2e^{-2} + \frac{5}{12} - 2e^{-3/2}}{\frac{5}{12}} \approx 0.5786$

7.16 $P(B) = 1 - \int_{\sqrt{\frac{5}{3}}}^{2}\int_{50}^{30x^2} \frac{1}{70}dydx \approx 0.6710$.

7.17 The idea of the proof is given by the comments after Example 7.7: If the range space R is not rectangular let T be the smallest rectangle containing R. Then a point $(x_0, y_0) \in T \backslash R$ satisfies $0 = f(x_0, y_0) \ne g(x_0)h(y_0) > 0$.

7.19 Show that the functions in eqs. (7.13) and (7.14) satisfy all requirements for a one-dimensional pdf.

7.20 $F(x, y) = \sum_{s=1}^{x}\sum_{t=1}^{y}\frac{1}{2^{s+t}} = \frac{1}{2^{x+y}} - \frac{1}{2^x} - \frac{1}{2^y} + 1$,

$G(x) = F(x, \infty) = 1 - \frac{1}{2^x}$, $H(y) = F(\infty, y) = 1 - \frac{1}{2^y}$.

7.21 (i) $S(x,y) = \int_{x}^{\sqrt{1-y}}\int_{y}^{1-s^2} \frac{5}{4}(s^2 + t)dtds = \sqrt{1-y}\left(\frac{1}{2} - \frac{y}{6} - \frac{y^2}{3}\right) + \frac{x}{8}(x^4 + 5y^2 - 5) + \frac{5}{12}yx^3$

(ii) $S(x,y) = \int_{-\sqrt{1-y}}^{\sqrt{1-y}}\int_{y}^{1-s^2} \frac{5}{4}(s^2 + t)dtds = \sqrt{1-y}\left(1 - \frac{y}{3} - \frac{2}{3}y^2\right)$.

7.22 $F(x,y) + S(x,y) = xy + (1-x)(1-y) < 1$ for $0 < x, y < 1$.

Chapter 8

8.1 $X :=$ number of tossings,

$$P(X = k) = \left(\frac{5}{6}\right)^{k-1}\frac{1}{6}, \; E(X) = \frac{1}{6}\sum_{k=1}^{\infty}k\left(\frac{5}{6}\right)^{k-1} = 6.$$

8.2 Any element assigned randomly to one of the positions 1, ..., n is a fixed point with probability $1/n$. Thus, in assigning n elements to the positions 1, ..., n, the expected number of fixed points is $n \cdot \frac{1}{n} = 1$.

This holds by linearity of the expectation, in spite of the dependence between the considered variables.

8.3 $E(X) = 0$ because of symmetry, $E(|X|) = \frac{1}{a}$.

8.4 $C \approx 0.4438$, $E \approx 3.3157$, $V \approx 1.0787$.

8.5 (i) $C = \frac{3}{25} \cdot 10^{-6}$, $E(X) = 60$, $E(X^2) = 4,000$, $V(X) = 400$,

(ii) expected profit: $a + bE(X) + cE(X^2) = a + 60b + 4000c$

8.6 $p := P(X \le 150) = 1 - e^{-3/4} \approx 0.5267$, expected profit: $-3p + 5(1-p) \approx 0.779$.

8.7 $g(x) = \frac{1}{\sqrt{2\pi}}e^{-x^2/2}$, $h(y) = \frac{1}{\sqrt{2\pi}}e^{-y^2/2}$

Thus $f(x, y) = g(x)h(y)$, i.e. X and Y are independent. $E(X) = E(Y) = 0$ implies $E(XY) = E(X)E(Y) = 0$.

8.8 (i) Joint pdf of (U, V):

$$g(u, v) = \frac{u}{\sqrt{u^2 - v^2}} \text{ for } 0 \le v \le 1, \; 0 \le u^2 - v^2 \le 1,$$

pdf of U: $h(u) = \int_0^u \frac{u}{\sqrt{u^2-v^2}}dv = \frac{\pi}{2}u$ for $0 \le u \le 1$,

$$h(u) = \int_{\sqrt{u^2-1}}^{1} \frac{u}{\sqrt{u^2 - v^2}}dv = u\left[\arcsin(1/u) - \arcsin\left(\frac{\sqrt{u^2-1}}{u}\right)\right] \text{ for } 1 \le u \le \sqrt{2}.$$

(ii) $E(U) = \int_0^{\sqrt{2}} h(u)du = \frac{\sqrt{2}}{3} + \frac{1}{4}\ln(1+\sqrt{2}) - \frac{1}{12}\ln(\sqrt{2}-1) \approx 0.765196$

Alternatively, $E(U) = \int_0^1\int_0^1 \sqrt{x^2+y^2}dydx \approx 0.765196$.

8.9 The expected distance is approximately 0.52.

8.10 The expected distance is

$$D_n := \frac{1}{n^4}\sum_{i=1}^{n}\sum_{j=1}^{n}\sum_{k=1}^{n}\sum_{l=1}^{n}\sqrt{\left(\frac{i}{n}-\frac{k}{n}\right)^2 + \left(\frac{j}{n}-\frac{l}{n}\right)^2}$$

$$= \frac{1}{n^5}\sum_{i=1}^{n}\sum_{j=1}^{n}\sum_{k=1}^{n}\sum_{l=1}^{n}\sqrt{(i-k)^2 + (j-l)^2}.$$

In particular, $D_{10} = 0.518687$, $D_{100} = 0.521380$ and $D_{200} = 0.521399$. Evidently D_n converges to the expected distance in the continuous case for $n \to \infty$ (see Exercise 8.9).

8.11 (i) pdf of $X + Y$: $h(u) = \frac{u}{a^2}$ for $0 \le u \le a$, $h(u) = \frac{2a-u}{a^2}$ for $a \le u \le 2a$,

pdf of $X - Y$: $h(u) = \frac{a-|u|}{a^2}$ for $-a \le u \le a$.

8.12 $E(Z) = 0.3\mu_A + 0.7\mu_B = 163$, $V(Z) = 0.3^2\sigma_A^2 + 0.7^2\sigma_B^2 = 90.81$.

8.13 $E(Z) = \mu$, $V(Z) = \frac{\sigma^2}{n}$

8.14 $E(U) = -5$, $V(U) = 7$, $E(T) = 79$, $V(T) = 830$, $E(W) = -84$, $V(W) = 837$.

8.15 $(a-b)^2/12$

8.16 (i) $\mu_k' = \frac{k!}{a^k}$, $\mu_5 = \frac{44}{a^5}$, $\mu_{[4]} = \frac{2}{a^4}\left(-3a^3 + 11a^2 - 18a + 12\right)$

(ii) $\mu_k' = \frac{a^{k+1} - b^{k+1}}{(k+1)(a-b)}$ $\quad \mu_k = \frac{(a-b)^k + (b-a)^k}{2^{k+1}(k+1)}$.

8.17 $\alpha_3 = \frac{E[(X-60)^3]}{20^3} = -\frac{2}{7}$, $\alpha_4 = \frac{E[(X-60)^4]}{20^4} = \frac{33}{14}$.

8.18 $\mu = E(X) = \frac{400}{7}$, $\sigma^2 = V(X) = \frac{15,000}{49}$, upper limit: $\frac{\sigma^2}{30^2} \approx 0.3401$, exact probability: ≈ 0.0815.

8.19 Apply Jensen's inequality to the convex function $h(x) = \frac{1}{x} + x^2 + \ln(x)$.

8.20 $E(XY) = \frac{1}{12}$, upper limit: $\sqrt{E(X^2)E(Y^2)} = \sqrt{\frac{1}{12} \cdot \frac{1}{3}} = \frac{1}{6}$.

8.21 $E(XY) = 8/63$, $E(X) = 2/7$, $V(X) = 17/49$, $E(Y) = 2/7$, $V(Y) = 20/441$, Cov$(X, Y) = 20/441$, $\rho \approx 0.3616$.

8.22 $E(XY) = 25/33$, $E(X) = 5/6$, $V(X) = 175/396$, $E(Y) = 1$, $V(Y) = 5/11$, COV$(X, Y) = -5/66$, $\rho = \frac{-1}{\sqrt{35}} \approx -0.1690$.

Chapter 9

9.2 $\left(\sum_{k=25}^{\infty} e^{-30} \frac{30^k}{k!}\right)^{10} \approx 0.1807$.

9.3 $\sum_{k=60}^{\infty} e^{-50} \frac{30^k}{k!} \approx 0.0923$.

9.4 (i) $\lambda_{\text{week}} = 20\frac{7}{30} = \frac{14}{3}$, $e^{-14/3}\frac{(14/3)^6}{6!} \approx 0.1349$,

(ii) $\lambda_{\text{day}} = \frac{2}{3}$, $\sum_{k=3}^{\infty} e^{-2/3}\frac{(2/3)^k}{k!} \approx 0.0302$.

9.7 A cost over \$100,000 occurs if and only if $X \ge 9$, and the latter occurs if and only if the first eight trials are failures. Hence the required probability is $P(X \ge 9) = q^8 = 0.85^8 = 0.2725$.

9.9 Setting the derivative with respect to p equal to zero yields $p = 1/k$.

9.10 (i) $X > k$ occurs if and only if the first k trials are failures, thus $P(X > k) = q_1 \cdots q_k$ and $P(X \le k) = 1 - q_1 \cdots q_k$.

(ii) $P(X = k) = q_1 \ldots q_{k-1} p_k = \frac{1}{2^{k(k-1)/2}} \left(1 - \frac{1}{2^k}\right)$

$$E(X) = \sum_{k=1}^{\infty} kP(X = k) \approx 1.6416, \quad V(X) \approx 0.5485.$$

9.11 0.2162

9.12 0.0473

9.13 $p = 1/10, r = 5$

9.17 (i) We ask for the probability that all persons are chosen from the 430 "non-Italian," which is $\dfrac{\binom{430}{20}}{\binom{500}{20}} \approx 0.0460.$

(ii) We consider three groups of persons: 80 French, 70 Italian and 350 others. The required probability is therefore

$$\frac{1}{\binom{500}{20}} \sum_{j=5}^{20} \sum_{i=4}^{20} \binom{350}{20-i-j}\binom{80}{j}\binom{70}{i} \approx 0.0418. \text{ Recall that } \binom{a}{b} = 0 \text{ for}$$

$b < 0$ or $b > a$. Therefore, various possibilities exist to choose the limits of the summation indices.

(iii) Let i denote the number of Austrian and j the difference between the numbers of German and Austrian. The required probability can therefore be written as

$$\frac{1}{\binom{500}{20}} \sum_{j=0}^{20} \sum_{i=0}^{20} \binom{200}{i+j}\binom{150}{i}\binom{150}{20-2i-j} \approx 0.7538.$$

9.18 (i) Since any number of the die occurs with the same probability 1/6 we obtain the result

$$\frac{5!}{2!2!}\left(\frac{1}{6}\right)^5 = \left(\frac{30}{6^5}\right) \approx 0.0039.$$

(ii) With the notation $p(x_1, \ldots, x_6) = P(X_1 = x_1, \ldots, X_6 = x_6)$ the required probability is

$f(1, 0, 0, 0, 2, 2) + f(0, 1, 0, 0, 2, 2) + f(0, 0, 1, 0, 2, 2) + f(0, 0, 0, 1, 2, 2)$

$+ f(0, 0, 0, 0, 2, 3) + f(0, 0, 0, 0, 3, 2) = 4\binom{4}{1, 2, 2}\left(\frac{1}{6}\right)^5 + 2\binom{5}{1, 2, 2}\left(\frac{1}{6}\right)^5$

$\approx 0.018.$

9.19 (i) $\left(\frac{430}{500}\right)^{20} \approx 0.0490$.

(ii) $\sum_{i=4}^{15} \sum_{j=5}^{20-i} \binom{20}{20-i-j,\ i,\ j} \left(\frac{350}{500}\right)^{20-i-j} \left(\frac{80}{500}\right)^{j} \left(\frac{70}{500}\right)^{i} \approx 0.0381$.

(iii) $\sum_{i=0}^{9} \sum_{j=0}^{20-2i} \binom{20}{i+j,\ i,\ 20-2i-j} \left(\frac{200}{500}\right)^{i+j} \left(\frac{150}{500}\right)^{i} \left(\frac{150}{500}\right)^{20-2i-j} \approx 0.7495$.

9.20 Let $X_1, \ldots, X_4$ denote the number of steps to the right, to the left, upward and downward, respectively.

(i) The required probability is $P(X_1 - X_2 = x, X_3 - X_4 = y)$.

(ii) In order to calculate the event $\{X_1 - X_2 = 5, X_3 - X_4 = 3\}$ we denote the value of X_1 by i and the value of X_3 by j. If this event occurs, it holds $X_2 = i - 5$ and $X_4 = j - 3$. Since $n = 20$ movements were performed, the r.v.'s must also satisfy $X_1 + \ldots + X_4 = 20 \Rightarrow 2i - 5 + 2j - 3 = 20 \Rightarrow j = 14 - i$. In summary, the event of interest occurs when $X_1 = i$, $X_2 = i - 5$, $X_3 = 14 - i$ and $X_4 = 11 - i$. The required probability is therefore

$$\sum_{i=5}^{11} P(X_1 = i, X_2 = i - 5, X_3 = 14 - i, X_4 = 11 - i)$$

$$= \sum_{i=5}^{11} \binom{20}{i,\ i-5,\ 14-i,\ 11-i} \left(\frac{1}{4}\right)^{20} \approx 0.005921.$$

Chapter 10

10.1 (i) 0.7150, 0.3821, 0.0668

(ii) 3.57

(iii) $\mu = 66.5$, the corresponding probability is 85.29%.

10.2 $\sigma = 8$.

10.3 $c = 3.65$.

10.4 0.1500.

10.5 $\mu \approx 85$, $\sigma \approx 20$.

10.7 $P(X > \frac{1}{a}) = e^{-1}$.

10.8 $E(X) = 0$, $V(X) = E(X^2) = \frac{2}{a^2}$

$F(x) = \frac{1}{2} e^{ax}$ for $x \le 0$ and $1 - \frac{1}{2} e^{-ax}$ for $x > 0$.

10.9 $C = a e^a$, $E(X) = \frac{1+at}{a}$.

10.12 The mode exists for $r > 1$. Then it is given by

$$\frac{r-1}{a}.$$

10.13 $\int_{-c}^{c} \int_{-\sqrt{c^2-x^2}}^{\sqrt{c^2-x^2}} \frac{1}{2\pi} e^{-(x^2-y^2)/2} dy dx = 1 - e^{-c^2/2}$.

10.14 mode $= n - 2$.

328 ━━ Appendix

10.16 Let F and G denote the cdf's of X and $Y = X^2$. Thus

$$G(y) = P(Y \leq y) = P(-\sqrt{y} \leq x \leq \sqrt{y}) = F(\sqrt{y}) - F(-\sqrt{y})$$

(see Section 6.1), and differentiating gives the pdf of Y as

$$g(y) = f(\sqrt{y}) \frac{1}{2\sqrt{y}} - f(\sqrt{y}) \frac{-1}{2\sqrt{y}},$$

where f is the standardized normal density. Hence

$$g(y) = y^{-1/2} \cdot \frac{e^{-y/2}}{\sqrt{2\pi}} = y^{-1/2} \cdot \frac{e^{-y/2}}{\sqrt{2}\Gamma(1/2)},$$

which is the density of χ_1^2, see eq. (1.9).

10.17 The r.v. $Y = X/6$ has distribution $N(0,1)$, thus $P(4 < X^2 < 9) = P(1/9 < Y^2 < 1/4) = \int_{1/9}^{1/4} z^{-1/2} \frac{e^{-z/2}}{\sqrt{2\pi}} dz \approx 0.1218$ since $Z = Y^2$ has distribution χ_1^2.

10.18 (i) 0.6398 and (ii) 0.2676.

10.19 mode $= \frac{a-1}{a+\beta-2}$.

10.20 If $a = \beta$, then the pdf is symmetric about 0.5.

10.21 $a = 2$, $\beta = 3$.

10.23 (i) For f the marginal density of X is

$$f_X = \int_{-\infty}^{\infty} f(x, y) dy = \frac{1}{\sqrt{3\pi}} \exp\left(-\frac{x^2}{2}\right) \int_{-\infty}^{\infty} \exp\left(-\frac{2}{3}\left(y - \frac{x}{2}\right)^2\right) dy$$

$$= \frac{1}{\sqrt{3\pi}} \exp\left(-\frac{x^2}{2}\right) \sqrt{\frac{3}{2}\pi} = \frac{1}{\sqrt{2\pi}} \exp\left(-\frac{x^2}{2}\right).$$

(ii) $\frac{2+a}{6}$ for h.

10.24 Set $\mu = b = 0$ and $\Sigma = I$ in Theorem 10.7.

10.25 The expectations do not change if ρ changes. The covariance matrices of the image distribution are $\begin{pmatrix} 0.01192 & 0.01604 \\ 0.01604 & 0.0217 \end{pmatrix}$ and $\begin{pmatrix} 0.01 & 0.0134 \\ 0.0134 & 0.0181 \end{pmatrix}$ for $\rho = 0.2$ and $\rho = 0$, respectively.

10.26 0.1856.

10.27 $\mu = \begin{pmatrix} 1/2 \\ 1 \end{pmatrix}$, $\Sigma = \begin{pmatrix} 7/4 & \sqrt{7/4} \\ \sqrt{7/4} & 1 \end{pmatrix}$.

10.28 $N(36,769)$.

10.30 (i) 0.0839 and (ii) 0.0891.

10.33 (i) 0.4418, (ii) 0.4153 and (iii) 0.2963.

10.34 $\text{Cov}(X, Y) = \frac{\rho}{a\beta}$.

10.35 $P(Y \geq 2X) = \frac{1}{2} - \frac{1}{8} \sqrt{\frac{2\pi}{\alpha\beta\theta}} \left[\mathrm{erf}\left(\frac{\alpha + 2\beta}{2\sqrt{2\alpha\beta\theta}} \right) - 1 \right] (\alpha - 2\beta) \exp\left(\frac{(\alpha + 2\beta)^2}{8\alpha\beta\theta} \right)$.

10.36 (i) $K^{-1} = \frac{1}{c} \Gamma\left(0, \frac{ab}{c} \right) \exp\left(\frac{ab}{c} \right)$,

    (ii) $\frac{K\,e^{-ax}}{b+cx}$ and $\frac{K\,e^{-by}}{a+cy}$,

    (iii) $E(X) = \frac{b}{c}\left(\frac{K}{ab} - 1 \right)$, $E(Y) = \frac{a}{c}\left(\frac{K}{ab} - 1 \right)$.

Chapter 11

11.1 Substituting p for λ/n in the mgf of the binomial distribution yields

$$\left[\frac{\lambda}{n} e^t + \left(1 - \frac{\lambda}{n} \right) \right]^n = \left[1 + \frac{\alpha}{n} \right]^n,$$

    where $\alpha = \lambda(e^t - 1)$. For $n \to \infty$, the latter converges to e^α which is the mgf of the Poisson distribution.

11.2 $M(t) = \left(\frac{\alpha}{\alpha - t} \right)^3$, $M^{(k)}(t) = \frac{(k+2)!\,\alpha^3}{2(\alpha - t)^{k+3}}$, $M^{(k)}(0) = \frac{(k+2)!}{2\alpha^k}$.

11.3   The distribution of Y is χ_1^2, yielding

$$M_Y(t) = \int\limits_0^\infty e^{ty} \frac{1}{\sqrt{2\pi y}} e^{-y/2} dy = \frac{1}{\sqrt{1 - 2t}}.$$

11.4   (i) $M(t) = 1 + t\sqrt{\frac{\pi}{2}} \exp\left(\frac{t^2}{2} \right) \left[1 + \mathrm{erf}\left(\frac{t}{\sqrt{2}} \right) \right]$,

    where $\mathrm{erf}(x) = \frac{2}{\sqrt{\pi}} \int_0^x e^{-t^2/2} dt$ denotes the *error function*.

    (ii) $M(t) = t\sqrt{\frac{\pi}{2}} + (t^2 + 1) \exp\left(\frac{t^2}{2} \right) \left[1 + \mathit{erf}\left(\frac{t}{\sqrt{2}} \right) \right]$,

11.5 $M(t) = \frac{6}{\pi^2} \mathrm{polylog}(2, e^t)$,

    where polylog $(a, z) = \sum_{n=1}^\infty \frac{z^n}{n^a}$ denotes the polylogarithm,

$$M'(t) = -\frac{6}{\pi^2} \ln\left(1 - e^t \right), \quad M''(t) = \frac{6e^t}{\pi^2(1 - e^t)},$$

$$M'''(t) = \frac{6e^t}{\pi^2(1 - e^t)^2}.$$

11.6 $M(t) = \frac{1}{2} \left[\frac{a}{a-t} + \frac{a}{a+t} \right]$,

$$M^{(k)}(t) = \frac{ak!}{2}\left[\frac{1}{(a-t)^{k+1}} + \frac{(-1)^k}{(a+t)^{k+1}}\right],$$

$$M^{(k)}(0) = \begin{cases} \dfrac{k!}{a^k} & \text{for } k \text{ even,} \\ 0 & \text{for } k \text{ odd.} \end{cases}$$

11.7 $M(t) = \frac{e^t(2t+ta-2a) - 2t + ta + 2a}{2t^2},$

G(t) is obtained by substituting t for $\ln(t)$.

11.8 $M_Y(t) = E\left(\exp(t\,X^2)\right) = \int_0^\infty \exp(t\,x^2)ae^{-ax}dx$

$$= -\frac{a}{2}\sqrt{-\frac{\pi}{t}}\exp\left(-\frac{a^2}{4t}\right)\operatorname{erf}\left(\frac{a}{2\sqrt{-t}}\right) \text{ for } a > 0, t < 0.$$

11.9 The required mgf is $\lambda\, M_f(t) + (1-\lambda)M_g(t)$.

11.11 $G(t) = \left(\frac{tp}{1-tp}\right)^r$, $G'(t) = G(t)\frac{r}{t(1-qt)}$, $G''(t) = G(t)\frac{r(r+2t-2tp-1)}{t^2(1-qt)^2}$,

$G'(1) = \frac{r}{p}$, $G''(1) = \frac{r(r+q-p)}{p^2}$.

11.12 From Example 11.18, we get

$$G^{(k)}(1) = \frac{k!}{p}\left(\frac{q}{p}\right)^{k-1}.$$

11.16

$$M^{(k)}(t) = \frac{\Gamma(k+1,\ -bt) - \Gamma(k+1,\ -at)}{(b-a)(-1)^k t^{k+1}},$$

where $\Gamma()$ denotes the incomplete gamma function.

Chapter 12

12.2 $E(X) = \frac{3}{40\,\lambda^4} - \frac{1}{4\,\lambda^2} + \frac{3}{8}$, $V(X) = \frac{1}{5} - (E(X))^2$,

λ	2	4	6	8	10
$a_3(\lambda)$	−0.1793	0.1049	0.2344	0.2919	0.3212
$a_4(\lambda)$	2.300	2.269	2.272	3.205	2.192

12.3 $h(x) = 2f(x)G(\lambda x),$

$$E(X) = \sqrt{\frac{2\lambda^2}{\pi\,(\lambda^2+1)}},$$

$$V(X) = \frac{\lambda^2 (\pi - 2) + \pi}{\pi (\lambda^2 + 1)},$$

$$\mu_3 = \sqrt{\frac{\lambda^2 + 1}{\lambda^2}} \frac{(4 - \pi) \lambda^4 \sqrt{2}}{\pi^{3/2} (\lambda^2 + 1)^2},$$

$$\mu_4 = \frac{\lambda^4 (3 \pi^2 - 4 \pi - 12) + \lambda^2 \pi (6\pi - 12) + 3 \pi^2}{\pi^2 (\lambda^2 + 1)^2}.$$

The coefficients $a_3(\lambda)$ and $a_4(\lambda)$ are calculated by eqs. (8.5) and (8.6). $a_3(0) = 0$, $a_3(\infty) \approx 0.9953$, $a_4(0) = 3$, $a_4(\infty) \approx 3.8692$.

12.4 The mode is given by $\mathrm{Mo} = 1 - \left(\frac{\alpha - 1}{\alpha \beta - 1}\right)^{1/\alpha}$, $\alpha \beta \neq 1$, $\alpha \beta \neq 0$.

12.5 $g(x) = \frac{e^{-x^2/2}}{\sqrt{2\pi}} \left(1 + \mathrm{erf}\left(\frac{x}{\sqrt{2}}\right)\right)$, $\mathrm{Mo} = 0.5061$, $E(X) = \frac{1}{\sqrt{\pi}}$, $V(X) = 1 - \frac{1}{\pi}$,

$$a_3 = \frac{4 - \pi}{2 (\pi - 1)^{3/2}} \approx 0.1369, \quad a_4 = \frac{3\pi^2 - 4\pi - 3}{(\pi - 1)^2} \approx 3.0617.$$

12.6 $F(x) = \frac{x \sqrt{2} + \sqrt{4 + 2x^2}}{2 \sqrt{4 + 2x^2}}$, $g(x) = \frac{x\sqrt{2} + \sqrt{4 + 2x^2}}{\sqrt{2} (2 + x^2)^2}$, $\mathrm{Mo} = \frac{\sqrt{30}}{15} \approx 0.3651$,

$$E(X) = \frac{\sqrt{2} \pi}{4} \approx 1.1107.$$

12.7 If the graphs of two densities f_1 and f_2 intersect in a point (x_0, y_0) then it holds for any $\lambda \in [0, 1]$:

$$f(x_0) = \lambda f_1(x_0) + (1 - \lambda)f_2(x_0) = \lambda y_0 + (1 - \lambda)y_0 = y_0,$$

i.e., for any $\lambda \in [0, 1]$ the graph of the mixture density passes through the point (x_0, y_0).

12.8 $F(x) = \frac{3}{20} \mathrm{erf}\left(\frac{x}{\sqrt{2} \sigma}\right) + \frac{1}{4} \mathrm{erf}\left(\frac{x-2}{\sqrt{2} \sigma}\right) + \frac{1}{10} \mathrm{erf}\left(\frac{x-4}{\sqrt{2} \sigma}\right) - \frac{1}{2}$.

σ	Modes
0.6	0.013819, 1.998391, 3.978446
0.8	1.977112
1.2	1.819468

12.9 $F(x) = \begin{cases} \frac{1}{2}\left(1 - \lambda + \lambda\, e^x + \mathrm{erf}\left(\frac{\sqrt{2}}{4}(x-3)\right)\right) (1-\lambda) & \text{for } x < 0, \\ \frac{1}{2}\left(1 + \lambda - e^{-x} + \mathrm{erf}\left(\frac{\sqrt{2}}{4}(x-3)\right)\right) (1-\lambda) & \text{for } x > 0. \end{cases}$

r	$E(X^r)$
1	$3 - 3\lambda$
2	$13 - 11\lambda$
3	$63 - 63\lambda$
4	$345 - 321\lambda$

12.10 The moments about the origin are

$$F(x) = \begin{cases} -2 \cdot \dfrac{k \cdot (k-2) \cdot \dots \cdot 1 \cdot a}{2+a^2} & \text{for } k \text{ odd,} \\[2ex] \dfrac{(k+1)a^2+2}{2+a^2}(k-1) \cdot (k-3) \cdot \dots \cdot 1 & \text{for } k \text{ even.} \end{cases}$$

12.11 $F(x) = \dfrac{a\,(2-ax)e^{-x^2/2}}{\sqrt{2\pi}\,(2+a^2)} + \dfrac{1+\text{erf}\left(\frac{x}{\sqrt{2}}\right)}{2}$.

12.12 $f(x) = \dfrac{1+k\,x^3}{a^3+6\,k}\,a^4\,e^{-ax}$, $E(X^r) = \dfrac{a^3\,r!+k(r+3)!}{a^r\,(a^3+6\,k)}$,

$F(x) = 1 - \dfrac{1}{e^{ax}} - \dfrac{akx\,(6+3ax+a^2x^2)}{e^{ax}\,(a^3+6k)}$.

12.13 $C = 2+a^2-a\,\sqrt{\pi}$, $E(X^r) = \dfrac{1}{C}\left[2\,\Gamma\!\left(\frac{r+2}{2}\right) - 2a\,\Gamma\!\left(\frac{r+3}{2}\right) + a^2\,\Gamma\!\left(\frac{r+4}{2}\right)\right]$,

$F(x) = \dfrac{2+a^2-a\sqrt{\pi}\,\text{erf}(x)}{C} - \dfrac{2+a^2-2ax+a^2\,x^2}{C \cdot e^{x^2}}$.

12.14 $g(y) = \dfrac{1}{2\sqrt{y}}\,ae^{-a\sqrt{y}}$, $E(Y^r) = \dfrac{(2\,r)!}{a^{2r}}$, $G(y) = 1-e^{-a\sqrt{y}}$.

12.15 $g(y) = \dfrac{ab}{y^{b+1}}\,\exp\left(-ay^{-b}\right)$, $E(Y^r) = a^{r/b}\,\Gamma\!\left(\frac{b-r}{b}\right)$ for $r < b$,

$G(y) = \exp\left(-ay^{-b}\right)$, $\text{Mo} = \left(\dfrac{ab}{b+1}\right)^{1/b}$.

12.16 $f_u(u) = \dfrac{1}{\sqrt{2\pi}\,u}\,\exp\left(-\dfrac{(\ln(u))^2}{2}\right)$ for $u > 0$, $f_v(v) = \dfrac{1}{2\,\sqrt{2\pi}\,v}\,\exp\left(-\dfrac{(\ln(v))^2}{8}\right)$ for $v > 0$,

$$\rho_{u,v} = \dfrac{e^{\sqrt{3}}-1}{\sqrt{(e-1)(e^4-1)}} = 0.4848, \quad \rho_{x,y} = \sqrt{3}/2 = 0.8660,$$

$P(U \le 1,\ V \le 1) = P(U \ge 1,\ V \ge 1) = 5/12$.

12.17 With the notation of Theorem 7.1, we have $U = X/Y$, $V = X$, thus $X = UV$, $Y = V$ and

$J = \begin{vmatrix} v & u \\ 0 & 1 \end{vmatrix} = v$, $g(u,v) = \varphi\,(uv)\dfrac{1}{2a}\,|v|$. We obtain the density

$h(u) = \int_{-a}^{a} g(u,v)\,dv = \dfrac{1}{\sqrt{2\pi}\,a\,u^2}\left(1-e^{-a^2u^2/2}\right)$ and the distribution function

$H(u) = \dfrac{1}{2} + \dfrac{1}{\sqrt{2\pi}\,a\,u}\left(e^{-a^2u^2/2}-1\right) + \dfrac{1}{2}\,\text{erf}\left(\dfrac{a\,u}{\sqrt{2}}\right)$.

12.18 $g(u,v) = \int_0^\infty ae^{-auv}b\,e^{-buv}\,v\,dv = \dfrac{ab}{(au+b)^2} = \dfrac{\frac{a}{b}}{\left(\frac{a}{b}u+1\right)^2}$. By setting $c = a/b$, one unnecessary parameter is eliminated.

12.19 $h(u) = \int_{-\infty}^{u} \varphi(v)\,a\,e^{-a(u-v)}\,dv = \dfrac{a}{2}e^{-au+a^2/2}\left(1-\text{erf}\left(\frac{a-u}{\sqrt{2}}\right)\right)$, see eq. (7.25).

12.20 $U = \dfrac{X}{Y^{1/q}}$, $V = Y^{1/q}$, $V = X$, thus $X = UV$, $Y = V^q$ and $J = \begin{vmatrix} v & u \\ 0 & qv^{q-1} \end{vmatrix} = qv^q$, $g(u,v) = \varphi(uv)\,qv^q$.

12.21 For $q > r$ we get

$$E(X^r) = \begin{cases} 0 & \text{for } r \text{ odd,} \\ \dfrac{(r-1)(r-3)\cdot\ldots\cdot 1\cdot q}{q-r} & \text{for } r \text{ even,} \end{cases}$$

$$\alpha_4 = \frac{E(X^4)}{(E(X^2))^2} = \frac{\frac{3q}{q-4}}{\left(\frac{q}{q-2}\right)^2} = \frac{3\,(q-2)^2}{q\,(q-4)} \quad \text{for } q > 4.$$

12.22 $h(u) = \int_0^1 a\, e^{-auv} q\, v^q dv = q\frac{\Gamma(q+1)-\Gamma(q+1,\ au)}{a^q\, u^{q+1}}$ for $u \ge 1$, where $\Gamma(z,\ x)$ is the incomplete gamma function, see eqs. (1.12) and (1.13).

Chapter 13

13.1 $\frac{(1/6)\cdot(5/6)}{n\cdot 0.02^2} \le 0.01 \Rightarrow n \ge 34{,}723.$

13.2 $\varepsilon = 1/15$ gives the lower limit $\frac{0.8\cdot 0.2}{150\cdot(1/15)^2} = 0.24.$

13.3 $P(|X_n - X| = 1) = \frac{1}{n}$, $P(|X_n - X| = 0) = 1 - \frac{1}{n}$.

Hence $P(|X_n - X| > \varepsilon) = \frac{1}{n}$ for $0 < \varepsilon < 1$ and 0 for $\varepsilon > 1$.

13.5 Exact probability: 0.6050, approximation: 0.6016

13.6 $P(T_{tot} \ge 52\cdot 60) = 1 - \Phi\left(\frac{3{,}120 - 3{,}000}{\sqrt{5{,}000}}\right) \approx 0.0446.$

13.7 (i) $E(W) = 135\cdot 80 + 105\cdot 70 + 60\cdot 40 = 20{,}550.$
(ii) $V(W) = 135\cdot 100 + 105\cdot 64 + 60\cdot 25 = 21{,}720.$

(iii) $\Phi\left(\frac{21{,}000 - 20{,}550}{\sqrt{21{,}720}}\right) \approx \Phi(3.05) \approx 0.9998.$

13.8 (i) $1 - \Phi\left(\frac{800 - 10\cdot 75}{\sqrt{10}\cdot 8.5}\right) \approx 1 - \Phi(1.86) \approx 0.0314.$

(ii) $1 - \Phi\left(\frac{800 - 11\cdot 75}{\sqrt{11}\cdot 8.5}\right) \approx 1 - \Phi(-0.89) \approx 0.8133$

(iii) $1 - \Phi\left(\frac{800 - 10\mu}{\sqrt{10}\cdot 8.5}\right) = 0.1 \Rightarrow \frac{800 - 10\mu}{\sqrt{10}\cdot 8.5} \approx 1.28 \Rightarrow \mu \approx 76.56.$

13.9 (i) $\Phi\left(\frac{50 - 45}{\sqrt{15}\cdot 1.2}\right) \approx \Phi(1.08) \approx 0.8599.$

(ii) $\Phi\left(\frac{Q - 45}{\sqrt{15}\cdot 1.2}\right) = 0.99 \Rightarrow \frac{Q - 45}{\sqrt{15}\cdot 1.2} \approx 2.33 \Rightarrow Q \approx 55.8.$

13.10 $\Phi\left(\frac{53 - 50}{\sqrt{2}}\right) - \Phi\left(\frac{48 - 50}{\sqrt{2}}\right) \approx \Phi(2.12) - \Phi(-1.14) \approx 0.9830 - 0.1271 = 0.8559.$

13.11 $\Phi\left(\frac{110 - 120}{\sqrt{200}}\right) \approx \Phi(-0.71) \approx 0.2389.$

13.12 Exact probability:

$$\sum_{k=61}^{\infty} \binom{k-1}{49} 0.7^{50} \cdot 0.3^{k-50} \approx 0.9861.$$

With

$$\mu = \frac{r}{p} = \frac{50}{0.7} \approx 71.43, \ \sigma^2 = \frac{r(1-p)}{p^2} \approx 30.61,$$

eq. (13.9) yields the approximate probability 0.9759.

The number X of productions up to the rth perfect item (success) is a sum of independent r.v.'s $X = X_1 + \cdots + X_r$, where X_i is the number of productions after the $(r-1)$th success to and including the rth success.

Chapter 14

14.1 (a) $P(165 \le \bar{X} \le 172) = \phi\left(\frac{172-170}{\sqrt{1.5}}\right) - \phi\left(\frac{165-170}{\sqrt{1.5}}\right) \approx 0.9484.$

(b) $P\left(200 \le S^2 \le 250\right) = P\left(\frac{149}{225} 200 \le \frac{149}{225} S^2 \le \frac{149}{225} 250\right) = \int_{1,192/9}^{1,490/9} \frac{1}{2^{n/2}\Gamma(n/2)} x^{n/2-1} e^{-x/2} \, dx.$

14.2 (a) $f(x) = p(1-p)^{x-1}$, $F(x) = 1 - (1-p)^x$, $F_j(x) = \sum_{k=j}^{n} \binom{n}{k} [1-(1-p)^x]^k (1-p) x^{(n-k)}.$

(b) $F_1(x) = 1 - (1 - F(x))^n = 1 - (1-p)^{xn}$ (geometric distribution with parameter $1-(1-p)^n$):

$$F_n(x) = F(x)^n = \left[1-(1-p)^x\right]^n.$$

14.3 $f_1(x) = F_1(x) - F_1(x-1) = (1-p)^{(x-1)n} - (1-p)^{xn} = (1-p)^{(x-1)n}\left(1-(1-p)^n\right).$

$f_n(x) = F_n(x) - F_n(x-1) = \left[1-(1-p)^x\right]^n - \left[1-(1-p)^{x-1}\right]^n.$

14.4 (a) $F_n(x) = F(x)^n$ $n = 10$.

$$F(25) = \sum_{k=0}^{25} e^{-20} \frac{20^k}{20!} \approx 0.887815.$$

Thus

$$F_{10}(25) = F(25)^{10} \approx 0.3042.$$

(b) $F_{n-1}(x) = \sum_{k=n-1}^{n} \binom{n}{k} F(x)^k (1-F(x))^{n-k} = F(x)^{n-1}[n-(n-1)F(x)].$

Hence,

$$F_9(25) = F(25)^9 (10 - 9F(25)) \approx 0.6887.$$

14.5 (b)
$$h(i, j) = \frac{1}{N^n} \qquad\qquad\qquad \text{for } i=j,$$
$$h(i, j) = \left(\frac{i-i+1}{N}\right)^n - 2\left(\frac{i-i}{N}\right)^n + \left(\frac{i-i-1}{N}\right)^n \quad \text{for } i<j.$$

(c) $f(r) = \frac{1}{N^{n-1}}$ for $r=0$,
$$f(r) = (N-r)\left[\left(\frac{r+1}{N}\right)^n - 2\left(\frac{r}{N}\right)^n + \left(\frac{r-1}{N}\right)^n\right] \qquad \text{for } r=1, \ldots, N-1,$$

(d) (ii) $\frac{h(4, 6)}{f(2)} = \frac{1}{4}$, $\frac{h(1, 6)}{1-\left(\frac{5}{6}\right)^{10}} \approx 0.8281$, $\frac{h(2, 4)}{\left(\frac{4}{6}\right)^{10} - \left(\frac{3}{6}\right)^{10}} \approx 0.0576$.

14.6 $F(x) = 1 - \exp\left(\frac{-x^2}{2s^2}\right)$,
$$f_{(n)}(x) = n\, f(x)\, F(x)^{n-1} = n\frac{x}{s^2}\exp\left(\frac{-x^2}{2s^2}\right)\left[1-\exp\left(\frac{-x^2}{2s^2}\right)\right]^{n-1}$$
$$f_{(n-1)}(x) = n(n-1)f(x)F(x)^{n-2}(1-F(x))$$
$$= n(n-1)\frac{x}{s^2}\exp\left(\frac{-x^2}{s^2}\right)\left[1-\exp\left(\frac{-x^2}{2s^2}\right)\right]^{n-2}.$$

14.7 (a) 188.67 cm,
(c) $(\Phi(2))^{20} \approx 0.6311$,
(d) $20\,\phi(2)^{19}\,(1 - \Phi(2)) + \Phi(2)^{20} \approx 0.9250$.

14.8 (a) $h(k, m) = n(n-1)\frac{km}{s^4}\exp\left(-\frac{k^2+m^2}{2s^2}\right)\left[\exp\left(\frac{-k^2}{s^2}\right) - \exp\left(\frac{-m^2}{2s^2}\right)\right]^{n-2}$,
(b) $P(M \geq 10K) = \int_{k=0}^{\infty}\int_{m=10k}^{\infty} h(k, m)dm\; dk \approx 0.4970$,
$$P(M \geq K^2) = \int_{k=0}^{\infty}\int_{m=k^2}^{\infty} h(k, m)dm\; dk \approx 0.9243.$$

14.10 $E(K) = \frac{1}{na} = \frac{1}{65}$, $E(M) \approx 0.6360$, $E(R) \approx 0.6206$.
14.11 (i) 8.07 and (ii) 0.110.
14.12 $E(X_{(n)}) = 97.50$, $E(X_{(n-1)}) = 93.55$.
14.13 The density of the range is
$$f_{(R)}(r) = 2\int_{-\infty}^{\infty} \varphi(x)\varphi(x+r)dx = \frac{1}{\sqrt{\pi}\sigma}\exp\left(-\frac{r}{4\sigma^2}\right) = 2\frac{1}{\sqrt{2\pi}\sigma_0}\exp\left(-\frac{r^2}{2\sigma_0^2}\right).$$

where $\sigma_0 = \sqrt{2}\sigma$ (half-normal distribution).

14.14 0.171.
14.15 1.73. As a possible cuboid one can choose
$$\left\{(x,y,z)^T \mid -1.5 \leq x, z \leq 1.5,\; 0.5 \leq y \leq 3.5\right\}.$$

14.16 (a)

	Normal	Maxwell	Exponential
Q_1	68.255	2.753	0.7192
Q_2	75	3.845	1.7329
Q_3	81.745	5.067	3.4657
$E(X_{(1)})$	66.533	2.608	0.8323
$E(X_{(2)})$	74.999	3.909	2.0861
$E(X_{(3)})$	83.464	5.451	4.6025
$E(X_{(11)})$	68.119	2.740	0.7258
$E(X_{(22)})$	75.010	3.851	1.7602
$E(X_{(33)})$	81.894	5.100	3.5575
$E(X_{(38)})$	68.213	2.749	0.7192
$E(X_{(76)})$	74.999	3.847	1.7397
$E(X_{(114)})$	81.786	5.076	3.4941

For the uniform distribution all expectations correspond to the respective quartiles, one can prove easily by means of eq. (14.9) that $E(X_{(i)}) = Q_i = i/4$ for $i = 1,2,3$.

Chapter 15

15.1 (i) $\bar{x} = 83.45$, $\hat{\sigma}^2 \approx 149.21$.
   (ii) 141.75.

15.2 (i) $E(X) = L$ for all values of a.
   (ii) Minimizing the function $V(a) = V(aL_1) + (1 - a)L_2) = a^2V(L_1) + (1 - a)^2V(L_2)$ yields
   $$a = \frac{V(L_2)}{V(L_1) + V(L_2)}.$$

15.3 For any $i = 1, \ldots, n - 1$ it holds
   $E(X_i - X_{i+1})^2 = E(X_i^2 - 2X_iX_{i+1} + X_{i+1}^2) = E(X^2) - 2E(X)^2 + E(X)^2 = 2V(X)$. Thus, $\sum_{i=1}^{n-1}$
   $(X_i - X_{i+1})^2$ has the expectation $2(n - 1)V(X)$, i.e. C must be chosen as $C = \frac{1}{2(n-1)}$.

15.5 $\hat{r} \approx 0.1278$, $\hat{a} \approx 0.0208$.
   The right side of eq. (15.11) is positive due to the inequality between arithmetic and geometric mean.

15.6 (i) $\hat{\beta} = -1 - \frac{n}{\ln(x_1 \ldots x_n)}$ and (ii) 0.3051.

15.7 $\hat{a} = \frac{1}{\bar{t} - t_0}$.

15.8 $\hat{a} = \max(|x_1|, \ldots, |x_n|)$.

15.9 $\hat{a} = \frac{n}{x_1^b + \cdots + x_n^b}$.

15.10 (i) $\hat{b} = \frac{1}{2n}\sum_{i=1}^{n} x_i^2$,
   (ii) Since $E(X^2) = 2b$, we get $E\left(\frac{1}{2n}\sum_{i=1}^{n} X_i^2\right) = \frac{1}{2n}nE(X^2) = b$

$$V\left(\frac{1}{2n}\sum_{i=1}^{n}X_i^2\right) = \frac{1}{4n^2}nV(X^2) = \frac{1}{4n}\left(E(X^4) - E(X^2)^2\right)$$

$$= \frac{1}{4n}(8b^2 - 4b^2) = \frac{b^2}{n}.$$

For $n \to \infty$ the latter converges to zero.

15.11 For the specified values of K and n we encountered $\bar{\theta} = 25.0144$ and $V = 6.1637$.

15.12 $\frac{\partial}{\partial\mu}\ln\left[\frac{1}{\sqrt{2\pi}\sigma}\exp\left(-\frac{1}{2}\left(\frac{x-\mu}{2}\right)^2\right)\right] = \frac{\partial}{\partial\mu}\left(-\frac{1}{2\sigma^2}(X-\mu)^2\right) = \frac{X-\mu}{\sigma^2}$,

thus

$$E\left(\frac{X-\mu}{\sigma^2}\right) = \frac{V(X)}{\sigma^4} = \frac{1}{\sigma^2} \Rightarrow t = \frac{n}{\sigma^2}.$$

15.13 $\hat{p} = \frac{\bar{x}-S^2}{\bar{x}}$, $\hat{n} = \frac{\bar{x}}{\hat{p}}$.

15.14 $E(X) = \frac{k}{\lambda}$, $E(X^2) = \frac{k(k+1)}{\lambda^2}$, $\hat{k} = \frac{\bar{x}^2}{m_2 - \bar{x}^2}$, $\hat{\lambda} = \frac{\hat{k}}{\bar{x}}$.

15.15 (i) The ML estimate is solution of

$$\ln(x_1)x_1^b + \cdots + \ln(x_n)x_n^b - \frac{n}{b} = \ln(x_1 \ldots x_n).$$

The moment estimate is solution of

$$\bar{x} = \frac{\Gamma(1/b)}{b}.$$

(ii) The estimator is consistent but biased, so it tends to "overestimate" the parameter value.

15.16 $\hat{a} = \sqrt{\frac{\pi}{8}}\bar{x}$.

From Theorem 14.1, it follows that the estimator is unbiased and consistent.

15.17 $\hat{c} = \bar{x} - \sqrt{m_2 - \bar{x}^2} = \bar{x} = \sqrt{\sum_{i=1}^{n}(x_i - \bar{x})^2}$,

$$\hat{a} = \frac{1}{\bar{x} - c}.$$

Note that the problem becomes easier if the parameter a is substituted by $1/\beta$. The root gives the (unbiased) sample variance.

15.18 $\hat{a} = -1.725$, $\hat{b} = 48.83$.

15.19 $E(\hat{b}) = E(\bar{y} - \hat{a}\bar{x}) = E(\bar{y}) - a\bar{x} = b$,
$V(\hat{b}) = V(\bar{y} - \hat{a}\bar{x}) = V(\bar{y}) + \bar{x}^2V(\hat{a}) = V(\bar{y}) + \frac{\bar{x}^2\sigma^2}{S_{xx}}$.
Now the second formula in eq. (15.29) follows from the fact that

$$V(\bar{y}) = V\left(\frac{1}{n}\sum_{i=1}^{n}(ax_i + b + \varepsilon_i)\right) = \frac{1}{n^2}\sum_{i=1}^{n}V(\varepsilon_i) = \frac{\sigma^2}{n}.$$

15.20 $\hat{b} = \bar{y} - \hat{a}\bar{x} = \frac{1}{n}\sum_{i=1}^{n}y_i - \bar{x}\frac{S_{xy}}{S_{xx}} = \frac{1}{n}\sum_{i=1}^{n}y_i - \sum_{i=1}^{n}y_i\bar{x}\frac{x_i - \bar{x}}{S_{xx}}.$

The estimates $\hat{a}$ and $\hat{b}$ are in fact the best linear unbiased estimates (see Definition 15.5).

15.21 $E(\hat{a}) = \frac{1}{\sum_{i=1}^{n}x_i^2}\sum_{i=1}^{n}x_iE(y_i) = \frac{1}{\sum_{i=1}^{n}x_i^2}\sum_{i=1}^{n}x_iax_i = a.$

15.22 Use the model $S = \frac{a}{2}T^2 + \varepsilon$ and minimize $\sum_{i=1}^{n}\left(at_i^2 - s_i\right)^2.$

15.25 $\hat{a} \approx -1.958,\ \hat{b} \approx 1.199,\ \hat{c} \approx 1.517.$

15.26 [86.32, 93.68].

15.27 (i) [0.4086, 0.4664].

(ii) $n \approx 6{,}660.$

15.28 $L = \frac{9 \cdot 26.044}{21.666} \approx 10.82,\ U = \frac{(n-1)s^2}{G_{n-1}^{-1}(\alpha)} = \frac{9 \cdot 26.044}{2.088} \approx 112.26.$

15.30 Setting $P(\mu \le \bar{X} + \varepsilon) = \alpha$ yields $\varepsilon = \frac{\sigma}{\sqrt{n}}\Phi^{-1}(1-\alpha)$ (compare with eq. (14.42)). The limits in the intervals are $U = \bar{X} + \varepsilon$ and $L = \bar{X} - \varepsilon$, respectively.

15.32 Setting $P(\mu \le \bar{X} + \varepsilon) = 1 - \alpha$ yields $\varepsilon = \frac{\sigma}{\sqrt{n}}H_{n-1}^{-1}(1-\alpha)$. The lower confidence interval can then be written as $]-\infty,\ \bar{X} + \varepsilon]$. Analogously the upper confidence interval is obtained as $[\bar{X} - \varepsilon,\ \infty[$.

Chapter 16

16.1 From (16.8) it follows that $\Phi\left(\frac{t-\mu_0}{\sigma/\sqrt{n}}\right) = 1 - \alpha$, thus $\frac{t-\mu_0}{\sigma/\sqrt{n}} = \Phi^{-1}(1-\alpha)$, yielding (16.9).

16.2 $\varepsilon \approx 3.12,\ t \approx 172.85.$

The acceptance region expands and the critical region becomes smaller, i.e. the "tolerance" with respect to acceptance of H_0 increases.

16.3 (i) $\alpha = 1 - \Phi\left(\frac{t-170}{20/\sqrt{50}}\right) = \Phi\left(\frac{170-t}{20/\sqrt{50}}\right),\ \beta = \Phi\left(\frac{t-180}{30/\sqrt{50}}\right),$

(iii) $t = 174.65,\ \beta \approx 0.104,$

(iv) $t = 173.02,\ \alpha \approx 0.143.$

(v) $\alpha = \beta \Rightarrow \frac{170-t}{20/\sqrt{50}} = \frac{t-180}{30/\sqrt{50}} \Rightarrow t = 174 \Rightarrow \alpha = \beta \approx 0.079.$

(vi) $t = 174.48,\ \alpha = 0.0566,\ \beta = 0.0966.$

16.4 (i) The critical value is $t = 8 + \frac{1.2}{\sqrt{50}}\Phi^{-1}(0.99) \approx 8.3948$. Since $\bar{x} = 8.5 > t$, the null hypotheses must be rejected.

(ii) Now we get $t \approx 8.5264 > \bar{x}$, thus H_0 is accepted.

16.5 (i) $\hat{p} = \frac{192}{300} = 0.64,\ \varepsilon = \sqrt{\frac{0.7 \cdot 0.3}{300}}\Phi^{-1}(0.99) \approx 0.6415.$

Thus $\hat{p}$ is contained in the acceptance region $[0.7 - \varepsilon,\ 0.7 + \varepsilon] = [0.6385, 0.7615]$ and H_0 is accepted.

(ii) $t = 0.7 - \sqrt{\frac{0.7 \cdot 0.3}{300}} \Phi^{-1}(0.98) \approx 0.6457$.

Now $\hat{p}$ is contained in the rejection region

$[-\infty, t]$ and H_0 is rejected.

16.6 One must choose n such that $\beta(\mu)$ in (15.17) equals 0.07. It follows $n = 2656$.

16.7 (i) 0.072, (ii) 0.036.

16.8 (i) $P(X \geq 532) = \frac{1}{21,000} \sum_{k=532}^{1,000} \binom{1,000}{k} \approx 0.0231$.

The exact p-value is thus $2 \cdot 0.0231 = 0.0462$.

(ii) The approximate value for $P(X \geq 532)$ is

$$P(\hat{p} \geq 0.532) = 1 - \Phi\left(\frac{0.532 - 0.5}{\sqrt{1/4,000}}\right) \approx 0.0215.$$

16.9 (i) 0.0519 for the unilateral and 0.1038 for the bilateral test.

(ii) n is given by $0.1 = \sqrt{\frac{0.7929}{n}} \cdot 2.1448 \Rightarrow n = 365$,

(iii) $\varepsilon = 0.451$, $t \approx 5.622$.

16.10 (i) $V = \frac{14 \cdot 0.7929}{1} = 11.10$ is not contained in the rejection region $[23.68, \infty [$, hence H_0 is accepted.

(ii) $V = \frac{14 \cdot 0.7929}{2} = 5.5$ is contained in the rejection region $[0, 6.57[$, hence H_0 is rejected.

16.11 (i) $\bar{x} \leq 7.67$ for $\alpha = 5\%$ and $\bar{x} \leq 7.49$ for $\alpha = 1\%$,

(ii) The right test decision is taken if

$$\bar{X} - \bar{Y} \leq \Phi^{-1}(0.05)\sigma, \text{ where } \sigma^2 = \frac{1.2^2}{60} + \frac{1.5^2}{50} = 0.069.$$

Since $\bar{X} - \bar{Y} \sim N(-0.5, \sigma^2)$, this event occurs with probability $\Phi\left(\frac{\Phi^{-1}(0.05)\sigma + 0.5}{\sigma}\right) \approx 0.6\,020$.

(iii) One must choose n such that $\Phi\left(\Phi^{-1}(0.05) + \frac{0.5}{\sqrt{\frac{1.2^2 + 1.5^2}{n}}}\right) \approx 0.95$.

It follows that $n = 160$.

16.12 (i) 0.9074,

(ii) Z_0 is contained in the acceptance region $[-1.96, 1.96]$, hence H_0 is accepted.

16.13 $\Delta := |\bar{X} - \bar{Y}|$ must be larger than the value given by $\frac{\Delta}{\sqrt{\frac{14}{10} + \frac{20}{15}}} = \varepsilon = 2.0639$. It follows that $\Delta > 4.16$.

16.16 (i) For s_2^2 outside of $[2.80, 69.93]$.

(ii) For $s_2^2 > 53.02$.

16.17 (ii) $4.586 \cdot 10^{-7}$.

16.19 The modified value of the statistic is $\chi_0^2 \approx 26.27$. As in Example 15.16, this value is much larger than the critical value, which is now given by $G_4^{-1}(0.99) \approx 13.28$.

16.20 The observed value of the statistic $\chi_0^2 \approx 7.01$ is smaller than the critical value $G_6^{-1}(0.98) \approx 15.03$.

Hence, the independency hypothesis cannot be rejected.

Chapter 17

17.1 $\bar{x} = 88.033$, $V = 2863.30$, Mo $= 35$, $(Q_1, Q_2, Q_3) = (35, 94, 137)$,
$(D_1, \ldots, D_9) = (9.5, 30.2, 38, 69.6, 94, 113, 127.4, 139.4, 160)$,
7-quantiles: $(Q_1, \ldots, Q_6) = (18.71, 35, 74.29, 108, 129.29, 157.86)$.

17.2 The number $\frac{i}{4}(n+1)$ is not an integer for i=1,2,3. Thus, $m = \frac{in}{4}$, $t = \frac{i}{4}$. Hence, the quartile Q_i lies between $x_{in/4}$ and $x_{in/4+1}$. The proof for the deciles is analogous.

17.3 $\bar{x} = 65.61$, $V = 84.77$, Mo $= 68.57$,
$(Q_1, Q_2, Q_3) = (59.25, 66.5, 72.36)$,
$(D_1, \ldots, D_9) = (51.875, 57, 61, 64, 66.5, 68.75, 71.111, 73.611, 77.5)$.

17.4 (i) The estimated mean is 43 kg.

(ii) The exact mean is 42.7 kg, so the error in estimation is 0.3 kg.

(iii) In general, the approximation is good, when the data are approximately uniformly distributed within classes. Theoretically, the maximum error occurs, e.g., when all values are identical to the lower class limits. Then the absolute error is $h/2$, where h is the amplitude of the classes.

17.5 The mode determined by the Czuber formula is Mo $= 3.78$, and the theoretical mode (maximum point of the density) is 3.5:

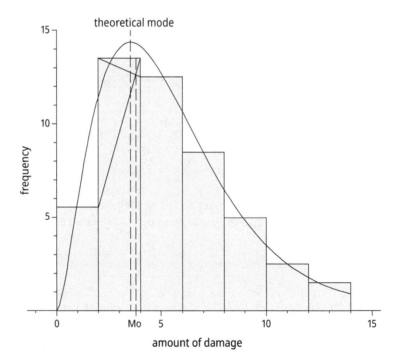

Note that the y-scale of the histogram is of "density type," i.e., the height of a bar is such that its area corresponds to the frequency of the class.

17.6 $\bar{x} = 2.24$, Mo = 2, $\tilde{x} = 2$.

17.7 $F(x) = \begin{cases} 0 & \text{for } x \leq 140 \\ \dfrac{1}{60}x - \dfrac{7}{3} & \text{for } 140 < x \leq 150 \\ \dfrac{2}{75}x - \dfrac{23}{6} & \text{for } 150 < x \leq 160 \\ \dfrac{7}{300}x - \dfrac{33}{10} & \text{for } 160 < x \leq 170 \\ \dfrac{1}{50}x - \dfrac{41}{15} & \text{for } 170 < x \leq 180 \\ \dfrac{1}{75}x - \dfrac{23}{15} & \text{for } 180 < x \leq 190 \\ 1 & \text{for } x > 190 \end{cases}$.

Geometric illustration:

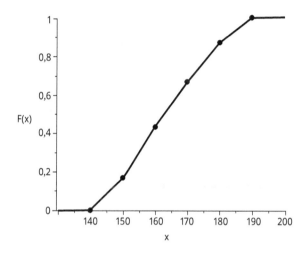

F is a piecewise linear function. The density f is the derivative of F, which exists in points inside of the classes, and f is a stair function. The quantile function is the inverse function F^{-1} of F.

17.8   The expression $\frac{1}{n}\sum_{i=1}^{n}|x_i - \bar{x}|^p$ generalizes (17.8) and (17.9). Expression (17.9) is obtained for $p = 1$ and (17.8) for $p = 2$. The higher the value p, the more severely high deviations of individual data from the mean are "penalized."

17.9   In fact, the measures (17.6)–(17.10) are identical for the two data sets, but (17.11)–(17.15) assign a significantly lower value to the second data set than to the first.

Measure	Set x_i	Set y_i
(17.6)	100	
(17.7)	100	
(17.8)	1666.7	
(17.9)	33.33	

(continued)

Measure	Set x_i	Set y_i
(17.10)	40.82	
(17.11)	0.5	0.01
(17.12)	0.5	0.01
(17.13)	25	0.3333
(17.14)	0.6124	0.0082
(17.15)	0.5	0.0067

17.10 We get $V = \frac{1}{n} \left(\sum_i x_i^2 \cdot F_i - 2\bar{x} \sum_i x_i \cdot F_i + \bar{x}^2 \sum_i F_i \right)$

$= \frac{1}{n} \sum_i x_i^2 - 2\bar{x}^2 + \bar{x}^2$ from which the assertion follows. There is no analogous transformation for eq. (17.17).

17.11 (i): $V = 169.33$, $\sigma = 13.01$.

(ii): $V = 183.81$, $\sigma = 13.56$.

17:12 (i): $a_3 = -0.0754$, $a_4 = 1.6645$.

(ii): $a_3 = 0.1081$, $a_4 = 2.1882$.

(iii): $a_3 = 0.6373$, $a_4 = 2.8576$

17:13 $a_3 = 0.1394, a_4 = 2.1055$, $\bar{x} = 163.6$, $\sigma = 12.33$, $Mo = 163.33$

$D_1 = 146.25$, $Q_1 = 153.75$, $Q_2 = 163.33$, $Q_3 = 172.78$, $D_9 = 181.67$,

$P_1 = 0$, $P_2 = 0.0649$, $G_{skew} = -0.0072$, $K = 0.2686$.

17:14 $P_1 = \frac{1}{\sqrt{r}}$. For the indicated values of r in Fig. 8.3 it holds approximately $P_1 = a_3/2$.

17:15 The quantile function is $Q(y) = \left(\frac{-\ln(1-y)}{a} \right)^{\frac{1}{b}}$.

For example, the first quartil is $Q(1/4) = \left(\frac{\ln(4/3)}{a} \right)^{\frac{1}{b}}$,

$$G_{skew} = \frac{(\ln(4/3))^{1/b} + (\ln(4))^{1/b} - 2 \, (\ln(2))^{1/b}}{(\ln(4))^{1/b} - (\ln(4/3))^{1/b}},$$

$$K = \frac{1}{2} \frac{(\ln(4))^{1/b} - \left(\ln\left(\frac{4}{3}\right)\right)^{1/b}}{(\ln(10))^{1/b} - \left(\ln\left(\frac{10}{9}\right)\right)^{1/b}}.$$

17.16 The considered shape coefficients are defined as $a_3 = \frac{m_3}{\sigma^3}$ and $a_4 = \frac{m_4}{\sigma^4}$ where σ and the m_i are defined by (17.16) and (17.21), respectively. By multiplying F_i by $c>0$, the numbers σ, m_3 and m_4 do not change. The same numbers do not change, if a constant c is added to all x_i. If the vector x is multiplied by c, then σ is multiplied by c and m_i is multiplied by c^i.

17.17 The assertion follows by substituting y_i for $c x_i$ in (17.30).

17.18 $\rho = -0.7198$, $\hat{a} = -4.801$, $\hat{b} = 96.91$, $SS_{tot} = 10052.5$, $SS_{res} = 4843.5$, $SS_{reg} = 5{,}209$, $R^2 = 0.5182$. The last value indicates that the linear model fit is not good.

17.19 $\rho(n) = \frac{\sqrt{7}}{5\,(n^2-1)} \frac{9n^4 + 15n^3 - 5n^2 - 15n - 4}{\sqrt{27n^4 + 84n^3 + 69n^2 - 8}}$. This function is monotone decreasing and convex with $\rho(10) = 0.9284$ and

$$\rho(\infty) = \frac{\sqrt{21}}{5} = 0.9165.$$

17.20 $P(T \geq c) = \int_c^\infty t_{18}\, dt = 0.05$ results in $c = 1.7341$.

Thus, $1.7341 = \rho\sqrt{\frac{18}{1-\rho^2}}$, implying $\rho = 0.3783$.

17.21 $\rho = 0.9982$, $\hat{a} = 12152$, $\hat{b} = 48934$, $SS_{tot} = 3.372 \cdot 10^{10}$, $SS_{res} = 1.246 \cdot 10^8$, $SS_{reg} = 3.360 \cdot 10^{10}$ and $R^2 = 0.9963$.
It is interesting to observe that the linear regression model fits the data much better than the exponential. Nevertheless, in general the latter is more appropriate for growth processes. For the considered example the exponential model is less suitable due to the change in the growth rate (see Example 17.12).

17.22 $\rho(c) = \dfrac{9,110 - 82\ c}{\sqrt{9.22 \cdot 10^7 - 1.55 \cdot 10^6 c + 3.59 \cdot 10^4 c^2}}$,

$\hat{a} = 1.14 - 0.0103\ c$, $\hat{b} = -127.4 + 1.94\ c$.

Table of distributions

No.	Name	Prob. fct./pdf	Parameters	E(X)	V(X)
1	Bernoulli	$f(k)=p^k(1-p)^{1-k}$	$0<p<1; k=0,1$	p	$p(1-p)$
2	Beta	$f(x)=\dfrac{x^{a-1}(1-x)^{b-1}}{B(x,y)}$	$a,b>0, 0\le x\le 1$	$\dfrac{a}{a+b}$	$\dfrac{ab}{(a+b+1)(a+b)^2}$
3	Binomial	$f(k)=\dbinom{n}{k}p^k(1-p)^{n-k}$	$0<p<1, n\in\mathbb{N}\ k=0,1,\dots,n$	np	$np(1-p)$
4	Binomial, negative	$f(k)=\dbinom{r+k-1}{k}p^r(1-p)^k$	$0<p<1, r>0, k=0,1,\dots$	$\dfrac{r(1-p)}{p}$	$\dfrac{r(1-p)}{p^2}$
5	Cauchy	$f(x)=\dfrac{1}{\pi\beta\left[1+\left(\frac{x-\alpha}{\beta}\right)^2\right]}$	$\beta>0, \alpha\in\mathbb{R}, x\in\mathbb{R}$	Not defined	Not defined
6	Chi	$f(x)=\dfrac{2^{1-n/2}x^{n-1}\exp(-x^2/2)}{\Gamma(n/2)}$	$x>0, n\in\mathbb{N}$	$\dfrac{\sqrt{2}\,\Gamma\left(\frac{n+1}{2}\right)}{\Gamma\left(\frac{n}{2}\right)}$	$n-2\left(\dfrac{\Gamma\left(\frac{n+1}{2}\right)}{\Gamma\left(\frac{n}{2}\right)}\right)^2$
7	Chi-square	$f(x)=\dfrac{x^{k/2-1}e^{-x/2}}{2^{k/2}\Gamma\left(\frac{k}{2}\right)}$	$x>0, k\in\mathbb{N}$	k	$2k$
8	Erlang	$f(x)=\dfrac{\lambda^k x^{k-1}e^{-\lambda x}}{(k-1)!}$	$\lambda>0, x>0, k\in\mathbb{N}$	$\dfrac{k}{\lambda}$	$\dfrac{k}{\lambda^2}$
9	Exponential	$f(x)=\lambda e^{-\lambda x}$	$\lambda>0, x>0$	$\dfrac{1}{\lambda}$	$\dfrac{1}{\lambda^2}$
10	Exponential, double (Laplace)	$f(x)=\dfrac{1}{2\beta}\exp\left(-\dfrac{|x-\alpha|}{\beta}\right)$	$\beta>0, \alpha, x\in\mathbb{R}$	α	$2\beta^2$

		$f(x)$	conditions	mean	variance
11	f-Distribution (Snedecor)	$f(x) = \dfrac{\Gamma\left(\frac{m+n}{2}\right)}{\Gamma\left(\frac{m}{2}\right)\Gamma\left(\frac{n}{2}\right)}\left(\dfrac{m}{n}\right)^{m/2}\cdot\dfrac{x^{(m-2)/2}}{\left(1+\frac{m}{n}x\right)^{(m+2)/2}}$	$m, n \in \mathbb{N}, x>0$	$\dfrac{n}{n-2}$ for $n>2$	$\dfrac{2n^2(m+n-2)}{m(n-2)^2(n-4)}$ for $n>4$
12	Gamma	$f(x) = \dfrac{\lambda^r}{\Gamma(r)}x^{r-1}\exp(-\lambda x)$	$\lambda, r, x>0$	$\dfrac{r}{\lambda}$	$\dfrac{r}{\lambda^2}$
13	Geometric	$f(k) = p(1-p)^k$	$0<p<1,$ $k=0,1,\dots$	$\dfrac{1-p}{p}$	$\dfrac{1-p}{p^2}$
14	Gumbel	$f(x) = \dfrac{1}{\beta}\exp\left(-\dfrac{x-a}{\beta}\right)\cdot\exp\left(-\exp\left(-\dfrac{x-a}{\beta}\right)\right)$	$\beta>0, a, x \in \mathbb{R}$	$a+\beta\gamma,\ \gamma\approx 0.5772$ (Euler's constant)	$\dfrac{\pi^2\beta^2}{6}$
15	Hypergeometric	$f(k) = \dfrac{\binom{a}{k}\binom{b}{n-k}}{\binom{a+b}{n}}$	$n, a, b \in \mathbb{N}, k=0,\dots,n$	$\dfrac{a\,n}{a+b}$	$\dfrac{(a+b-n)\,abn}{(a+b)^2(a+b-1)}$
16	Logistic	$f(x) = \dfrac{\exp\left(\frac{x-a}{\beta}\right)}{\beta\left(1+\exp\left(\frac{x-a}{\beta}\right)\right)^2}$	$\beta>0, a, x \in \mathbb{R}$	a	$\dfrac{\pi^2\beta^2}{3}$
17	Lognormal	$f(x) = \dfrac{1}{x\sqrt{2\pi}\sigma}\cdot\exp\left(-\dfrac{(\ln(x)-\mu)^2}{2\sigma^2}\right)$	$\mu \in \mathbb{R}, \sigma, x>0$	$e^{\mu+\sigma^2/2}$	$\exp(2\mu+2\sigma^2) - \exp(2\mu+\sigma^2)$
18	Maxwell-Boltzmann	$f(x) = \dfrac{4x^2}{\sqrt{\pi}}\beta^{3/2}e^{-\beta x^2}$	$\beta, x>0$	$\dfrac{2}{\sqrt{\pi}\beta}$	$\dfrac{1}{\beta}\left(\dfrac{3}{2}-\dfrac{\pi}{4}\right)$
19	Normal	$f(x) = \dfrac{1}{\sqrt{2\pi}\sigma}\exp\left(-\dfrac{(x-\mu)^2}{2\sigma^2}\right)$	$\mu, x \in \mathbb{R}, \sigma>0$	μ	σ^2
20	Pareto	$f(x) = \dfrac{\theta c^\theta}{x^{\theta+1}}$	$x>c>0, \theta>0$	$\dfrac{c\theta}{\theta-1}$ for $\theta>1$	$\dfrac{c^2\theta}{(\theta-1)^2(\theta-2)}$ for $\theta>2$
21	Poisson	$f(k) = \dfrac{e^{-\lambda}\lambda^k}{k!}$	$\lambda>0, k=0,1,\dots$	λ	λ

(continued)

(continued)

No.	Name	Prob. fct./pdf	Parameters	E(X)	V(X)
22	Rayleigh	$f(x)=\dfrac{x\,\exp\left(-\dfrac{x^2}{2s^2}\right)}{s^2}$	$x>0, s>0$	$s\sqrt{\frac{\pi}{2}}$	$\left(2-\frac{\pi}{2}\right)s^2$
23	t-Distribution	$f(x)=\dfrac{\Gamma\left(\frac{k+1}{2}\right)}{\Gamma\left(\frac{k}{2}\right)\sqrt{k\pi}\left(1+\frac{x^2}{k}\right)^{(k+1)/2}}$	$k>0, x\in\mathbb{R}$	$\mu=0$ for $k>1$	$\frac{k}{k-2}$ for $k>2$
24	Uniform (discrete)	$f(k)=1/n$	$k=1,\dots,n$	$\frac{n+1}{2}$	$\frac{n^2-1}{12}$
25	Uniform (continuous)	$f(x)=\frac{1}{b-a}$	$a<x<b$	$\frac{a+b}{2}$	$\frac{(b-a)^2}{12}$
26	Weibull	$f(x)=abx^{b-1}\exp\left(-ax^b\right)$	$a,b,x>0$	$a^{-1/b}\cdot\Gamma(1+1/b)$	$a^{-2/b}\cdot\left[\Gamma(1+2/b)-\Gamma^2(1+1/b)\right]$

Standard normal distribution

For a positive z, the table shows the value

$$\Phi(z) = \int_{-\infty}^{z} \frac{1}{\sqrt{2\pi}} e^{-s^2/2} ds = P(Z \le z).$$

For example, $\Phi(1.43) = 0.9236$.

For negative values of z one can make use of the relation

$$\Phi(z) = 1 - \Phi(-z).$$

z	0	1	2	3	4	5	6	7	8	9
0.0	0.5000	0.5040	0.5080	0.5120	0.5160	0.5199	0.5239	0.5279	0.5319	0.5359
0.1	0.5398	0.5438	0.5478	0.5517	0.5557	0.5596	0.5636	0.5675	0.5714	0.5753
0.2	0.5793	0.5832	0.5871	0.5910	0.5948	0.5987	0.6026	0.6064	0.6103	0.6141
0.3	0.6179	0.6217	0.6255	0.6293	0.6331	0.6368	0.6406	0.6443	0.6480	0.6517
0.4	0.6554	0.6591	0.6628	0.6664	0.6700	0.6736	0.6772	0.6808	0.6844	0.6879
0.5	0.6915	0.6950	0.6985	0.7019	0.7054	0.7088	0.7123	0.7157	0.7190	0.7224
0.6	0.7257	0.7291	0.7324	0.7357	0.7389	0.7422	0.7454	0.7486	0.7517	0.7549
0.7	0.7580	0.7611	0.7642	0.7673	0.7703	0.7734	0.7764	0.7794	0.7823	0.7852
0.8	0.7881	0.7910	0.7939	0.7967	0.7995	0.8023	0.8051	0.8078	0.8106	0.8133
0.9	0.8159	0.8186	0.8212	0.8238	0.8264	0.8289	0.8315	0.8340	0.8365	0.8389
1.0	0.8413	0.8438	0.8461	0.8485	0.8508	0.8531	0.8554	0.8577	0.8599	0.8621
1.1	0.8643	0.8665	0.8686	0.8708	0.8729	0.8749	0.8770	0.8790.	0.8810	0.8830
1.2	0.8849	0.8869	0.8888	0.8907	0.8925	0.8944	0.8962	0.8980	0.8997	0.9015
1.3	0.9032	0.9049	0.9066	0.9082	0.9099	0.9115	0.9131	0.9147	0.9162	0.9177
1.4	0.9192	0.9207	0.9222	0.9236	0.9251	0.9265	0.9278	0.9292	0.9306	0.9319
1.5	0.9332	0.9345	0.9357	0.9370	0.9382	0.9394	0.9406	0.9418	0.9430	0.9441
1.6	0.9452	0.9463	0.9474	0.9484	0.94951	0.9505	0.9515	0.9525	0.9535	0.9545
1.7	0.9554	0.9564	0.9573	0.9582	0.9591	0.9599	0.9608	0.9616	0.9625	0.9633
1.8	0.9643	0.9648	0.9656	0.9664	0.9671	0.9678	0.9686	0.9693	0.9700	0.9706
1.9	0.9713	0.9719	0.9726	0.9732	0.9738	0.9744	0.9750	0.9756	0.9762	0.9767
2.0	0.9772	0.9778	0.9783	0.9788	0.9793	0.9798	0.9803	0.9808	0.9812	0.9817
2.1	0.9821	0.9826	0.9830	0.9834	0.9838	0.9842	0.9846	0.9850	0.9854	0.9S57
2.2	0.9861	0.9864	0.9868	0.9871	0.9874	0.9878	0.9881	0.9884	0.9887	0.9890
2.3	0.9893	0.9896	0.9898	0.9901	0.9904	0.9906	0.9909	0.9911	0.9913	0.9916
2.4	0.9918	0.9920	0.9922	0.9925	0.9927	0.9929	0.9931	0.9932	0.9934	0.9936
2.5	0.9938	0.9940	0.9941	0.9943	0.9945	0.9946	0.9948	0.9949	0.9951	0.9952
2.6	0.9953	0.9955	0.9956	0.9957	0.9959	0.9960	0.9961	0.9962	0.9963	0.9964
2.7	0.9965	0.9966	0.9967	0.9968	0.9969	0.9970	0.9971	0.9972	0.9973	0.9974
2.8	0.9974	0.9975	0.9976	0.9977	0.9977	0.9978	0.9979	0.9979	0.9980	0.9981
2.9	0.9981	0.9982	0.9982	0.9983	0.9984	0.9984	0.9985	0.9985	0.9986	0.9986
3.0	0.9987	0.9990	0.9993	0.9995	0.9997	0.9998	0.9998	0.9999	0.9999	1.0000

Student's t-distribution

For a given probability p and k degrees of freedom the table shows the value c, satisfying

$$\int_{-\infty}^{c} \frac{\Gamma\left(\frac{k+1}{2}\right)}{\Gamma\left(\frac{k}{2}\right)\sqrt{\pi k}} \left(1 + \frac{t^2}{k}\right)^{-(k+2)/2} dt = p.$$

p / k	0.75	0.90	0.95	0.975	0.99	0.995	p / k	0.75	0.90	0.95	0.975	0.99	0.995
1	1.0000	3.0777	6.3138	12.7062	31.8207	63.6574	46	0.6799	1.3002	1.6787	2.0129	2.4102	2.6870
2	0.8165	1.8856	2.9200	4.3027	6.9646	9.9248	47	0.6797	1.2998	1.6779	2.0117	2.4083	2.6846
3	0.7649	1.6377	2.3534	3.1824	4.5407	5.8409	48	0.6796	1.2994	1.6772	2.0106	2.4066	2.6822
4	0.7407	1.5332	2.1318	2.7764	3.7469	4.6041	49	0.6795	1.2991	1.6766	2.0096	2.4049	2.6800
5	0.7267	1.4759	2.0150	2.5706	3.3649	4.0322	50	0.6794	1.2987	1.6759	2.0086	2.4033	2.6778
6	0.7176	1.4398	1.9432	2.4469	3.1427	3.7074	51	0.6793	1.2984	1.6753	2.0076	2.4017	2.6757
7	0.7111	1.4149	1.8946	2.3846	2.9980	3.4995	52	0.6792	1.2980	1.6747	2.0066	2.4002	2.6737
8	0.7064	1.3968	1.8595	2.3060	2.8965	3.3554	53	0.6791	1.2977	1.6741	2.0057	2.3988	2.6718
9	0.7027	1.3830	1.8331	2.2622	2.8214	3.2498	54	0.6791	1.2974	1.6736	2.0049	2.3974	2.6700
10	0.6998	1.3722	1.8125	2.2281	2.7638	3.1693	55	0.6790	1.2971	1.6730	2.0040	2.3961	2.6682
11	0.6974	1.3634	1.7959	2.2010	2.7181	3.1058	56	0.6789	1.2969	1.6725	2.0032	2.3948	2.6665
12	0.6955	1.3562	1.7823	2.1788	2.6810	3.0545	57	0.6788	1.2966	1.6720	2.0025	2.3936	2.6649
13	0.6938	1.3502	1.7709	2.1604	2.6503	3.0123	58	0.6787	1.2963	1.6716	2.0017	2.3924	2.6633
14	0.6924	1.3450	1.7613	2.1448	2.6245	2.9768	59	0.6787	1.2961	1.6711	2.0010	2.3912	2.6618
15	0.6912	1.3406	1.7531	2.1315	2.6025	2.9467	60	0.6786	1.2958	1.6706	2.0003	2.3901	2.6603

k	0.75	0.90	0.95	0.975	0.99	0.995
16	0.6901	1.3368	1.7459	2.1199	2.5835	2.9208
17	0.6892	1.3334	1.7396	2.1098	2.5669	2.8982
18	0.6884	1.3304	1.7341	2.1009	2.5524	2.8784
19	0.6876	1.3277	1.7291	2.0930	2.5395	2.8609
20	0.6870	1.3253	1.7247	2.0860	2.5280	2.8453
21	0.6864	1.3232	1.7207	2.0796	2.5177	2.8314
22	0.6858	1.3212	1.7171	2.0739	2.5083	2.8188
23	0.6853	1.3195	1.7139	2.0687	2.4999	2.8073
24	0.6848	1.3178	1.7109	2.0639	2.4922	2.7969
25	0.6844	1.3163	1.7081	2.0595	2.4851	2.7874

k	0.75	0.90	0.95	0.975	0.99	0.995
26	0.6840	1.3150	1.7056	2.0555	2.4786	2.7787
27	0.6837	1.3137	1.7033	2.0518	2.4727	2.7707
28	0.6834	1.3125	1.7011	2.0484	2.4671	2.7633
29	0.6830	1.3114	1.6991	2.0452	2.4620	2.7564
30	0.6828	1.3104	1.6973	2.0423	2.4573	2.7500
31	0.6825	1.3095	1.6955	2.0395	2.4528	2.7440
32	0.6822	1.3086	1.6939	2.0369	2.4487	2.7385
33	0.6820	1.3077	1.6924	2.0345	2.4448	2.7333
34	0.6818	1.3070	1.6909	2.0322	2.4411	2.7284
35	0.6816	1.3062	1.6896	2.0301	2.4377	2.7238
36	0.6814	1.3055	1.6883	2.0281	2.4345	2.7195
37	0.6812	1.3049	1.6871	2.0262	2.4314	2.7154
38	0.6810	1.3042	1.6860	2.0244	2.4286	2.7116
39	0.6808	1.3036	1.6849	2.0227	2.4258	2.7079

k	0.75	0.90	0.95	0.975	0.99	0.995
61	0.6785	1.2956	1.6702	1.9996	2.3890	2.6589
62	0.6785	1.2954	1.6698	1.9990	2.3880	2.6575
63	0.6784	1.2951	1.6694	1.9983	2.3870	2.6561
64	0.6783	1.2949	1.6690	1.9977	2.3860	2.6549
65	0.6783	1.2947	1.6686	1.9971	2.3851	2.6536
66	0.6702	1.2945	1.6683	1.9966	2.3842	2.6524
67	0.6782	1.2943	1.6679	1.9960	2.3833	2.6512
68	0.6781	1.2941	1.6676	1.9955	2.3824	2.6501
69	0.6781	1.2939	1.6672	1.9949	2.3816	2.6490
70	0.6780	1.2938	1.6669	1.9944	2.3808	2.6479

k	0.75	0.90	0.95	0.975	0.99	0.995
71	0.6780	1.2936	1.6666	1.9939	2.3800	2.6469
72	0.6779	1.2934	1.6663	1.9935	2.3793	2.6459
73	0.6779	1.2933	1.6660	1.9930	2.3785	2.6449
74	0.6778	1.2931	1.6657	1.9925	2.3778	2.6439
75	0.6778	1.2929	1.6654	1.9921	2.3771	2.6430
76	0.6777	1.2928	1.6652	1.9917	2.3764	2.6421
77	0.6777	1.2926	1.6649	1.9913	2.3758	2.6412
78	0.6776	1.2925	1.6646	1.9908	2.3751	2.6403
79	0.6776	1.2924	1.6644	1.9905	2.3745	2.6395
80	0.6776	1.2922	1.6641	1.9901	2.3739	2.6387
81	0.6775	1.2921	1.6639	1.9897	2.3733	2.6379
82	0.6775	1.2920	1.6636	1.9893	2.3727	2.6371
83	0.6775	1.2918	1.6634	1.9890	2.3721	2.6364
84	0.6774	1.2917	1.6632	1.9886	2.3716	2.6356

(continued)

(continued)

p	0.75	0.90	0.95	0.975	0.99	0.995	p	0.75	0.90	0.95	0.975	0.99	0.995
k							k						
40	0.6807	1.3031	1.6839	2.0211	2.4233	2.7045	85	0.6774	1.2916	1.6630	1.9883	2.3710	2.6349
41	0.6805	1.3025	1.6829	2.0195	2.4208	2.7012	86	0.6774	1.2915	1.6628	1.9879	2.3705	2.6342
42	0.6804	1.3020	1.6820	2.0181	2.4185	2.6981	87	0.6773	1.2914	1.6626	1.9876	2.3700	2.6335
43	0.6802	1.3016	1.6811	2.0167	2.4163	2.6951	88	0.6773	1.2912	1.6624	1.9873	2.369	2.6329
44	0.6801	1.3011	1.6802	2.0154	2.4141	2.6923	89	0.6773	1.2911	1.6622	1.9870	2.369	2.6322
45	0.6800	1.3006	1.6794	2.0141	2.4121	2.6896	90	0.6772	1.2910	1.6620	1.9867	2.3685	2.6316

Chi-square distribution

For a given probability p and n degrees of freedom the table shows the value c, satisfying

$$\int_0^c \frac{x^{n/2-1} e^{-x/2}}{2^{n/2}\Gamma\left(\frac{n}{2}\right)}\, dx = p.$$

p	0.005	0.01	0.025	0.05	0.10	0.25	p	0.75	0.90	0.95	0.975	0.99	0.995
n							n						
1	–	–	0.001	0.004	0.016	0.102	1	1.323	2.706	3.841	5.024	6.635	7.879
2	0.010	0.020	0.051	0.103	0.211	0.575	2	2.773	4.605	5.991	7.378	9.210	10.597
3	0.072	0.115	0.216	0.352	0.584	1.213	3	4.108	6.251	7.815	9.348	11.345	12.838
4	0.207	0.297	0.484	0.711	1.064	1.923	4	5.385	7.779	9.488	11.143	13.277	14.860
5	0.412	0.554	0.831	1.145	1.610	2.675	5	6.626	9.236	11.071	12.833	15.086	16.750

df												
6	0.676	0.872	1.237	1.635	2.204	3.455	7.841	10.645	12.592	14.449	16.812	18.548
7	0.989	1.239	1.690	2.167	2.833	4.255	9.037	12.017	14.067	16.013	18.475	20.278
8	1.344	1.64	2.180	2.733	3.490	5.071	10.219	13.362	15.507	17.535	20.090	21.955
9	1.735	2.088	2.700	3.325	4.168	5.899	11.389	14.684	16.919	19.023	21.666	23.589
10	2.156	2.558	3.247	3.940	4.865	6.737	12.549	15.987	18.307	20.483	23.209	25.188
11	2.603	3.053	3.816	4.575	5.578	7.584	13.701	17.275	19.675	21.920	24.725	26.757
12	3.074	3.571	4.404	5.226	6.304	8.438	14.845	18.549	21.026	23.337	26.217	28.299
13	3.565	4.107	5.009	5.892	7.042	9.299	15.984	19.812	22.362	24.736	27.688	29.819
14	4.075	4.660	5.629	6.571	7.790	10.165	17.117	21.064	23.685	26.119	29.141	31.319
15	4.601	5.229	6.262	7.261	8.547	11.037	18.245	22.307	24.996	27.488	30.578	32.801
16	5.142	5.812	6.908	7.962	9.312	11.912	19.369	23.542	26.296	28.845	32.000	34.267
17	5.697	6.408	7.564	8.672	10.085	12.792	20.489	24.769	27.587	30.191	33.409	35.718
18	6.265	7.015	8.231	9.390	10.865	13.675	21.605	25.989	28.869	31.526	34.805	37.156
19	6.844	7.633	8.907	10.117	11.651	14.562	22.718	27.204	30.144	32.852	36.191	38.582
20	7.434	8.260	9.591	10.851	12.443	15.452	23.828	28.412	31.410	34.170	37.566	39.997
21	8.034	8.897	10.283	11.591	13.240	16.344	24.935	29.615	32.671	35.479	38.932	41.401
22	8.643	9.542	10.982	12.338	14.042	17.240	26.039	30.813	33.924	36.781	40.289	42.796
23	9.260	10.196	11.689	13.091	14.848	18.137	27.141	32.007	35.172	38.076	41.638	44.101
24	9.886	10.856	12.401	13.848	15.659	19.037	28.241	33.196	36.415	39.364	42.980	45.559
25	10.520	11.524	13.120	14.611	16.473	19.939	29.339	34.382	37.652	40.646	44.314	46.928

n	0.005	0.01	0.025	0.05	0.10	0.25	0.75	0.90	0.95	0.975	0.99	0.995
26	11.160	12.198	13.844	15.379	17.292	20.843	30.435	35.563	38.885	41.923	45.642	48.290
27	11.808	12.879	14.573	16.151	18.114	21.749	31.528	36.741	40.113	43.194	46.963	49.645
28	12.461	13.565	15.308	16.928	18.939	22.657	32.620	37.916	41.337	44.461	48.278	50.993
29	13.121	14.257	16.047	17.708	19.768	23.567	33.711	39.087	42.557	45.722	49.508	52.336
30	13.787	14.954	16.791	18.493	20.599	24.478	34.800	40.256	43.773	46.970	50.892	53.672
31	14.458	15.655	17.539	19.281	21.434	25.390	35.887	41.422	44.985	48.232	52.191	55.003
32	15.134	16.362	18.291	20.072	22.271	26.304	36.973	42.585	46.194	49.480	53.486	56.328
33	15.815	17.074	19.047	20.867	23.110	27.219	38.058	43.745	47.400	50.725	54.776	57.648
34	16.501	17.789	19.806	21.664	23.952	28.136	39.141	44.903	48.602	51.966	56.061	58.964
35	17.192	18.509	20.569	22.465	24.797	29.054	40.223	46.059	49.802	53.203	57.342	60.275
36	17.887	19.233	21.336	23.269	25.643	29.973	41.304	47.212	50.998	54.437	58.619	61.581
37	18.586	19.960	22.106	24.075	26.492	30.893	42.383	48.363	52.192	55.663	59.092	62.883
38	19.289	20.691	22.878	24.884	27.343	31.815	43.462	49.513	53.384	56.896	61.162	64.181
39	19.996	21.426	23.654	25.695	28.196	32.737	44.539	50.660	54.572	58.120	62.428	65.476
40	20.707	22.164	24.433	26.509	29.051	33.660	45.616	51.805	55.758	59.342	63.691	66.766
41	21.421	22.906	25.215	27.326	29.907	34.585	46.692	52.949	56.942	60.561	64.950	68.053
42	22.138	23.650	25.999	28.144	30.765	35.510	47.766	54.090	58.124	61.777	66.200	69.336
43	22.859	24.398	26.785	28.965	31.625	36.436	48.840	55.230	59.304	62.990	67.459	70.616
44	23.584	25.148	27.575	29.787	32.487	37.363	49.913	56.369	60.481	64.201	68.710	71.893
45	24.311	25.901	28.366	30.612	33.350	38.291	50.985	57.505	61.656	65.410	69.957	73.166

References

This book should be comprehensible without the aid of further reading. Either way, we cite a few further textbooks and encyclopedias that enable the reader to continue the study in different directions. Research papers are only cited in Chapter 12, which is related to a currently very active topic.

Textbooks

Dwass, Meyer: Probability and Statistics: An Undergraduate Course. W.A. Benjamin, Inc., New York, 1970.
Feller, W.: An introduction to Probability Theory and Its Applications, J. Wiley & Sons, Inc., New York, 1968.
Rohatgi, V. K, Saleh, A.K. and Md Ehsanes: An introduction to probability and statistics, J.Wiley&Son, Inc.,
    New York, 2001.
Ross, S. M.: Introduction to probability models, Elsevier, Amsterdam, New York, 2007.

Encyclopedias

Johnson, N.L., Kemp, A. and Kotz, S.: Univariate discrete distributions, Hoboken, Wiley-Interscience, c2005.
Johnson, N.L., Kotz, S. and Balakrishnan, N.: Continuous univariate distributions, New York, Wiley, 1994.
Wimmer, G; Altmann, G.: *Thesaurus of univariate discrete probability distributions*. Essen/Germany, STAMM
    publisher, 1999.

Research papers

Azzalini, A.: A class of distributions which includes the normal ones. Scandinavian Journal of Statistics, 12,
    171–178, 1989.
Elal-Olivero, D.: Alpha-skew-normal distributions, Proyecciones (Antofagasta) 29(3), 224–20, 2010.
Shah, S., Hazarika, P.J., Chakraborty, S.: A Generalized alpha-beta-skew normal distribution with
    applications, Annals of Data Science 10, 1127–1155, 2023.
Zörnig, P.: On Generalized Slash Distributions: Representation by Hypergeometric Functions, MDPI Stats
    2(3), 371–387, 2019.

https://doi.org/10.1515/9783111332277-019

Index

https://doi.org/10.1515/9783111332277-020